Third Edition

ION MOBILITY
SPECTROMETRY

Third Edition

ION MOBILITY SPECTROMETRY

G.A. Eiceman • Z. Karpas • H.H. Hill, Jr.

CRC Press
Taylor & Francis Group
Boca Raton   London   New York

CRC Press is an imprint of the
Taylor & Francis Group, an **informa** business

CRC Press
Taylor & Francis Group
6000 Broken Sound Parkway NW, Suite 300
Boca Raton, FL 33487-2742

First issued in paperback 2016

Version Date: 20130927

ISBN 13: 978-1-138-19948-4 (pbk)
ISBN 13: 978-1-4398-5997-1 (hbk)

Library of Congress Cataloging-in-Publication Data

Eiceman, Gary Alan.
    Ion mobility spectrometry / G.A. Eiceman, Z. Karpas, and H.H. Hill, Jr. -- 3rd ed.
        p. cm.
    "A CRC title."
    Includes bibliographical references and index.
    ISBN 978-1-4398-5997-1
    1. Ion mobility spectroscopy. I. Karpas, Zeev. II. Hill, H. H. (Herbert H.) III. Title.

QD96.P62E33 2013
543'.65--dc23   2012023859

Visit the Taylor & Francis Web site at
http://www.taylorandfrancis.com

and the CRC Press Web site at
http://www.crcpress.com

Contents

Preface

At the publication of our second monograph *Ion Mobility Spectrometry* nearly a decade ago, a transformation of ion mobility spectrometry (IMS) was already in a nascent stage, with new comprehensions of principles, innovations in technology, and broadening of applications. Demonstrated and emerging uses of IMS in clinical practices, pharmaceutical explorations, and biomolecular investigations eventually complimented an already-large presence of this method in military establishments for measuring chemical warfare agents, in security organizations for detecting explosives, and in drug enforcement agencies for identifying illicit drugs. While new mobility methods with asymmetric electrical fields and the miniaturization of analyzers were distinctive of our previous edition, the combination of ion mobility with mass spectrometry has been foremost in the developments in recent years. A decade ago, every publication concerning IMS could be read, and in some instances anticipated, owing to a relatively small number of well-acquainted researchers and users. This is no longer easily done with true globalization of the method and commercial availability of instruments.

In the present monograph, innovations or advances in all aspects of IMS are treated in a fresh, thorough, and wholly new format comprised of 19 chapters in four sections Overview, Technology, Fundamentals, and Applications. A fifth section is a single chapter on concluding remarks. Section I is comprised of chapters introducing the definitions, theory, and practice of IMS and summarizing the rich history of IMS from the beginnings of study of ions to the present standing of commercial and scholarly activities. These have been sharpened and reorganized compared to the prior editions of this book. In the second section, the technology of IMS is presented as viewed from a perspective of a measurement. Thus, the order follows from inlet (Chapter 3) through ion formation (Chapter 4), ion injection (Chapter 5), electric fields and drift tube structures (Chapter 6), to detectors (Chapter 7). The section ends with discussion of the end results of a measurement, the mobility spectrum (Chapter 8) and the transformative trend of ion mobility–mass spectrometry (Chapter 9). This organization is intended to aid readers by structuring discussions and treating each facet of IMS with a new and advanced level of detail.

In Section III, the central element of an IMS measurement, the mobility of ions, is discussed from the perspective of meaning (Chapter 10) and influences of the experimental parameters (Chapter 11). An exciting aspect of this edition is the section on applications; seven Chapters (12 to 18) are needed to present the wide breadth and depth of activities. Mobility-based methods are no longer restricted to volatile substances; indeed, the large benefits of this technology—simplicity, convenience, and low cost of technology—along with the advantages of measurements (high speed, distinctive spectral features, and operation in ambient pressure with thermalized ions) have become recognized as meritorious for use throughout a large range of the human experience.

The book is meant to serve specialists in the field of IMS who are interested in details of recent developments and to aid researchers and engineers who want a comprehensive overview of this technology. This text may also be considered useful as a foundation for graduate-level university courses. For those new to IMS, we trust you will find this blend of chemistry, physics, and engineering informative, and we hope it will inspire you to innovation and activity in a field of exploration still rich with opportunity to discover and advance. We welcome you to this third edition with our sincere hope that our labors here will be helpful and interesting.

G. A. Eiceman
Las Cruces, NM

H. H. Hill, Jr.
Pullman, WA

Z. Karpas
Omer, Israel

1 Introduction to Ion Mobility Spectrometry

1.1 BACKGROUND

1.1.1 A DEFINITION OF ION MOBILITY SPECTROMETRY

The term *ion mobility spectrometry* (IMS) refers to the principles, methods, and instrumentation for characterizing substances from the speed of swarms (defined as ensembles of gaseous ions) derived from a substance, in an electric field and through a supporting gas atmosphere.[1,2] This simple definition encompasses all combinations of pressure, flow, and composition of the gases, strength and control of electric fields, and methods of forming ions from samples.[3-7] Proliferation in mobility methods during the past decade was accompanied by variations in styles, geometries, and dimensions of mobility analyzers. While reasons for the rise of such diversity are complex, a consequence of this creativity is an expanded range of measurement possibilities that are improved significantly over descriptions in the 2005 second edition of this monograph.[2] An attempt to summarize these advances from this period of high activity, based on principles of mobility by pressure and electric field, is shown in Figure 1.1.

For ion mobility measurement as found in analyzers now widely distributed at military establishments and in airports,[8] an ion swarm is injected into a drift region using an ion shutter, which establishes the time base for the measurement (Figures 1.2a and 1.2b). In the drift region, the ion swarm moves through purified air at ambient pressure in a voltage gradient or electric field (E in V/cm) as shown in Figure 1.2b, with drift velocities v_d determined from the time (ms) needed for a swarm to traverse the distance (d, cm) between the ion shutter and detector (i.e., the drift time t_d):

$$v_d = d/t_d \qquad (1.1)$$

The length of drift regions for modern in-field IMS analyzers is typically 4 to 20 cm. An ion swarm moving 6 cm with a drift time of 15 ms in an electric field of 200 V/cm has a drift velocity of 4 m/s. The normalization of drift velocity to E produces the mobility coefficient K, which is swarm velocity per unit field (cm^2/Vs):

$$K = v_d/E \qquad (1.2)$$

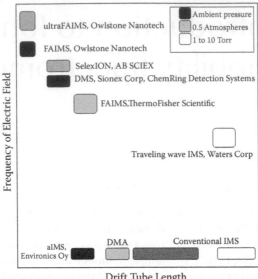

Drift Tube Length

FIGURE 1.1 A proliferation of mobility methods occurred during the past decade, ranging over dimensions, pressures, and arrangements of electric field (constant or time dependent). Pressure of measurements is given in the fill pattern of boxes for individual methods.

and these same ions have a mobility coefficient of 2.4 cm²/Vs at a pressure of 660 torr and temperature of 25°C. Drift velocity is affected by temperature (T) and pressure (P) inside the drift tube, and K is commonly normalized to 273 K and 760 torr, producing a reduced mobility coefficient K_o:

$$K_o = K * \left(273/T\right)\left(P/760\right) \qquad (1.3)$$

which here is 2.0 cm²/Vs. The relationship between drift velocity and electric field (Equation 1.2) is valid for an ion swarm at thermal energies (or thermalized) measured in a constant composition of gas atmosphere, pressure, and temperature. Ions are thermalized when the energy gained by ions between collisions with the supporting gas atmosphere is low compared to thermal energy; at ambient pressure, this is found when electric fields are about 600 V/cm and less. The energy of ions under this condition is $(3/2)kT$, where k is the Boltzmann constant and T is the gas temperature. Notably, ion mobility measurements pertain only to ion swarms and not to individual ions, for which the speeds can be comparatively large; for example, the median speed between collisions for N_2^+ at ambient pressure at 25°C is about 450 m/s.

Describing ion mobilities with values of K_o is a common and useful practice yet subtly complex and can be somewhat confusing. Mobility formulas were derived from studies of ions in nonclustering atmospheres, often at low pressures of 1 to 10 torr. Reduced mobility values may accurately account for ion molecule processes for measurements in low pressures with nonclustering gases such as He; however, normalization for pressure and temperature alone cannot account for changes observed in all

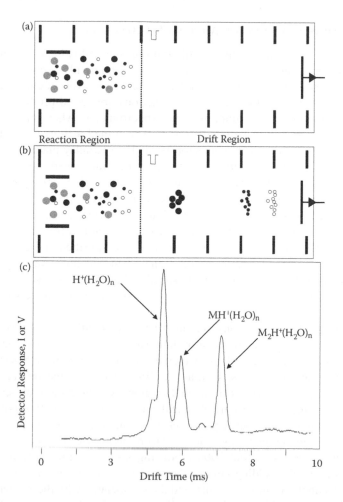

FIGURE 1.2 Schematic of drift tube for ion mobility spectrometry. (a) The drift tube is comprised of a reaction region and drift region, both under an electric field gradient. Two types of neutral sample molecules (small and large grey symbols) are introduced into the ion source region. (b) Sample molecules are ionized (small and large black symbols). Ions are injected using an ion shutter into the drift region and are separated according to differences in ion mobility. (c) A positive polarity mobility spectrum for 2-pentanone in air. The reactant ion peak is apparent at 4.45 ms. The protonated monomer and the proton-bound dimer appear at 5.075 and 6.225 ms, respectively.

mobility measurements, particularly those at ambient pressure in polarizable gases. Such changes arise from the influences of gas temperature, pressure, and gas composition on ion identities, collision cross sections, and hence on K_o values. Mobility measurements are exquisitely sensitive to ion identities, including adducted polar neutrals in gases such as nitrogen or air,[9] and there is today no direct accounting for such changes in K_o. When experimental parameters are well controlled or stable, values for K_o can be understood as, and provide a measure of, mass and shape of ions. The theory underlying the mobility of ions in gases is treated in detail in Chapter 10.

1.1.2 A MEASUREMENT BY ION MOBILITY SPECTROMETRY

A measurement in all experimental configurations of mobility begins when gas ions are formed from components in a sample. This is commonly accomplished using gas phase reactions between analyte molecules and a reservoir of charge called the reactant or reagent ions. Ions from these reactions are called product ions, characterized for mobility in common mobility spectrometers when injected into the drift region (Figure 1.2), where drift time is determined as ions reach the end of the drift tube and collide with a detector, typically a Faraday plate. Ions colliding on the plate are neutralized, drawing current that is transduced into voltage in an amplifier with some gain; for example, currents of 0.1 to 10 nA are typically transformed to volts in the range of 0.1 to 10 V. A distinction of IMS, in contrast to mass spectrometry (MS), is that ions are formed and characterized in a supporting gas atmosphere, also called a buffer gas or the drift gas, that is refreshed continuously. A main practical purpose of this gas is to maintain a purified and constant atmosphere for collision-based movement of the ion swarm. The linear velocity of the gas is typically 0.05 m/s, compared to swarm speeds of 2 to 5 m/s, and variations in flow do not noticeably alter a measurement, except in extremes or when pressures are low (e.g., 1 to 5 torr).

A plot of the detector response versus drift time is called a mobility spectrum (Figure 1.2c), which contains all the information available in a mobility measurement. This includes drift time; peak shape, which is a measure of drift tube performance; and secondary spectral details, such as baseline distortion, that provide information on ion–molecule reactions in the drift region. Although mobility spectra may also contain fragment ions that are chemical class specific,[10–12] methods of chemical ionization are largely low-energy processes, resulting in spectra with simple profiles even for large and complex molecules. Measurements with ion mobility methods that have emerged since about 2000 producing spectra significantly unlike those in Figure 1.2c while yielding comparable information on ion swarms in electric fields. The technology, measurements, and data associated with these relatively new methods, including field-dependent mobility, aspirator-style designs, and traveling wave techniques, are treated in Chapters 6, 10, and 11.

1.1.3 THE FORMATION OF IONS FROM SAMPLE THROUGH GAS PHASE CHEMICAL REACTIONS

Ions are formed in mobility spectrometers mostly through chemical reactions between sample molecules and a reservoir of ions known as reactant ions. These reactions are affected by the properties of particular molecules; thus, the very first step in response adds a layer of selectivity in addition to the mobility characterization. Reactant ions, which are formed through beta emitters in air at ambient pressure and in the absence of a regent gas, are in positive polarity $H^+(H_2O)_n$ and in negative polarity $O_2^-(H_2O)_n$.[13–17] Sample or analyte molecules M are ionized in positive polarity through collisions with hydrated protons, forming a cluster ion (Equation 1.4), which is stabilized through the displacement of adducted water, yielding a product ion, a protonated monomer:

$$M + H^+(H_2O)_n \quad \leftrightarrow \quad MH^+(H_2O)_n^* \quad \leftrightarrow \quad MH^+(H_2O)_{n-x} + xH_2O \qquad (1.4)$$

Sample neutral + Reactant ion   Cluster ion   Product ion

Protonated monomer + Water

Should the vapor concentration of M in the reaction region increase further, a second product ion can be formed as another sample neutral attaches to the protonated monomer, displacing a water molecule and yielding a proton-bound dimer $M_2H^+–(H_2O)_{n-x}$:

$$MH^+(H_2O)_n + M \quad \leftrightarrow \quad M_2H^+(H_2O)_{n-x} + xH_2O \qquad (1.5)$$

Protonated monomer + Sample   Proton bound dimer + water

While formation of proton-bound trimers and tetramers may occur in the vapor-rich region of the ion source, these ions have short lifetimes when extracted into the purified atmosphere of a drift region and are rarely observed in mobility spectra at ambient temperatures or above.

Product ions are formed in negative polarity by association of a molecule and an oxygen anion as in Equation 1.6:

$$M + O_2^-(H_2O)_n \quad \leftrightarrow \quad MO_2^-(H_2O)_n^* \quad \leftrightarrow \quad MO_2^-(H_2O)_{n-x} + xH_2O \qquad (1.6)$$

Sample + Negative   Cluster ion   Product ion + water

reactant ion

As with the analog cluster ion in positive polarity, this cluster ion can be stabilized to a product ion, here $MO_2^-(H_2O)_{n-x}$, by displacement of a water molecule.

These and other ion reactions in gas atmospheres at ambient pressure are understood to occur under thermalized conditions without activation barriers, where the kinetics of ion formation is governed by collision frequencies adjusted for the strength of ion dipole interactions. Based on chemical structure, compounds may exhibit a type of preferred response in a given polarity. A consequence of high collision frequencies and relatively long residence times in reaction regions makes IMS a trace detector (subpicogram limits of detection) with some control of selectivity in ion formation and ion mobility. Quantitative treatment of these reactions is given in Chapters 4 and 11.

Selectivity and versatility of measurements have been enhanced through the use of a range of ion sources operated at ambient pressure and easily combined with drift tubes. Techniques to measure samples as liquids and solids, not only gases, have been incorporated significantly into IMS technology during the past decade and have transformed IMS from a vapor analyzer with niche applications to a measurement technique broadly applicable to semivolatile or nonvolatile substances. Indeed, a reason for the increase in visibility of mobility spectrometry as an analytical method is the combination of electrospray ionization (ESI) with drift tubes and the combination of mobility analyzers with mass spectrometers. This combination of ESI-IMS-MS with drift tubes at ambient pressure and at reduced pressures of

1 to 10 torr is part of an endeavor worldwide to characterize and understand biological systems. Inorganic compounds and metals are also easily accessible now using ESI or laser desorption. Consequently, mobility spectrometry today can be seen as a general concept in analytical measurement sciences.

1.1.4 THE MOBILITY OF IONS IN AN ELECTRIC FIELD THROUGH GASES

The central question in a mobility measurement is the relationship between ion swarm velocity and the chemical identity of the ions in the swarm, associated largely with the reduced mobility coefficient. Early efforts to relate ion structure or identity to mobility coefficients arose mainly from studies of mono- or diatomic ions in pure gases at subambient pressure[18] and led to models for K (as in Equation 1.7)[19]:

$$K = \frac{3e(2\pi)^{1/2}(1+\alpha)}{16N(\mu k T_{eff})^{1/2}\Omega_D(T_{eff})} \tag{1.7}$$

where e is the electron charge; N is the number density of neutral gas molecules at the measurement; α is a correction factor; μ is the reduced mass of ion and gas of the supporting atmosphere; T_{eff} is the effective temperature of the ion determined by thermal energy and the energy acquired in the electric field; and $\Omega_D(T_{eff})$ is the effective collision cross section of the ion at the temperature of the supporting atmosphere.

The mass and shape of an ion strongly affect the mobility coefficient through the cross section for collision as shown in Figure 1.3. Difficulties in Equation 1.7 occur since drift velocities are sensitive to the exact identity of ions in a swarm, which is affected by the composition of the gas atmosphere and by changes in temperature. Thus, the relationship between K and Ω_D is inaccurately described or incomplete for large organic ions where Ω_D is influenced by polar neutrals or moisture in the drift region. There is no comprehensive model that includes ion-neutral associations in polarizable gas atmospheres and the role of temperature in controlling Ω_D and thus K. The distribution of charge within a large organic ion also affects observed mobility values, once again through formation of clusters, and this is not included in current formulas for mobility. Consequently, Equation 1.7 and other formulas for mobility are incomplete descriptions of mobility measurements in air at ambient pressure, and the correlation of ion structure, collision cross section Ω_D, and K over a range of experimental conditions should be considered an underdeveloped topic in IMS.

The previous discussion is intended to introduce briefly elementary aspects of IMS and to emphasize that IMS should be understood best as two sequential processes:

a. formation of ions that are representative of a sample and
b. the determination of these ions for mobilities in an electric field.

Both of these can be discussed and treated separately, although the final analytical result must be seen as a sum of both events. Although the measurement of mobility for an ion swarm at ambient pressure is founded on principles of physics and chemistry,

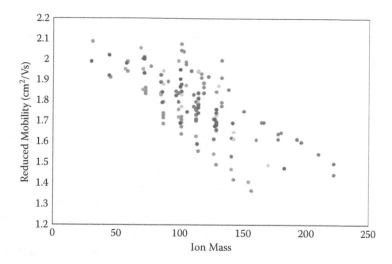

FIGURE 1.3 Plots of reduced mobility versus mass for volatile organic compounds. When all parameters of instrument and chemistry are controlled, the mobility coefficient is governed by size-to-charge and reduced mass of the ion in the supporting atmosphere and is an exquisitely sensitive measure of the structure and behavior of ions in a supporting gas atmosphere. One of the key facets of IMS is the influence of structure shape and size on mobility. Mobility coefficients are influenced by ion mass and a linear relationship exists within a homologous series; however, ions of the same mass but different functional groups, or even different geometrical arrangement of the same functional groups (isomers), often exhibit different K_o values, reflecting the influence of shape and size on mobility.

the details of which are nontrivial, the practice of IMS is comparatively simple, fast, economic, and robust with significant value in chemical analyses.

1.2 METHODS OF ION MOBILITY SPECTROMETRY

Discoveries and developments already under way prior to the 2005 second edition of this book,[2] including ESI-IMS and emerging nontraditional methods of mobility measurement, have greatly altered the practice, commercial development, and applications of IMS. These have led to an era that has seen abundant innovations, applications, and technology. These developments, summarized in the following material, reveal that there is overlap and synergism, even if not recognized initially, and that the sum of all efforts has broadened and expanded IMS as a general concept.

1.2.1 NEW TECHNOLOGIES, CONFIGURATIONS, AND METHODS FOR ION MOBILITY SPECTROMETRY

1.2.1.1 Field Asymmetric IMS, Differential Mobility Spectrometry, Ion Drift Spectrometry

A journal article in 1993 heralded a new method for measurements of ion mobility in which ions do not move under thermalized conditions at ambient pressure.[20] This method is based on ions undergoing changes in mobility coefficients under strong field conditions (at constant N), so that mobility should be understood for most ions as

$$K(E/N) = K_o\left(1 + \alpha(E/N)\right) \tag{1.8}$$

where α is a function describing the dependence of mobility on the ratio of electric field strength to neutral density. In this new method, termed variously ion drift spectrometry, field ion spectrometry, field asymmetric IMS (FAIMS), or differential mobility spectrometry (DMS), ions are carried by gas flow through a gap with conducting surfaces. These drift tubes can be either curved from concentric cylinders[21] or flat from parallel plates.[22] An electric field is applied to this gap, using an asymmetric waveform of 20,000 V/cm or greater and −1,000 V/cm, and ion swarms move with the electric fields according to Equations 1.1 to 1.3 and 1.8. The waveform is designed so that the integrals of these two regions are equal, and ions with mobility coefficients that are independent of E, even at high fields, will pass line of sight through the gap of the analyzer (see Figure 1.4a) and reach the detector. In contrast, ions with a dependence of K_o on E undergo a net displacement toward a surface with repeated exposure of the ion swarm to the periodic changes in direction and strength of the electric field. The magnitude of displacement depends on the differences in mobility at field extremes, and a comparatively low direct current (DC) voltage can be added to the plates to superimpose a DC field that allows control or compensation of ion motion toward a plate. Ions restored to the center of the gap will be passed to the detector, and a sweep of this voltage, often 10 to 40 V or fields of 100 to 500 V/cm, provides a measure of all ions in the analyzer for a given waveform. These methods can be seen as mobility filters rather than spectrometers, and separations are based on differences in mobility, hence the name differential mobility spectrometry, an alternative to FAIMS. Comprehensive treatments of this method can be found in a review[6] and a book.[7]

The complex mixture of ions sometimes produced with electrospray ion sources, particularly with biological or metabolic samples, invites ion filtering before mass analysis, and methods of FAIMS and DMS nicely match the requirements potentially for small and inexpensive ion filters to reduce "chemical noise" in MS determinations.[21,23,24] An ESI-FAIMS instrument was commercialized first in 2001 by Ionanalytics and continued in 2005 by ThermoScientific as an inlet to their single-quadrupole mass spectrometer.[25] In 2010 to 2011, AB SCIEX in Toronto, Ontario, Canada, introduced a small planar DMS attachment to their mass spectrometers under the name SelexION Technology.[26] In both the FAIMS device from ThermoScientific and the DMS SelexION Technology, the intentions for analytical measurements are comparable, although the DMS technology is small (Figure 1.5a). Even smaller is the μFAIMS analyzer of Owlstone Nanotech of Oxford, England, with silicon-etched channels in place of the gaps of other analyzers.[27] Their analyzer is scaled in micrometers rather than millimeters (Figure 1.5c), where ions pass a distance of 300 μm through a gap 30-μm wide experiencing 50-MHz waveforms, where electric fields can reach 80 townsends (at ambient pressure about 100,000 V/cm). Ion residence times caused ions to be thoroughly declustered, exhibiting behavior largely associated with a negative alpha function. Electric fields are so strong that

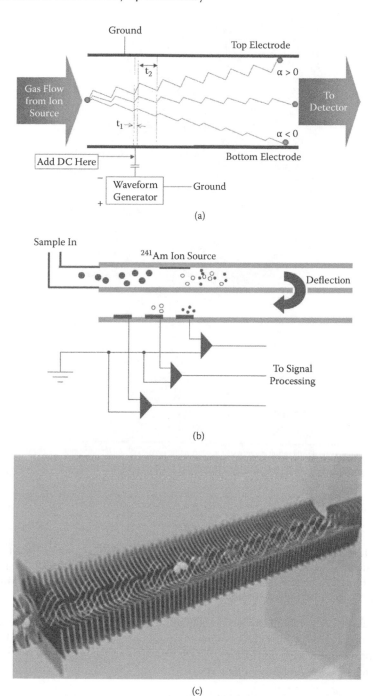

FIGURE 1.4 Schematics of methods for (a) differential mobility (or FAIMS), (b) aspirator, and (c) traveling wave methods of mobility measurements. (From Alistar Wallace, Waters Corp. With permission.)

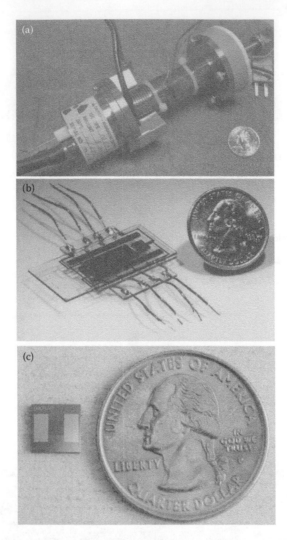

FIGURE 1.5 A trend in the past decade has been miniaturization, with the CAM reduced from handheld to pocket size. Original drift tubes for FAIMS (a) were reduced first to 5×13 mm (b) and then to etched silicon with 10 mm (c). Compliments of Ranaan Miller and Owlstone Nanotech for Figures 1.5b and 1.5c, respectively.

ions are electrically heated to dissociation and fragmentation,[28] providing chemical information in detail richer than commonly possible. An emphasis with μFAIMS, as previously made with FAIMS and DMS, has been to separate ions, particularly isobaric ions, by ion mobility before MS.

1.2.1.2 Aspirator or Flow-Field Filter Designs, Differential Mobility Analyzer

In the method of aspirator IMS (aIMS), ions are introduced into a flow of gas between two parallel plates with a constant electric field (Figure 1.4b).[29,30] The ions move through the laminar gas profile toward a collector plate, which is fitted with several detector (Faraday) plates at locations downflow from the point where ions are introduced in the drift tube. The mobility of ion swarms is measured using the ion impact distance on the detectors measured from the point of entry. This distance depends on drift velocities in the electric field and rate of gas flow. Ions with high mobility reach the detectors, being pushed only a small distance, while ions of low mobility reach the detector farthest from the ion inlet.

A variant of aIMS in which gases flow at high velocity through the analyzer is the differential mobility analyzer (DMA); high gas velocities and strong electric fields are designed to suppress diffusion of ion swarms and to provide improved resolving power.[31,32] Unlike aIMS, the voltage over the mobility region is swept, bringing ions to a single point—the aperture of the mass spectrometer—as a function of electric field strength in constant gas flow. Details of aIMS and DMA methods can be found in Chapter 6.

On the boundary of definitions of mobility spectrometry lie electric aerosol mobility analyzers, which are used to determine sizes and distributions of particulates or aerosols through mobility measures.[33] Significantly, an electric mobility analyzer was transformed for mobility measurements of large biological molecules.[34] Very large biomolecules or intact organisms were introduced into the flow and field, obtaining measurements of the biological particle size.

1.2.1.3 Traveling Wave Methods of IMS

Perhaps one of the most surprising technical and commercial developments in IMS during the past decade with a large impact on the acceptance and use of IMS was the introduction of traveling wave methods in combination with time-of-flight mass spectrometry (TOF-MS) for investigations of biological molecules.[35–38] In traveling wave methods, a set of rings is placed at low pressure, and as sample ions are introduced into the drift tube, the potential is raised on a ring, establishing an electric field and initiating swarm movement by mobility. The potential is then lowered as potential on a neighboring ring is raised; the process of lifting and dropping potential sequentially with some time delay establishes a "wave" as an electric field in the drift tube and moves ions in the direction of the wave, traveling down the tube. Once the first wave is initiated and is moving in the tube, a second and then subsequent waves occur. Ion swarms are propagated through the drift tube with a certain period or velocity associated with characteristics of the wave and mobility of ions. All ions eventually reach the end of the drift tube, and ion separation is based on time or the number of waves experienced by the swarm. Ion injection is through an ion trap before the drift tube to accumulate and enter ions into the traveling wave drift region. TOF-MS provides strong capabilities to the mobility analyzer, which has relatively low resolving power for an analyzer operated at low pressure, in this instance nitrogen. Nonetheless, this instrument has been transformative in providing researchers with a tool to separate large biomolecules by principles of ion mobility.

1.2.2 Tandem Drift Tubes in Mobility Spectrometry

Ion mobility spectrometers are sufficiently simple and inexpensive that several drift tubes can be combined for sequential measurements of ions or for characterization using principles of orthogonality. The first tandem IMS instrument was described in the late 1970s, and this development was accompanied by the construction of three IMS/IMS instruments by PCP, Incorporated, under contract for the U.S. Army.[39,40] A tandem DMS-IMS instrument was shown to exhibit some orthogonal character but was best described as sequential for small organic ions.[41] Although ion peak separation by orthogonal properties was intended with DMS-IMS, the sequential nature of the measurement was useful, providing a relatively constant resolving power over a relatively large range of ion masses. When separation by orthogonal principles in mobility methods is considered critical, the chemistry of ion–molecule interactions may be introduced to tandem instruments, and a DMSxDMS instrument has been described[42] in which ions from a first DMS analyzer are blended with a reagent gas before being characterized in a second DMS. In this, reagents are intended to provide the orthogonal mobility characterization through selectivity of ion cluster formation, charge stripping, or changed alpha functions in the DMS method. The components of DMS and IMS were reversed recently with an IMS/FAIMS/MS experiment; an IMS was used as a filter before the DMS through a two-ion-shutter method.[43]

Several teams have demonstrated tandem mobility with drift tubes of conventional linear field designs. A "kinetic" IMS has been used to study the rate of fragmentation or dissociation of gas ions at ambient pressure with mobility drift tubes.[44] In this instrument, product ions formed in the reaction region are separated by differences in mobility in a first drift region, arriving not at a detector but at a second set of ion shutters. Synchronization between the first and second ion shutters allows only a single peak or ion swarm to be isolated by mobility and then injected into a second drift region. In this second region, ion decomposition is followed using perturbations of baselines and distortion of peak shapes. Originally demonstrated using an ion mobility–mass spectrometer (IM–MS),[44] an IMSxIMS with a Faraday plate detector was used in ion decomposition studies.[45] In another approach to IMSxIMS, a tandem drift tube was planned for separation of ion swarms in one dimension, with ions of a certain mobility extracted at right angles into a second mobility region.[46] In principle, ion swarms at all distances throughout the first drift tube could be transferred into a second region of mobility, offering the possibility, not yet demonstrated, of full two-dimensional (2D) mobility measurements on a near-continuous basis. A tandem IMS instrument with linear or traditional drift tube structures has also been described for experiments with IMSxIMS-MS; the drift tubes are operated at low pressures in nonclustering gases.[47] Studies with biomolecules were made using ESI sources which emerged significantly in combination with mobility spectrometers during the 1990s.

1.2.3 MOBILITY MEASUREMENTS IN BIOMOLECULAR STUDIES, PHARMACEUTICAL RESEARCH, AND CLINICAL DIAGNOSIS

A surge of interest has occurred in the interest in and use of mobility methods to provide additional ion filtering between ESI sources and mass spectrometers and to explore gas phase ions of biomolecules for shape, size, and transformations. This was simulated by availability of commercial instruments, including the ThermoFisher FAIMS/MS, AB SCIEX SelexION, the Owlstone μFAIMS/MS, and

Waters Corporation Synapt. The samples are solutions of biological material, and the ion sources are uniformly EIS, nanospray ionization, or matrix-assisted laser desorption and ionization as described in Chapters 4, 9, 15, and 18. Studies with the Synapt traveling wave instrument have revealed details of biomolecular ions in the gas phase that are not available by MS alone or by other methods.[35–38] The full meaning of such studies and relevance for in vivo biomolecular activity is currently under discussion and debate;[48] nonetheless, IM–MS for explorations of biomolecules certainly has affected the visibility of mobility as a measurement method and the level of technology that has been advanced through pharmaceutical and medical concerns.

Electrospray ion sources initiated the current revolution in biological and medical uses of MS, yet the chemistry of these sources is complex, and the number of useful ions derived from a sample can be few amid a complex mixture of interfering ions. When placed between the ESI source and the mass spectrometer, FAIMS and DMS instruments can isolate ions of importance, significantly simplifying mass spectra and improving signal-to-noise ratios in quantitative measurements.[49–51] This was initiated with FAIMS instruments[23–25] and is now complimented with ultraFAIMS[27] and DMS.[24] A direct and near-immediate use of IMS in medical research has arisen with the development of clinical applications of IMS, as described in Section 1.3.

1.2.4 MINIATURE DRIFT TUBES AND PORTABLE ION MOBILITY ANALYZERS

1.2.4.1 Miniature and Pocket-Size Drift Tubes

Nearly all advances in miniaturization of ion mobility instruments were made by commercial teams in support of military or security needs for personal monitoring of those fighting wars. The lightweight chemical detector (LCD) has a pulsed corona discharge that was substituted for the previously common 10-mCi radioactive ^{63}Ni foil, as found in handheld instruments such as the Chemical Agent Monitor (CAM) or Rapid Alarm Identification Device (RAID). The LCD weighs 1.5 pounds, including batteries; it is $10.54 \times 17.93 \times 4.72$ cm and has an operating life of 75 h continuously with a single set of ordinary AA batteries. Another traditional design with small dimensions was introduced by Sandia National Laboratories as the MicroHound; it was uniquely fabricated from flexible ceramic rolled into a cylindrical shape. A DMS instrument appeared as a field-ready, rugged, handheld analyzer, JUNO, from General Dynamics Armament and Technical Products and is now the concern of Chemring Detection Systems, Incorporated. A handheld aspiration-style mobility drift tube, perhaps the simplest of all small mobility analyzers, was introduced as the Chem Pro 100i; it weighs 800 g and is used by militaries in Finland and the Netherlands and first responders in Norway.

A miniature configuration of a gas chromatograph (GC) with DMS, complete with vapor-sampling inlet with sorbent trap enrichment, was introduced as the Micro-analyzer by Sionex Corporation. The very small FAIMS drift tube of Owlstone Technologies was commercialized as the Lonestar portable gas analyzer with production of a variant, the Nexsense C Chemical Detection and Identification System for chemical agent detection as a joint development by SELEX Galileo and Owlstone Inc.

1.2.4.2 Handheld or Portable Explosive Analyzers

Although drift tubes in benchtop analyzers for trace levels of high explosives are relatively small, they could not be called miniature and must be heated to temperatures of 150°C or higher. Most handheld analyzers have been equipped with drift tubes that are unheated or warmed slightly above ambient temperatures to minimize demand for power and batteries. Thus, these room temperature analyzers cannot be extended directly to measurements of high explosives. In the past decade, manufacturers of trace explosive detectors based on ion mobility have developed handheld analyzers with benchtop capabilities, including thermal desorption of sample wipes. These instruments include heated surfaces and drift tubes within highly portable packages. Examples include the Sabre family of analyzers from Smiths Detection, the Mobile Trace series from Morpho Detection, the MicroHound from Sandia National Laboratories, and the Quantum Sniffer QS-H150 from Implant Sciences Corporation. This last instrument is unique, with a noncontact sampler and photoionization source instead of a beta emitter or corona discharge source. Three other trace explosive detectors were introduced in the 2000s and should be mentioned here: the EGIS Defender, a flash GC DMS from Thermo Electron Corporation; the Quantum Sniffer™ QS-B220 from Implant Sciences Corporation; and the benchtop-size DE-tector from Bruker Daltonics.

1.3 EMERGING PATTERNS IN THE DEVELOPMENT OF ION MOBILITY METHODS

The vitality of developments of mobility spectrometry, seen in the preceding section and evident in the Contents for this book, demonstrates the highly dynamic and creative world of ion mobility measurements today. The level of technology and understanding of ion mobility is improved in comparison to those extant for the second edition of *Ion Mobility Spectrometry*. There are nonetheless exciting developments emerging and recognizable in areas of applications, technology, and science. Communications of findings and highlights of these are described next.

1.3.1 APPLICATIONS

Promising and satisfying uses of ion mobility instruments are emerging in prototype configurations for routine clinical and diagnostic measurements. While some applications involved rather complex instrumentation for proteomics, others involved stand-alone handheld or desktop designs. Diagnosis of common human vaginal infections, based on the measurement of biogenic amines, has been performed with a stand-alone IMS,[52] and some devices are being deployed by gynecologists on an experimental basis and have received technical approval within the European community. The scope has been extended to detection of vaginal infections that affect reproduction in farm animals, and the initial results are promising.[53] Preliminary tests with domestic animals indicated that detection of the presence of biomarkers for cancer using a rapid testing procedure is feasible.

Another exciting clinic-based use of IMS, as a combined GC-IMS instrument, is the determination of metabolites in human breath.[54,55] Emphasis was given to sampling VOCs (volatile organic compounds) from breath to aid choices of therapy for respiratory infections, to monitor the course of treatment, and finally to probe for tumors. These efforts have received both acceptance and honors in the medical community in Europe.[56]

1.3.2 PHARMACEUTICAL RESEARCH

Distinctive applications of mobility methods during the past decade have been in the pharmaceutical industry; FAIMS has been employed for significant reduction of chemical noise in liquid chromatographic (LC)-ESI-MS measurements of drugs and drug metabolites.[23–26] Such applications, in which a metabolite was isolated from a complex mass spectrum using mobility selection, was encouraged with a commercial attachment to mass spectrometers from Waters Corporation. This aspect of mobility measurements should increase with the introduction of the SelexION technology for sample characterization by LC-ESI-DMS-MS instruments.

A use of IMS within the pharmaceutical industry, for which ultratrace detection limits were combined with known analytes in a minimal matrix, was the verification of cleanliness in production lines scheduled for change to the manufacture of new formulations.[57] The rapid analysis of samples, made on site, should have led to widespread adoption of this method.

1.3.3 TECHNOLOGY AND SCIENCE OF MOBILITY

Although multiple methods and dozens of configurations for measuring mobility exist today (Figure 1.1), the technology of drift tubes and the principles underlying the meaning of mobility are active topics in research laboratories worldwide. In the past decades, the search for alternatives to radioactive sources has been satisfied by pulsed corona discharges, ESI, photodischarge, and more than 20 other methods. Efforts to extend the resolving power of drift tubes have received interest and should remain as a central element in technology advances in mobility methods.[58,59] Nonetheless, the relatively low resolving power of mobility spectrometers can be seen alternately as an advantage where spectral profiles are relatively simple in those applications for which ion populations from a sample are chemically simple and charge is concentrated in one or a few ion species. In other instances when detailed structures of substances are sought or samples are complex and a comprehensive measurement is desired, resolving power of drift tubes may be seen as a limitation. Another emerging feature of drift tubes, with near-term and long-range impact, is replacement for the Faraday plate. This detector is based on microchannel technology and is said to multiply ion current by 10^4 at atmospheric pressure.[60]

During the past decade, a milestone occurred in technology and applications of mobility methods with the decommissioning of the Volatile Organic Analyzer (VOA). This instrument was the first GC/IMS instrument with an integrated analytical design that includes an ambient air sampler, enrichment or sorbent traps, high-performance capillary columns, and relatively small high-temperature drift tubes. This instrument demonstrated that mobility spectrometers could be engineered and built as relatively complex instrument packages and operated remotely in difficult environments.

At the level of science, the concept of mobility in fundamentally disclosing properties of an ion and the relationship between the collision cross section and the ion shape/structure is under discussion. While this may have little effect on the applications of IMS as given in Section IV, there is value in clarifying with precision these meanings. At the moment, mobility methods particularly at ambient pressure and the meaning of a mobility coefficient are still disproportionally influenced by the experimental parameters or perhaps the control or knowledge of these parameters. How the community of researchers will develop this topic is unclear and is yet another opportunity to bring mobility methods to an improved level of practice and value.

1.3.4 DISSEMINATION OF DISCOVERIES

Evidence that mobility spectrometry is no longer an emerging technology and that a broad acceptance and appreciation for mobility determinations have developed can be seen in the appearance of a scholarly nonprofit organization, the International Society for Ion Mobility Spectrometry (ISIMS) and a related publication from Springer-Verlag, *The International Journal for Ion Mobility Spectrometry*. The society meets annually at sites alternately in Europe and North America for a week of lectures, poster sessions, and technology exhibition. The journal includes all aspects of mobility, from chemistry to engineering to applications in all methods and measurement regimes. Perhaps a broader measure of the vitality of IMS today is the sheer number of mobility-related publications, as seen in Figure 1.6. In the 1980s and following, articles increased in number year by year with refinements in instruments and exploration by academic teams. The biological revolution occurred in the 1990s

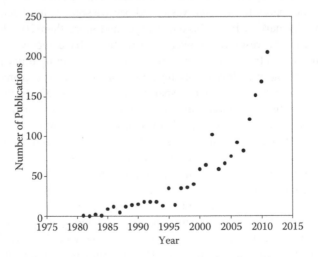

FIGURE 1.6   The rate of publishing in ion mobility spectrometry has increased reflecting the interest, discovery, and applications in mobility based methods.

and availability of instruments commercially began in the late 1990s and increased as anticipated in the second edition of this title.

1.4 SUMMARY COMMENTS

Exploration of response in IMS, after the formative years of discovery (Chapter 2), revealed that ionization chemistry for explosives in negative polarity and for nerve agents in positive polarity was favorable for analyzers in a range of venues. This initiated rapid development and acceptance of IMS analyzers as ultratrace detectors in military and security applications; this continues today with smaller and more versatile instruments.

The biological revolution and capabilities in forming gas phase ions directly from liquid and solid samples has expanded the usefulness of IMS. A mobility measurement can be obtained whenever a gas ion can be made from a substance. Ion mobility instruments have also enhanced mass spectrometers with improved specificity, reduced noise of measurement, and details on ion structures that complement mass measurements. This is now occurring with ion mobility–mass spectrometry (IM–MS) supplementing an already-active world of stand-alone mobility measurements now extending into the clinical, environmental, and industrial venues as described in Section IV of this volume. An encouraging development in the past decade concerning a wider recognition of IMS was the inclusion and mention of IMS in an analytical chemistry undergraduate textbook.[61] Finally, vibrant scientific activity seen in the number and quality of publications is now accompanied by energetic and dynamic commercial ventures as described in Chapter 2.

REFERENCES

1. Eiceman, G.A.; Karpas, Z., *Ion Mobility Spectrometry*, CRC Press, Boca Raton, FL, 1994.
2. Eiceman, G.A.; Karpas, Z., *Ion Mobility Spectrometry*, 2nd edition, CRC Press, Boca Raton, FL, 2005.
3. Borsdorf, H.; Eiceman, G.A., Ion mobility spectrometry: principles and applications, *Appl. Spectrosc. Rev.* 2006, 41, 323–375.
4. Borsdorf, H.; Mayer,T.; Zarejousheghani, M.; Eiceman, G.A., Recent developments in ion mobility spectrometry, *Appl. Spectrosc. Rev.* 2011, 46(6), 472–521.
5. Kanu, A.B.; Dwivedi, P.; Tam, M.; Matz, L.; Hill, H.H., Jr., Ion mobility–mass spectrometry, *J. Mass Spectrom.* 2008, 43, 1–22.
6. Kolakowski, B.M.; Mester Z., Review of applications of high-field asymmetric waveform ion mobility spectrometry (FAIMS) and differential mobility spectrometry (DMS), *Analyst* 2007, 132(9), 842–864.
7. Shvartsburg, A.A., *Differential Ion Mobility Spectrometry: Nonlinear Ion Transport and Fundamentals of FAIMS*, CRC Press, Taylor and Francis Group, Boca Raton, FL, 2009.
8. Mäkinen, M.A.; Anttalainen, O.A.; Sillanpää, M.E.T., Ion mobility spectrometry and its applications in detection of chemical warfare agents, *Anal. Chem.* 2010, 82, 9594–9600.
9. Berant, Z.; Karpas, Z.; Shahal, O., The effects of temperature and clustering on mobility of ions in CO_2, *J. Phys. Chem.* 1989, 93, 7529–7532.
10. Bell, S.E.; Nazarov, E.G.; Wang, Y.F.; Eiceman, G.A., Classification of ion mobility spectra by chemical moiety using neural networks with whole spectra at various concentrations, *Anal. Chim. Acta* 1999, 394, 121–133.

11. Bell, S.E.; Nazarov, E.G.; Wang, Y.F.; Rodriguez, J.E.; Eiceman, G.A., Neural network recognition of chemical class information in mobility spectra obtained at high temperatures, *Anal. Chem.* 2000, 72, 1192–1198.

12. Eiceman, G.A.; Nazarov, E.G.; Rodriguez, J.E., Chemical class information in ion mobility spectra at low and elevated temperatures, *Anal. Chim. Acta*, 2001, 433, 53–70.

13. Brosi, A.R.; Borkowski, C.J.; Conn, E.E.; Griess, J.C., Jr. Characteristics of Ni[59] and Ni[63], *Phys. Rev.* 1951, 81, 391–395.

14. Siu, K.W.M.; Aue, W.A., [63]Ni β range and backscattering in confined geometries, *Can. J. Chem.* 1987, 65(5), 1012–1024.

15. Kebarle, P.; Searles, S.K.; Zolla, A.; Scarborough, J.; Arshadi, M., The solvation of the hydrogen ion by water molecules in the gas phase. Heat and entropies of solvation of individual reaction: $H^+(H_2O)_{n-1} + H_2O = H^+(H_2O)_n$, *J. Am. Chem. Soc.* 1967, 89(25), 6393–6399.

16. Arshadi, M.; Kebarle, P., Hydration of OH^- and O^{2-} in the gas phase. Comparative solvation of OH^- by water and the hydrogen halides effects of acidity, *J. Phys. Chem.* 1970, 74(7), 1483–1485.

17. Kim, S.H.; Betty K.R.; Karasek, F.W., Mobility behavior and composition of hydrated positive reactant ions in plasma chromatography with nitrogen carrier gas, *Anal. Chem.* 1978, 50(14), 2006–2016.

18. McDaniel, E.W.; Mason, E.A., *The Mobility and Diffusion of Ions in Gases*, Wiley-Interscience, New York, 1973.

19. Revercomb, H.E.; Mason, E.A., Theory of plasma chromatography/gaseous electrophoresis: a review, *Anal. Chem.* 1975, 47, 970–983.

20. Buryakov, I.A.; Krylov, E.V.; Nazarov, E.G.; Rasulev, U.K., A new method of separation of multi-atomic ions by mobility at atmospheric pressure using a high-frequency amplitude-asymmetric strong electric field, *Int. J. Mass Spec. Ion Proc.* 1993, 128, 143–148.

21. Guevremont, R.; Purves, R.W., High field asymmetric waveform ion mobility spectrometry-mass spectrometry: an investigation of leucine enkephalin ions produced by electrospray ionization, *J. Am. Soc. Mass Spectrom.* 1999, 10, 492–501.

22. Miller, R.A.; Eiceman, G.A.; Nazarov, E.G., A micro-machined high-field asymmetric waveform-ion mobility spectrometer (FA-IMS), *Sensor Actuators B Chem.* 2000, 67, 300–306.

23. Kapron, J.T.; Jemal, M.; Duncan, G.; Kolakowski, B.; Purves, R., Removal of metabolite interference during liquid chromatography/tandem mass spectrometry using high-field asymmetric waveform ion mobility spectrometry, *Rapid Commun. Mass Spectrom.* 2005, 19(14), 1979–1983.

24. Levin, D.S.; Miller, R.A.; Nazarov, E.G., Vouros, P., Rapid separation and quantitative analysis of peptides using a new nanoelectrospray-differential mobility spectrometer–mass spectrometer system, *Anal. Chem.* 2006, 78(15), 5443–5452.

25. Thermo Electron Corporation, Press Release: Thermo Electron Acquires Provider of Novel Mass Spectrometry Ion Filtering Device, Thermo Electron Corporation, Waltham, MA, August 11, 2005.

26. AB SCIEX SelexION™ Technology. A new dimension in selectivity, brochure no. 2530311, 2011.

27. Brown, L.J.; Toutoungi, D.E.; Devenport, N.A.; Reynolds, J.C.; Kaur-Atwal, G.; Boyle, P.; Creaser CS., Miniaturized ultra high field asymmetric waveform ion mobility spectrometry combined with mass spectrometry for peptide analysis, *Anal. Chem.* 2010, 82(23), 9827–9834.

28. Wilks, A., A Consideration of Ion Chemistry Encountered on the Microsecond Separation Timescales of Ultra-High Field Ion Mobility Spectrometry, 20th annual conference, International Society for Ion Mobility Spectrometry, Edinburgh, Scotland, July 23–28, 2011.

29. Tammet, H., The Aspiration Method for the Determination of Atmospheric-Ion Spectra, Israel Program for Scientific Translations, Jerusalem, 1970; also see http://ael.physic.ut.ee/tammet/am/.

30. Tuovinen, K.; Paakkanen, H.; Hänninen, O., Determination of soman and VX degradation products by an aspiration ion mobility spectrometry, *Anal. Chim. Acta* 2001, 440, 151–159.

31. de la Mora, J.F.; de Juan, L.; Eichler, T.; Rosell, J., Differential mobility analysis of molecular ions and nanometer particles TrAC, *Trends Anal. Chem.* 1998, 17(6), 328–339.

32. Hogan, C.J., Jr.; de la Mora, J.F., Ion mobility measurements of non-denatured 12–150 kda proteins and protein multimers by tandem differential mobility analysis-mass spectrometry (DMA-MS), *J. Am. Soc. Mass Spectrom.* 2011, 22,158–172.

33. Bacher, G.; Szymanski, W.; Kaufman, S.; Zollner, P.; Blass, D.; Allmaier, G., Charge-reduced nanoelectrospray ionization combined with differential mobility analysis of peptides, proteins, glycoproteins, noncovalent protein complexes and viruses. *J. Mass Spectrom.* 2001, 36(9), 1038–1052.

34. Thomas, J.J.; Bothner, B.; Traina, J.; Benner, W.H.; Siuzdak, G., Electrospray ion mobility spectrometry of intact viruses, *Spectroscopy* 2004, 18, 31–36.

35. Pringle, S.D.; Giles, K.; Wildgoose, J.L.; Williams, J.P.; Slade, S.E.; Thalassinos, K.; Bateman, R.H.; Bowers, M.T.; Scrivens, J.H., An investigation of the mobility separation of some peptide and protein ions using a new hybrid quadrupole/travelling wave IMS/oa-TOF instrument, *Int. J. Mass Spectrom.* 2007, 261, 1–12.

36. Williams, J.P.; Bugarcic, T.; Habtemariam, A.; Giles, K.; Campuzano, I.; Rodger, P.M.; Sadler, P.J., Isomer separation and gas-phase configurations of organoruthenium anticancer complexes: ion mobility mass spectrometry and modeling, *J. Am. Soc. Mass Spectrom.* 2009, 20, 1119–1122.

37. Scarff, C.A.; Patel, V.J.; Thalassinos, K.; Scrivens, J.H., Probing hemoglobin structure by means of traveling-wave ion mobility mass spectrometry, *J. Am. Soc. Mass Spectrom.* 2009, 20, 625–631.

38. Shvartsburg, A.A.; Smith, R.D., Fundamentals of traveling wave ion mobility spectrometry, *Anal. Chem.* 2008, 80, 9689–9699.

39. Stimac, R.M.; Wernlund, R.F.; Cohen, M.J.; Lubman, D.M.; Harden, C.S., Initial studies on the operation and performance of the tandem ion mobility spectrometer, presented at the 1985 Pittsburgh Conference and Exposition on Analytical Chemistry and Applied Spectroscopy, Pittcon 1985, New Orleans, LA, March 1985.

40. Stimac, R.M.; Cohen, M.J.; Wernlund, R.F., Tandem Ion Mobility Spectrometer for Chemical Agent Detection, Monitoring and Alarm, Contractor Report on CRDEC contract DAAK11-84-C-0017, PCP, Inc., West Palm Beach, FL, May 1985, AD-B093495.

41. Eiceman, G.A.; Schmidt, H.; Rodriguez, J.E.; White, C.R.; Nazarov, E.G.; Krylov, E.V.; Miller, R.A.; Bowers, M.; Burchfield, D.; Niu, W.; Smith, E.; Leigh, N. Characterization of Positive and Negative Ions Simultaneously through Measures of K and ΔK by Tandem DMS-IMS, 14th International Symposium on Ion Mobility Spectrometry, Château de Maffliers, France, July 26, 2005.

42. Eiceman, G.A. Ion Preparation before Differential Mobility Spectrometry including DMS/DMS Analyzers, PittCon 2010, Orlando, FL, February 2010.

43. Pollard, M.J.; Hilton, C.K.; Li, H.; Kaplan, K.; Yost, R.A.; Hill, H.H., Ion mobility spectrometer-field asymmetric ion mobility spectrometer-mass spectrometry, *Int. J. Ion Mobil. Spectrom.* 2011, 14(1), 15–22.

44. Ewing, R.G.; Eiceman, G.A.; Harden, C.S.; Stone, J.A., The kinetics of the decompositions of the proton bound dimers of 1,4-dimethylpyridine and dimethyl methylphosphonate from atmospheric pressure ion mobility spectra, *Int. J. Mass Spectrom.* 2006, 255–256, 76–85.

45. An, X.; Stone, J.A.; Eiceman, G.A., Gas phase fragmentation of protonated esters in air at ambient pressure through ion heating by electric field in differential mobility spectrometry and by thermal bath in ion mobility spectrometry, *Int. J. Mass Spectrom.* 2011, 303(2–3), 181–190.

46. Wu, C., Multidimensional Ion Mobility Spectrometry Apparatus and Methods, Patent number 7576321; filing date December 29, 2006; issue date August 18, 2009; application number 11/618,430.

47. Koeniger, S.L.; Merenbloom, S.I.; Valentine, S.J.; Jarrold, M.F.; Udseth, H.R.; Smith, R.D.; Clemmer, D.E., An IMS–IMS analogue of MS–MS, *Anal. Chem.* 2006, 78(12), 4161–4174.

48. Ruotolo, B.T.; Robinson, C.V., Aspects of native proteins are retained in vacuum, *Current Opin. Chem. Biol.* 2006, 10, 402–408.

49. Klaassen, T.; Szwandt, S.; Kapron, J.T., Validated quantitation method for a peptide in rat serum using liquid chromatography/high-field asymmetric waveform ion mobility spectrometry, *Rapid Commun. Mass Spectrom.* 2009, 23, 2301–2306.

50. Xia, Y.Q.; Wu, S.T.; Jemal, M., LC-FAIMS-MS/MS for quantification of a peptide in plasma and evaluation of FAIMS global selectivity from plasma components, *Anal. Chem.* 2008, 80, 7137–7143.

51. Guddat, S.; Thevis, M.; Kapron, J.; Thomas, A.; Schänzer, W., Drug testing and analysis, *Adv. Sports Drug Testing*, 2009, 1(11–12), 545–553.

52. Chaim, W.; Karpas, Z.; Lorber, A., New technology for diagnosis of bacterial vaginosis, *Eur. J. Obstet. Gynecol. Reprod. Biol.* 2003, 111, 83–87.

53. Karpas, Z.; Marcus, S.; Golan, M., Method for the Diagnosis of Pathological Conditions in Animals, filing date June 18, 2009; application number 12/456,591; publication number U.S. 2009/0325191 A1.

54. Westhoff, M.; Litterst, P.; Freitag, L.; Urfer, W.; Bader, S.; Baumbach, J.I., Ion mobility spectrometry for the detection of volatile organic compounds in exhaled breath of patients with lung cancer: results of a pilot study, *Thorax* 2009, 64, 744–748.

55. Baumbach, J.I., Ion mobility spectrometry coupled with multi-capillary columns for metabolic profiling of human breath, *J. Breath Res.* 2009, 3, 034001.

56. Science Prize of the German Association of Pneumologists, 2006. Recipient: Jörg Ingo Baumbach.

57. Strege, M.A.; Kozerski, J.; Juarbe, N.; Mahoney, P., At-line quantitative ion mobility spectrometry for direct analysis of swabs for pharmaceutical manufacturing equipment cleaning verification, *Anal. Chem.* 2008, 80, 3040–3044.

58. Siems, W.F.; Wu, C.; Tarver, E.E.; Hill, H.H., Jr.; Larsen, P.R.; McMinn, D.G., Measuring the resolving power of ion mobility spectrometers, *Anal. Chem.* 1994, 66(23), 4195–4201.

59. Tolmachev, A.V.; Clowers, B.H.; Belov, M.E.; Smith, R.D., Coulombic effects in ion mobility spectrometry, *Anal. Chem.*, 2009, 81(12), 4778–4787.

60. Denson, S.; Denton, B.; Sperline, R.; Rodacy, P.; Gresham, C., Ion mobility spectrometry utilizing micro-Faraday finger array detector technology, *Int. J. Ion Mobil. Spectrom.* 2002, 5-3, 100–103.

61. Harris, D.C., *Quantitative Analysis*, 8th edition, Freeman, New York, 2010, pp. 518–519.

2 History of Ion Mobility Spectrometry

2.1 INTRODUCTION

Ion mobility spectrometry (IMS) arose from discoveries of the formation and behavior of ions in air and other gases at ambient pressure and from an interest in exploring the chemistry and properties of electrical discharges in gases. These were strongly linked to studies on lightning that began in the late 1700s[1] and to interests in the electricity throughout the 1800s.[2] Certain aspects of this history can still be seen today in IMS technology with corona discharges as ion sources. Similarly, ion–molecule reactions that occur in the troposphere can be associated with the chemical reactions that occur in a mobility spectrometer at ambient pressure. IMS has passed through decades of study in government, industry, and academic laboratories and applications in military establishments and commercial aviation security. Today, ion mobility spectrometers remain a preferred solution in these same venues, offering high speed, durability, and reliability. The attractive features of IMS methods for biomolecule measurements, providing details on molecular size and shape in addition to mass, have motivated the present period of vigorous innovation, expanded commercialization, and growing breadth of applications. Four major stages of development for IMS can be recognized, and each is described separately here.

2.2 THE FORMATIVE YEARS OF DISCOVERY (1895 TO 1960)

Studies to explore and understand the origins, behavior, and significance of ions in gases were initiated with sophisticated laboratory experiments beginning in the 1890s and reached a fairly mature level within a few decades (Table 2.1). Indeed, the principles and practices for nearly all the ion sources used in IMS today were described by 1900 and are neatly summarized in J. J. Thompson's 1903 monograph, *Conduction of Electricity through Gases.*[10] Under his direction, pioneering and sustained discoveries on ions, electrons, and gases were made in the Cavendish Laboratory in Cambridge, England. The mobility of ions in electric fields was recognized early as a measurable characteristic of gaseous ions,[5,8] and by the first decade of the 1900s, Langevin's sophisticated treatment of ion-neutral interactions contributed essential understandings in theory, instrumentation, and experimental findings for ions in gases at ambient or elevated pressures.[12,13] In the next two decades, technology and understandings of ion mobility were refined, and a substantial body of knowledge existed by 1938 according to Tyndall's monograph, *Mobility of Positive Ions in Gases.*[11] Little of the technology of this early era continued into the next

TABLE 2.1
Discovery of the Mobility of Ions in Gases to the Development of a Modern Analytical Method: Years 1895 to 1960

	Ions and Ionization of Gases	Reference
1895	Discovery of x-rays	3
1897	Studies of ionization of ambient air using x-rays	4
1898	Ionization of a radioactive ion source and mobility characterization of ions	5
	Gas ions formed from ultraviolet radiation	6
	Conductivity in flames from gaseous ions	7
1899	Mobility of 1.8-cm^2/Vs negative ions of corona discharge in dried air	8
	Ions produced from uranium radiation	9
1928	Monograph reviewing gaseous electricity with sections on the chemistry and mobility of ions in gases	10
1938	Monograph showing ions affected by temperature, pressure, moisture, and gas purity. Mobility was shown early in the study of ions in gases to be independent of the ratio of electric field to pressure becoming dependent on E/P above certain ion-characteristic values	11
	Technology and Principles of Ion Mobility	
1903	Langevin recognized attractive forces of charge on neutral molecule in effective	12
1905	collision cross sections and the influence on mobility from ion-molecule interactions	13
1908	Mobility determined as 1.37 and 1.80 cm^2/Vs formed in dry air	14
1912	Ion measured in atmosphere	15
1911	Effect of moisture on drift velocities vd with ions from x-rays	16
1913	Measurement of ion swarm velocities in dried gases	17
1929	Ion shutters with parallel wires for ion injection, first published mobility spectrum with drift time axis	18
	Kinetic study for rate of formation of negative ions by electron attachment	19
1933	Book summarizing Cavendish laboratory findings	20
1936	Bradbury two-shutter design with an early boxcar integrator	21
1940s	Autolycus, a diesel fume detection system	22, 23
1942	Study of mechanism in spark discharges in gases	24
1949	Lovelock's discovery that airborne vapors affect simple ionization detectors, establishing a link to ambient air monitoring; origins of electron capture detector	25, 26

era of IMS described in the next section. A notable exception was ion shutters with coplanar sets of parallel wire, known today as Bradbury–Nielson shutters[21]; these were described first by van de Graaff, who also produced the first mobility spectrum (Figure 2.1) with a drift time axis as presented today.[18]

A renewed and deepened interest in the chemistry of gas ions developed in the 1950s and 1960s largely from interests in ionization of air by radiation and in chemical reactions in the upper atmosphere portending exploration of space. Refinements and extensions of Langevin's descriptions on ion-molecule interactions were made during the 1950s and were undertaken in physics laboratories with ions in rare gases

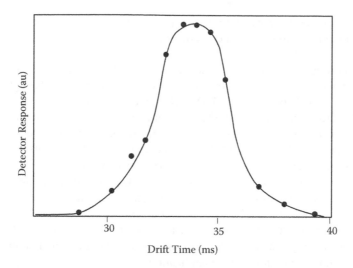

FIGURE 2.1 Adaptation of a 1929 mobility spectrum, the first with modern axes, by van de Graaff. He said: "It is the purpose of the present note to describe an improvement in the previous method giving a marked increased in both resolving power and absolute accuracy. This is obtained by the introduction of a new type of grid for producing the periodic 'shutter effect' of Fizeau." "Grids for these experiments may be conveniently constructed by first gridding a series of parallel slots in a thin glass plate, then completely silvering the surface of the glass and finally scraping off the silver where insulation is desired." (Adapted from van de Graaff, Mobility of ions in gases, *Nature*, 1929.)

at low pressures to avoid the complications of ion clusters. Studies with ions in air as a practical matter for chemical measurements can be attributed to Lovelock's studies with ionization detectors and response to airborne vapors.[25,26] His findings established a link between the presence of manmade substances in air and the response in ionizing detectors. This detector, which was termed the electron capture detector and popularized with gas, chromatographs originated with his efforts to measure breezes inside homes using a self-built vapor anemometer that also showed the direction of the breeze. While there is no record that his work influenced thinking in the conceptual development of IMS for chemical measurements, the similarity between ion chemistry inside an electron capture detector and that in an ion mobility spectrometer was recognized by the early 1970s.

2.3 ION MOBILITY SPECTROMETRY FOR CHEMICAL MEASUREMENTS (1960 TO 1990)

Efforts to use mobility coefficients of ions for *chemical measurements* was accompanied by a change in drift tube operation from low pressure to ambient pressure in the mid-1960s, a type of return to the early 1900s. This transition was followed rapidly and quietly by the transformation of large laboratory instruments into handheld military-grade analyzers by 1980. Continuously operating on-site explosive detectors were released by 1990, and ion mobility spectrometers today are central to commercial aviation security and military preparedness.[27-30] At least four developments of this period can be identified as distinctive and include

a. The introduction of IMS as a modern analytical method with commercial instrumentation. This occurred at Franklin GNO in 1970 through some transfer of technology from McDaniels at Georgia Tech.
b. The demonstration that mobility spectrometers exhibit response to a broad range of substances at ultratrace levels. This work came from one academic research team through a sustained publication record from 1970 to 1980.
c. The exploration of the reaction chemistry between ions and gases in air or other gases using mass spectrometry (MS) to mass identify ions. While some of these studies were made using ion sources at ambient pressure, others were made using high-pressure ion sources.
d. The advancement and acceptance of IMS as a rugged and reliable technology by military centers and security agencies pushing the method into life-threatening or mission-critical applications.

The chronology of these four achievements is interwoven and in some instances synergistic.

The development of a detector for chemical warfare agents using gas ions began in the mid-1960s through government contracts to Franklin GNO (Table 2.2) and with studies by C. Steve Harden at Edgewood Arsenal. Some details of the developments can be gleaned from titles and abstracts of contracts to PCP, Inc. in which drift tube construction and handling trace levels of substances were emphasized. The technology and practice of IMS by 1970 was deemed refined enough to introduce a commercial IMS instrument, the Beta VI (Figure 2.2)[31] and was accompanied by nearly a decade of study from Professor Francis W. Karasek at the University of Waterloo, Ontario, Canada. He maintained throughout the 1970s the only sustained program in IMS outside military and security establishments, and his team explored responses for a large range of substances. A small number of his publications in which mobility spectra and K_o values were reported are cited here.[32-38] His work established IMS as a detector with response over a broad range of chemical families with exquisitely low detection limits for that period.

During nearly this same period, mid-1960s to early 1970s, pioneering studies in gas phase ion molecule chemistry at ambient pressure with a MS were described for corona discharges by Shahin[39-41] and for the kinetics and thermodynamics of gas phase ion molecule reactions, often in air at ambient pressure, principally by Kebarle[42-44] and Meot-Ner (Mautner).[45-47] Although these may not have been recognized as vital to IMS by contemporaries, this period should be seen as a vigorous time of discovery during which the foundations of response in IMS were established. The knowledge from study investigations is essential in describing the appearance of mobility spectra from modern IMS analyzers. An abbreviated, although incomplete, list of contributors from this dynamic era of discovery includes Munson and Field,[48] Grimsrud,[49] Wentworth,[50] Ausloos,[51] Bowers,[52] and others.[53]

During this time and earlier, several teams were actively exploring the foundations of ion-neutral interactions using mobility spectrometers. These were made usually in

TABLE 2.2
Partial List of the Record of Patents (Annotated) and Publications Associated with the Commercialization of IMS in 1970 under the Name *Plasma Chromatography*

Government Contracts to Franklin GNO, the Predecessor of PCP, Incorporated

Title *Summary of Contract*	Contract No.	Agency	Date
"Experimental Investigation of Electron Attachment Characteristics of Certain Materials in Atmospheric Air" *Studies of dimethylhydrogenphosphite, triethylphospate, and Sarin with a pulsed D2 lamp at 50–100 torr in negative polarity. The drift tube was in a glass chamber.*	Nonr-4977(00)	Office of Naval Research	May 1965 March 1967
"Investigation of the Properties of Negative Ions Produced by the Interaction of Large Electronegative Gas Molecules with Free Electrons" *Studies of ionization of SF$_6$*	Nonr-4924(00)	Office of Naval Research	June 1965 July 1967
Performance Study of the PC Marking System *Study for personnel detection and chemicals for tracking and marking. This was the first all-metal stainless steel drift tube equipped with 250-μs shutter pulses. Two-shutter coplanar design with continuous UV lamp.*	FO8635- 67-C-0075	Air Force Armament Laboratory, Eglin Air Force Base	April 1967 June 1968
"PC Experimentation with Already Available PC Instrument: Design, Construct, and Laboratory Evaluate Modular Prototype PC" *New drift tube with wide rings, high-temperature operation (200°C), vibration rugged. Operated in truck and helicopter from batteries or generator.*	DAADO5- 69-C-0139	Land Warfare Laboratory, Aberdeen Proving Ground, U.S. Army	Nov 1968 May 1970

Patents Awarded to Franklin GNO on Ion Mobility Spectrometry

Title	Inventor(s)	U.S. Patent No.	Date Filed/ Granted
"Apparatus and Methods for Separating, Detecting, and Measuring Trace Gases with Enhanced Resolution"	Carroll, Cohen, Wernlund	3,626,180	Dec. 3, 1968 Dec. 7, 1971
"Apparatus and Methods for Separating, Concentrating, Detecting, and Measuring Trace Gases"	Carroll	3,668,383	Jan. 9, 1969 June 6, 1972
"Detecting a Trace Substance in a Sample Gas Comprising Reacting the Sample with Different Species of Reactant Ions"	Cohen	3,621,239	Jan. 28, 1969 Nov. 16, 1971

continued

TABLE 2.2 (continued)
**Partial List of the Record of Patents (Annotated) and Publications
Associated with the Commercialization of IMS in 1970 under the Name**
Plasma Chromatography

Patents Awarded to Franklin GNO on Ion Mobility Spectrometry

Title	Inventor(s)	U.S. Patent No.	Date Filed/ Granted
"Apparatus and Methods for Separating Electrons from Ions"	Carroll	3,629,574	Jan. 28, 1969 Dec. 21, 1971
"Gas Detecting Apparatus with Means to Record Detection Signals in Superposition for Improved Signal-to-Noise Ratios"	Wernlund	3,526,137	Feb. 11, 1969 Dec. 7, 1971
"Apparatus and Method for Improving the Sensitivity of Time of Flight Ion Analysis by Ion Bunching"	Cohen	3,626,182	Apr. 1, 1969 Dec. 7, 1971
"Time of Flight Ion Analysis with a Pulsed Ion Source Employing Ion-Molecule Reactions"	Cohen	3,593,018	Apr. 1, 1969 July 13, 1971

nonclustering atmospheres at reduced pressure, notably by Mason,[54] McDaniels,[55,56] and Compton.[57] Such investigators established both technology precedents for drift tubes in analytical ion mobility spectrometers and models of ion mobility that were highly refined, if only incomplete for ions in polar gases at ambient pressure. The development of average dipole orientation theory for the description of ion-neutral interactions was a significant advance over Langevin's model from the early 1900s for collision rates between ions and molecules.[58] This period of discovery beginning in the late 1950s, in retrospect, set in place the principles to describe the chemistry of ion formation and the mobility of ions in gases. The experimental regimes of such studies had only indirect importance for modern analytical ion mobility methods.

The largely unpublicized development programs in detecting chemical warfare agents by the U.K. Ministry of Defense and the U.S. Army during the 1960s and 1970s (principally Edgewood Arsenal for the U.S. Army and Porton Down for the U.K. military and their associated civilian companies) resulted in the introduction of the chemical agent monitor (CAM) in 1981 to 1983 (Table 2.3). This was an unprecedented technical achievement with superbly engineered handheld chemical instrumentation. Large-scale production of CAMs (Figure 2.3a) was followed with distribution among armed forces of the United Kingdom, the United States, and allied forces. These and nearly all IMS instruments in the following decade contained radioactive ion sources, often 10 mCi of ^{63}Ni, which provided a stable and predictable response without a demand for electrical power. The ion chemistry provided by ^{63}Ni was a favorable match to the ionization properties of chemical warfare agents, explosives, and narcotics.

A similarly impressive engineering effort occurred with detectors of trace levels of high explosives (Table 2.3). This arose in response to the 1985 bombing of an Air

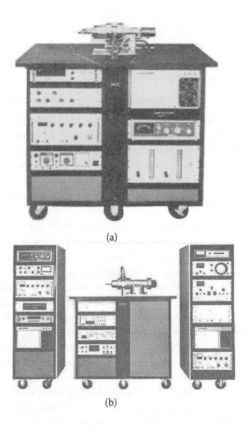

(a)

(b)

FIGURE 2.2 Photograph of the Beta VII plasma chromatograph from the early 1970s (a) and an Alpha II model IMS/MS instrument, also from 1970 (b). The Beta VI was a dual-shutter drift tube with a boxcar integrator for obtaining spectra. One of these was located in the laboratory of Prof. F. W. Karasek at the University of Waterloo, Ontario, Canada, and was used to establish the broad range of response by IMS during the 1970s. The Alpha II IMS/MS instrument or successors were employed significantly by Timothy W. Carr, then at IBM, and C. Steve Harden at Edgewood Arsenal, Maryland. (From commercial brochures ca. 1972 to 1980 from PCP, Inc.)

India flight originating in Canada first and then to the 1986 bombing of Pan Am 111 over Lockerbie, Scotland. In both incidents, hundreds of passengers and crew aboard the Boeing model 747 aircraft died when bombs brought onboard were detonated in flight. In 1990, Barringer Research Limited in Canada introduced a bench-scale, high-temperature IMS analyzer (IONSCAN; Figure 2.3b), initiating a decades-long reliance on this instrument and comparable analyzers from other vendors for security in international commercial air travel. Documents and reports from this period of development of CAM and IONSCAN are generally not available, and dissemination of results was not encouraged; thus, the chronology and content of Table 2.3 cannot be documented. There are nonetheless occasional open source documents or presentations that provide clues to some aspects of the development of IMS for explosives, chemical agents, and contraband narcotics.[59–62]

TABLE 2.3
Developments in Ion Mobility Spectrometry by Military and Security Establishments in 1965 to 2000

Year	Chemical Warfare Agent Detectors
1967 ff.	Studies begun in the United Kingdom on gas phase ion chemistry for chemical warfare agents (CWA) at Edgewood, MD, by C.S. Harden using mass spectrometry. In the United Kingdom at Porton Down in Salisbury, England, D. Blyth explored response to CWA on surfaces first, then on gas phase reactions. Contracts to Cohen for study of CWAs by IMS.
1969	Simple mobility-based filter-type detectors called the M-8A1 placed into service in the United States.
1970–1972	Ion mobility explored to improve specificity of M-8A at Edgewood. In the United Kingdom, Blyth develops DICE (detection by ion combination effect). Spangler explores IMS at U.S. Army Mobility Command.
1973–1974	Harden and Blyth exchange results and meet, agreeing to develop IMS jointly. Eventually, the responsibility of the U.S. military was to produce a continuous air monitor known as the ACADA (automatic chemical agent detector and alarm), and the United Kingdom was to develop a personal monitor, eventually called CAM, the chemical agent monitor, a handheld IMS analyzer.
1980	Introduction of CAM to U.K. forces and eventually the U.S. Army (1984).
1988	Environics Oy releases the M86, a single-measurement cell aspiration detector; the M90 was released in 1992.
1994	Bruker Saxonia Analytik GmbH begins development of RAID handheld instruments.
1995	Development of lightweight chemical detector (LCD) begins.
2002	LCD placed into service with U.K. forces and later (2006) with U.S. Army.

	Explosive Detectors
June 23, 1985	Air India Flight 182 on route Montreal-London-Delhi with a Boeing 747-237B *Emperor Kanishka* blown up at 31,000 feet over the Atlantic Ocean, killing 329 people. Canadian government initiates trace detection technology.
1987	Development of a high speed explosives detector using ion mobility spectrometry, A. D. Murray and Lucy Danylewich-May, Canada. Transportation Development Centre, Barringer Research Limited, Publisher: Barringer Research, Ltd., 1987, 44 pages.
1988	Development of the IMS Prototype Narcotics Detector, Contract No. 25ST.32032-7-3271, Barringer Research, Ltd., Final Report to Revenue Canada Customs and Excise, August 1988.
December 21, 1988	Pan American Airlines Flight 103, a Boeing 747-121 *Clipper Maid of the Seas*, destroyed by a bomb, killing all 243 passengers, 16 crew members, and 11 residents of Lockerbie, Scotland.
1990	Introduction of benchtop IMS analyzers for detection of trace levels of explosives: IONSCAN from Barringer Research Ltd. (now Smiths Detection) and later ITEMIZER from IonTrack (now Morpho Detection).

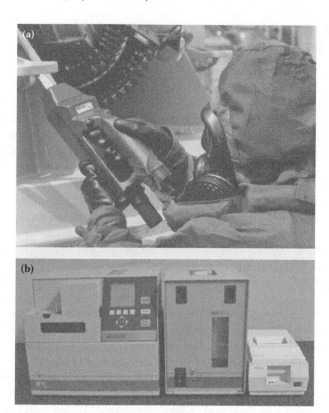

FIGURE 2.3 Two instruments that changed perceptions about and visibility of ion mobility spectrometry. In the top frame is a photograph of a soldier in a protective suit with a handheld military-style ion mobility spectrometer, the chemical agent monitor (CAM), from Graseby Dynamics, Limited, in Watford, England. In the bottom frame is a benchtop analyzer for trace detection of explosive residues, the model 400A IONSCAN, from Barringer Research in Toronto, Ontario, Canada. (Courtesy of Smiths Detection, both companies are now part of Smiths Detection.)

The enormous success and widespread acceptance of CAM and IONSCAN invited competition; soon, companies other than Graseby Dynamics (CAM) and Barringer Research (IONSCAN) were active in producing instruments, including Iontrack Instruments, Bruker Daltonics, and others. Most of these companies have undergone one or more reorganizations and a steady improvement in technology. Indeed, the original companies of CAM and IONSCAN are now consolidated at Smiths Detection. The production commercially of comparable instruments for military and security applications may have suggested to some that IMS had little value elsewhere in measurement science. Several developments in technology were soon to broaden the range of investigations by ion mobility significantly so that today the method can be considered universal in applicability. Where an ion can be made from a substance at ambient pressure and survives long enough to traverse the drift region, a mobility spectrum can be obtained to allow chemical characterizations or quantitative determinations.

2.4   MOBILITY METHODS BEYOND MILITARY AND SECURITY VENUES (1990 TO 2000)

An essential technical achievement for the further advancement of IMS as a defensible analytical method was the development of drift tubes, in which ion chemistry was reproducible and clustering of product ions with residual neutrals of sample was averted or controlled.[63] Such complications plagued early IMS instruments and led to disappointment with IMS in the 1970s.[64,65] These drawbacks were eliminated in this single innovation: drift tubes were pneumatically sealed and operated with a unidirectional flow of drift gas; gas flow was entered at the detector, passed through the entire drift tube and, vented after cleaning the ion source of residual sample neutrals. This ensured a constant gas composition in the drift region, free of sample vapors, which were rapidly vented from the source region along with drift gas. The residence times of neutrals in the reaction region were short and definite, and this allowed the response to return rapidly to original conditions after the sample was introduced into the drift tube. Thus, a mobility measurement was freed from complications from sample neutrals that could diffuse from the source to the drift region and persist in the instrument. These changes helped establish a mobility measurement at ambient pressure as broadly reliable even though not all instrument manufacturers adopted the unidirectional flow concept. While this innovation was achieved in the early 1980s,[63] mention is made here since the work can be seen as a defining step for a modern era in IMS.

A second defining development for expanded applications of IMS was the electrospray ionization (ESI) source, which was so powerfully employed previously in MS studies of biomolecules in the 1980s.[66] Electrospray was incorporated first into a mobility spectrometer by Dole et al., who demonstrated,[67] using mobility, the formation of gas ions directly from liquid samples using electrospray sources; however, their mobility spectra were broad, having limited value for ion characterization (Figure 2.4a). Although ESI permitted MS of large molecules,[68] the poor spectral profiles were attributed to clustering of ions with solvents and were solved by Shumate and Hill.[69,70] The addition of a counterflow of heated gas in a desolvation region, preceding the drift region, resulted in significantly improved peak shape and resolving power (Figure 2.4b). Their studies demonstrated that ionization of large molecules by ESI could be combined with IMS drift tubes at ambient pressure, extending ion mobility methods into biomolecule and pharmaceutical measurements. A next advance was the coupling of nanospray to an IMS drift tube at ambient pressure.[71] Even greater resolving power than these was obtained by Clemmer and Jarrold,[72] who employed long drift tubes with nonclustering gas atmospheres at 1 torr and with mass spectrometers as detectors. Other ion sources were added[73] to increase the utility of IMS in biomolecule studies further, as described in Chapter 4. Appreciation of mobility as a measurement method has grown as a result of impressive findings on and medical significance for the size and shape of biological molecules.

Finally, significant transformation of IMS technology occurred through one highly visible application, the use of IMS on the International Space Station as the Volatile Organic Analyzer,[74] a dual gas chromatograph-ion mobility spectrometer for analysis of air before and after so-called contingencies (Figure 2.5). Construction of IMS instruments as space flight-qualified instrumentation left no doubt that this

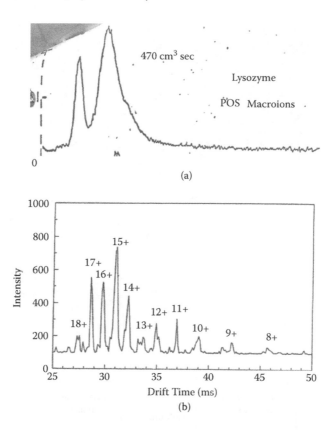

FIGURE 2.4 Mobility spectra from early ESI IMS measurements of large molecules including (a) lysozyme (Gieniec et al., Electrospray mass spectrometry, *Biomed. Mass Spectrom.* 1984. With permission.), and (b) Cytochrome c (Wittmer et al., Electrospray ionization–ion mobility spectometry, Anal. Chem. 1994. With permission.). The combination of ESI sources with IMS have become an essential feature of biological and biomedical studies with IMS–MS (or IMMS).

combination of methods of chemistry and physics could be engineered into chemical analyzers without equal for cost, size, power, and reliability. The launch of the VOA had been preceded by another space-based use of IMS to demonstrate the reliability of IMS technology in microgravity. This was a hydrazine-selective analyzer to detect hydrazine on astronaut's spacesuits while in the air lock of the U.S. shuttles after a space walk.[75] This was initiated from concerns that hydrazine, used in shuttle thrusters, would condense on astronauts' spacesuits during space walks if check valves froze leaking vapors. After return of astronauts to the inside of the shuttle, absorbed hydrazines could be volatilized and place a toxic chemical into the air environment of the shuttle. Handheld mobility spectrometers derived from CAM were modified with reagent gas chemistry to separate ammonia, hydrazine, and monomethylhydrazine by mobility.[76]

At the beginning of this century, nascent developments in ion mobility methods were recognizable with on-site instruments for use in clinical and laboratory settings.

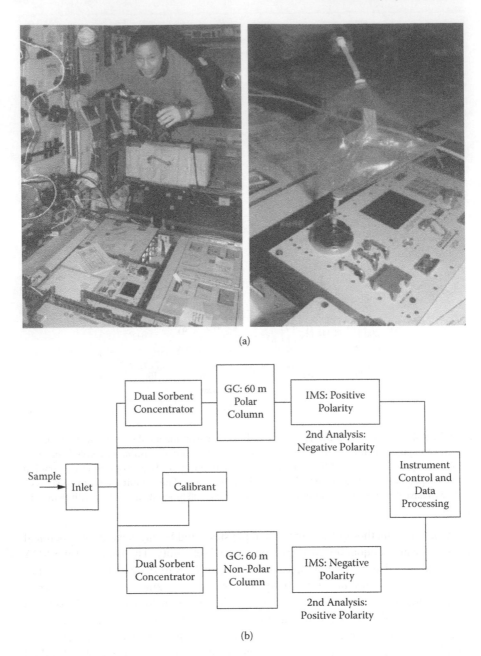

(a)

(b)

FIGURE 2.5 (a) Photographs of the Volatile Organic Analyzer (VOA) on board the International Space Station or ISS (top frames) where the GC IMS instrument was used to monitor the presence of contaminants in air during a ten year period. This was the first application of a GC IMS in a routine manner in a rugged environment. (b) The internal architecture of the VOA with dual instrument design. (Photograph ISS007e5845 from NASA, schematic from Dr. T. Limero, with approval from Dr. Edward Lu, pictured on-board ISS.)

These analyzers provided timely and valuable information on medical conditions or therapeutic progress. Details of these developments can be found throughout this edition of *Ion Mobility Spectrometry* and are especially highlighted in applications found in the chapters of Section IV. This flourishing of ion mobility was associated, accompanied, and stimulated by increased availably of instrumentation, including some instruments specifically designed to meet a growing demand in pharmaceutical and biomolecular research.

2.5 COMMERCIAL PRODUCTION OF MOBILITY-BASED ANALYZERS (2000 TO PRESENT)

Another striking feature of the past decade has been the level of commercialization of ion mobility methods and the variety of instrument designs. Eighteen companies now offer mobility-based instruments, as summarized in Table 2.4, eliminating one of the critical barriers to an enlarged presence of mobility measurements in modern analytical chemistry: convenient access to ion mobility technology. During this period, the historic, large-volume uses of IMS analyzers for military and security uses continued to dominate the total sale of unit items and attracted new companies; however, small startup ventures with new embodiments of ion mobility technology, new concepts, and new instruments have changed all aspects of the practice and applications of ion mobility. These companies have appeared worldwide, for example, in Israel, China, Finland, Germany, Spain, England, Canada, and the United States. Lastly, a few large instrument companies, with broad name recognition in analytical chemistry, began offering instruments, including a mobility component to compliment mass analyses; this development significantly altered the visibility of ion mobility within chemistry and other scientific communities.

The start of this new trend in commercialization may be attributed to the addition of an ESI source to cylindrical FAIMS (field asymmetric IMS), which was then combined with a mass spectrometer.[77] A relatively simple modification between the ESI and mass spectrometer simplified response to the complex ion mixtures created in ESI sources. This received an enthusiastic response and was commercialized as ESI FAIMS MS first by IonAnalytics Corporation (of Ottawa, Ontario, Canada) and then by ThermoScientific, which acquired IonAnalytics. They have promoted FAIMS as an ion filter with special emphasis on pharmaceutical and drug metabolite measurements by liquid chromatography with mass spectrometry.[78]

Another large instrument company, Waters Corporation, pioneered a new ion mobility method termed the traveling wave and marketed an ion mobility-time-of-flight (TOF)-MS instrument, the Synapt G1, and successors.[79–81] This instrument had a surprising and beneficial effect on the whole enterprise of IMS by introducing capabilities to investigators who otherwise were unlikely to build their own ion mobility MS (IM–MS) but who could purchase and use this IM–MS for discovery in biomolecule studies. Their efforts and the increased visibility of IMS as a measurement method should be highlighted as a milestone in the history of ion mobility methods. Although commercialization was energetic and targeted to biomolecular and pharmaceutical markets, activity was also vibrant, if not accelerating, in historic

TABLE 2.4
Alphabetical Summary of Commercial Activity in Ion Mobility Spectrometry between 2004 and 2011

Company Details	Recent History and Comments	Products
AB SCIEX	http://www.absciex.com/products DMS inlet for Sciex LC ESI MS/MS	SelexION™ Technology
Bruker Daltonics	http://www.bdal.com/ Recent development of explosive detectors with IMS	RAID family of handheld IMS-based analyzers, including µRAID; DE-tector™; stationary IMS instruments such as RAID-AFM (automated facility monitor)
Environics Oy	http://www.environics.fi/ Aspirator method for mobility with open-loop flow design	ChemPro family of instrument, including ChemPro 100i and stationary ChemProFX
Excellims	http://www.excellims.com/ Developing chiral IMS MS and multidimensional IMS	GA-2100 electrospray IMS CIMS (chiral ion mobility spectrometry)-MS Source of IMS components
G.A.S.	http://www.gas-dortmund.de/ Integrated instruments, including software; eight configurations of IMS instruments, including general GC IMS	FlavourSpec® BreathSpec® µIMS®-ODOR
Ramem S.A.	http://www.ioner.eu/ DMA method of IMS	High-resolution IMS
Implant Sciences Corporation	http://implantsciences.com/ Technology based on noncontact sampling with photoionization sources	Quantum Sniffer QS-H150 Quantum Sniffer™ QS-B220
Morpho Detection	http://www.morpho.com/detection/ Originated as Ion Track, Inc., then GE-Interlogix; now part of Safran Group	Itemiser® DX Itemiser® 3 Enhanced MobileTrace®
Nuctech Company Limited	http://www.nuctech.com/templates/ T_Second_EN/index.aspx?nodeid=148	TR1000 handheld explosive trace detector R1000NB handheld narcotics trace detector TR2000DB desktop trace detector
Owlstone Nanotech	http://www.owlstonenanotech.com/	Lonestar portable gas analyzer LC/FAIMS/mass spectrometry NEXSENSE C military CWA analyzer
Particle Measuring Systems, Inc.	http://www.pmeasuring.com/ particleCounter/molecularMonitors Clean air monitoring within industry	AirSentry® II family of instruments
3QBD, Ltd.	http://3qbd.com/English/ Formerly Q SCENT Ltd. Products for medical diagnostics of bacterial vaginosis	The VGTest

TABLE 2.4 (continued)
Alphabetical Summary of Commercial Activity in Ion Mobility Spectrometry between 2004 and 2011

Company Details	Recent History and Comments	Products
Scintrex Trace Corp.	http://www.scintrextrace.com/ Now part of Autoclear	E family of handheld explosive detectors and N family of handheld narcotic detectors EV 2500 and 3000 explosive vapor analyzers VE6000 for vehicle screening
Smiths Detection	http://www.smithsdetection.com/ Formed from Grasby Dynamics and Barrierger Research and Environmental Technologies Group	GID-3™ CWA detection system LCD (lightweight chemical detector) CAM (chemical agent monitor) MCAD (manportable chemical agent detector)
SIONEX (dissolved August 2010)	Microfabricated drift tubes for differential mobility spectrometry; microAnalyzer now produced in Draper Laboratory	microDMx™ sensor chip microAnalyzer Sionex Value Added Component SVAC™ DMS/IMS2
Thermo Fisher Scientific	http://www.thermoscientific.com/faims Origins with Ionanalytics, which adapted an MSA field ion spectrometer and innovated an original design and added ESI	FAIMS MS
ThermoScientific (Part II)	https://fscimage.fishersci.com/images/ D00700~.pdf Company arose from Thermidics, then Thermo Electron Corporation	EGIS Defender Explosives Trace Detection System
TSI	http://www.tsi.com/scanning-mobility-particle-sizer-spectrometers Scanning mobility particles by differential mobility methods	Scanning Mobility Particle Sizer Spectrometer 3034 Scanning Mobility Particle Sizer Spectrometer 3936
Waters Corporation	http://www.waters.com/waters/nav. htm?cid=514257&locale=en_US/	SYNAPT G2 family of traveling wave mobility spectrometers-mass spectrometers

uses of IMS for detecting chemical warfare agents and explosives. In each of these, catchwords of future versions of IMS analyzers were smaller, faster, and more selective than existing instruments.

The period can be described as broadly dynamic with reference to brands and businesses. Old names in this commercial scene have undergone sales or resales, including Iontrack, which was sold to GE Securities; GE Securities was sold to Morpho Detection of the Safran Group in 2006. Barringer Research Limited was acquired by Smiths Detection earlier in the 1990s, which remains a leading presence in the commercial world of IMS. Significantly, explosive detection was addressed anew by relative newcomers in the manufacture of IMS analyzers, including Implant Sciences in the United States and Nutech Company Limited in China, as shown in Table 2.4. As the threat list of explosives has grown, new versions of these reliable analyzers have been introduced. For example, the DT-500 from Smiths Detection is equipped with a twin drift tube and provides a response in both positive and negative polarity. Some significant commercial efforts that have either failed or not received broad acceptance, although not necessarily technical failures, are walk-through portals for explosive trace detection, building air monitors, ticket and document scanners, and purity verification in pharmaceutical production lines, although this last application appears to have some sustained interest.

Another feature of the past decade is the growing acceptance and refinement of aIMS (aspirator IMS). The aIMS has long been associated with ion mobility, dating at least to 1900,[82,83] and has merged in the past decade with mainstream IMS in both historic security and military applications with small instruments[84,85] and with large laboratory-based instruments known as differential mobility analyzers (DMAs).[86,87] The former are commercialized by Environics Oy in Finland, and the large versions are commercialized by SEADM, Parque Tecnológico de Boecillo, Valladolid, Spain. The Environics Oy handheld Chem Pro 100 and new 100i are open-loop instruments deployed with militaries in several nations. Efforts to improve the resolving power of these analyzers, as described in Chapter 9, suggest a future presence, perhaps a growing presence, for aIMS.

Another distinction of the 2000s was a miniature version of FAIMS,[88,89] developed and refined jointly at Draper Laboratory and New Mexico State University.[90] This led to a commercial venture in 2001 called Sionex, Incorporated (Table 2.4); the principal advantage of Sionex technology was size, with an analyzer region comprised of conductive plates only 5 mm wide by 13 mm long and separated by a 0.5-mm gap (see Figure 2.5b). This differential mobility spectrometry (DMS) technology was developed by General Dynamics, now ChemRing Detection Systems, into a chemical warfare agent detector called JUNO and featured packaging, simplicity of operation, and robustness of handheld, in-field ion mobility analyzers such as the CAM or LCD (lightweight chemical detector).[91] The DMS was combined early in the development program with a GC[92] and culminated with a completely portable GC-DMS with sample enrichment traps on the inlet of the GC, known as the micro-Analyzer.[93] The Sionex technology was also included in a replacement of the chemiluminescent detector in a commercial trace explosive analyzer, the EGIS, becoming a flash GC-DMS instrument called the Defender. The Defender was the first commercial DMS design based on modified gas atmospheres to adjust peak positions on the compensation voltage axis.[94] In 2010, Sionex Corporation was dissolved with its intellectual property sold, licensed, or returned to Draper Laboratory and recently ABI SCIEX (SelexION attachment to their MS/MS, derived from Sionex Corp.).[95]

An English company, Owlstone Limited, a subsidiary of Advance Nanotech, was formed in 2003 around a version of FAIMS or DMS with previously unprecedented dimensions (Chapter 1 and Chapter 9). This was termed nanofabrication, and Owlstone Nanotech impressively developed several vapor analyzers based on their version of FAIMS. The company has won several industry awards and teamed with Agilent Corporation for development of an inlet to the Agilent TOF-MS. In addition, it has entered an agreement with SELEX Galileo to address military needs for chemical sensing and with Pacific Northwest National Laboratory for technology development. A current emphasis is on combination of ESI, their μFAIMS, and MS.[96]

A German company, G.A.S. (Gesellschaft für analytische Sensorsysteme mbH) in Dortmund was launched in 1997 from ISAS (Institute for Analytical Sciences) and has produced some integrated GC-IMS instruments with conventional drift tubes, the only known versions of this technology. They are distinguished by two configurations of common technology platforms (FlavourSpec and BreathSpec) for specific applications, principally in vapor analysis from foodstuffs and in determination of volatile organic compounds as metabolites in breath.

In the United States, Excellims was launched in 2005 to commercialize ESI IMS technology significantly developed at Washington State University and offers today a stand-alone ESI-IMS instrument, an ESI-IMS-MS instrument, and parts for IMS under a research tool kit category. In Israel, a small startup called Q SCENT Limited, in 2000 rebranded as 3QBD Limited, has brought to market an IMS analyzer for clinical diagnosis of the most common vaginal infections through detection of biogenic amines.[97]

Spain has become home to two mobility groups during the past decade. The Sociedad Europea de Análisis Diferencial de Movilidad SL (SEADM) in Boecillo was founded in February 2005 to develop DMAs based on the work of its technical consultant and cofounder Professor Juan Fernandez de la Mora of Yale University. Efforts are largely directed to DMA/MS methods. In contrast, stand-alone DMA instruments are produced by RAMEM, SA, in Madrid. A stand-alone mobility analyzer is commercially available with multivariate analysis software; a default library includes standards and one of several nonradioactive ionization sources, including photoionization, corona discharge, and ESI.

The products and activities shown in Table 2.4 suggest competitive and vibrant commercial activity in IMS. Whereas market opportunities existed only in security and military venues a decade ago, biomolecular research, pharmaceutical development, and metabolite discovery have become second-generation applications of IMS instruments and methods. Such instruments have opened mobility MS to an enlarged user community, making mobility no longer esoteric. While IMS is no longer an obscure method known only to a few individuals or teams, the method has not reached a level of widespread utility or familiarity that stands alongside other analytical methods such as atomic absorption spectrometry or LC. Ion mobility spectrometry is not yet taught as part of the standard analytical chemistry curriculum in academic institutions, except for a small number of universities. Furthermore, a large part of the research and development of novel applications is carried out by commercial entities and not necessarily in academia.

2.6 THE SOCIETY FOR ION MOBILITY SPECTROMETRY AND JOURNAL

In 1992, the U.S. Army sponsored a first meeting for investigators and practitioners of IMS with an intention for international representation (Figure 2.6). Attendance at this

FIGURE 2.6   Participants at the First International Workshop in Ion Mobility Spectrometry held June 1992 in Mescalero, New Mexico.

workshop was by invitation, with the intent that those at the meeting present results from their activities in IMS. This meeting had been preceded by three instances when researchers or users of IMS had met in symposia at scientific meetings and one caucus of scientists/engineers at Snowbird, Utah, which resulted in a call for an expanded meeting. These meetings led to the formation of the International Society for Ion Mobility Spectrometry (ISIMS), registered first in Germany as a scholarly society in 1997.

The constitution of ISIMS defines that the society is open to those using IMS or with a strong interest in IMS as an analytical technique or who are interested in ambient pressure gas phase ion–molecule chemistry. The society promotes the use of IMS and ambient pressure gas phase ion–molecule chemistry as an analytical technique; provides the opportunity for development and training through short courses on IMS theory and practice; provides opportunity for free exchange of ideas and information on IMS through yearly conferences, poster sessions, vendor exhibits, and publications; and encourages the spirit of unity and cooperation among society members to advance ISIMS objectives.

In 2010, ISIMS was recognized as a nonprofit organization by the Internal Revenue Service in the United States and also registered in New Mexico. This eases practical economic matters by assisting in organizing conferences and managing society's structure, including the Web site and the ISIMS journal. These meetings today are attended by over 100 from nations worldwide, compared to a beginning of 25 attendees in 1992 (Table 2.5).

The *International Journal for Ion Mobility Spectrometry* is the official publication of ISIMS and publishes fully peer-reviewed research and application-oriented papers, critical reviews, technical notes, and topical issues. The journal is abstracted or indexed in Chemical Abstracts Service (CAS), Compendex, Google Scholar, Online Computer Library Center (OCLC), SCOPUS, and Summon by Serial Solutions. Articles can include presentations in all aspects of the mobility, reaction chemistry, and behavior of ions in a supporting gaseous atmosphere and in an electrical field. Also included are biomolecular studies, mobility at comparatively low pressure, instrumentation for field and process monitoring, and measurements of samples for

TABLE 2.5
History of Locations of Annual Meetings of the International Society of Ion Mobility Spectrometry

Year	Location	Organizer and Supporting Institution
1992[a]	Mescalero, New Mexico, USA	Gary Eiceman and C. Steve Harden
1993	Quebec City, Quebec, Canada	Pierre Pilon and Andre Lawrence
1994	Galveston, Texas, USA	Tom Limero and Jay Cross
1995	Cambridge, England, UK	Alan Brittain
1996	Jackson Hole, Wyoming, USA	Herbert H. Hill and Dave Atkinson
1997	The Bastei, Germany	Jörg Baumbach and Joachim Stach
1998	Jekyll Island, South Carolina, USA	C. Steve Harden
1999	Buxton, England, UK	C. L. Paul Thomas and Hilary Bollan
2000	Halifax, Nova Scotia, Canada	Pierre Pilon and Andre Lawrence
2001	Wernigerode, Harz, Germany	Jörg Baumbach and Joachim Stach
2002	San Antonio, Texas, USA	Angie Detulleo and Tom Limero
2003	Umeå, Sweden	Sune Nyholm
2004	Gatlinburg, Tennessee, USA	Jun Xi
2005	Château de Maffliers, Paris, France	Christine Fuche
2006	Honolulu, Hawaii, USA	C. Steve Harden
2007	Mikkeli, Finland	Osmo Anttalainen
2008	Ottawa, Canada	Pierre Pilon
2009	Thun, Switzerland	Herbert Hill, Marc Gonin, and Katrin Fuhrer
2010	Albuquerque, New Mexico, USA	Gary, Mary, and Abigail Eiceman
2011	Edinburgh, Scotland, UK	Hilary Bollan
2012	Orlando, Florida, USA	C. Steve Harden
2013	Boppard, Germany	Wolfgang Vautz

[a] The first meeting was funded by the U.S. Army Chemical Research, Development, and Engineering Center, Aberdeen Proving Grounds, Maryland, under contract no. DAAL03-91-C-0034, TCN number 92-057 (DO. No. 0127), Scientific Services Program.

industry, the environment, medicine, and security. Coverage extends to theoretical and experimental findings from fundamental and applied studies, touching on relevant aspects of chemistry, physics, geology, life sciences, and engineering. Papers and notes on novel instrument design, related new applications, and their validation can also be found among the articles.

REFERENCES

1. Franklin, B., A letter of Benjamin Franklin, Esq., to Mr. Peter Collinson, F. R. S. concerning an electric kite, *Trans.* 1751–1752, 47, 565–567.
2. Charles V. Walker, Ed., *Proceedings of the London Electrical Society during the Sessions 1841–2 and 1842–3*, Simplkin, Marshall, London, 1843.
3. Röntgen, W.C., On a new kind of rays, *Nature* 1895, 53, 274–276; also see *Science* 1896, 3, 726.

4. Rutherford, E., The velocity and rate of recombination of the ions of gases exposed to röntgen radiation, *Philos. Mag.* 1897, 44, 422–440.
5. Curie, M.S., Rayons émis par les composés de l'uranium et du thorium, *Comptes Rendus* 1898, 126, 1101–1103.
6. Rutherford, E., The discharge of electrification by ultraviolet light, *Proc. Camb. Philos. Soc.* 1898, 9, 401–416.
7. McClelland, J.A., On the conductivity of the hot gases from flames, Philos. Mag. 1898, 46, 29–42.
8. Chattock, A.P., On the velocity and mass of ions in the electric wind in air, *Philos. Mag.* 1899, 48, 401–420.
9. Rutherford, E., Uranium radiation and the electrical conduction produced by it, *Philos. Mag.* 1899, 47, 109–163.
10. Thomson, J.J., *Conduction of Electricity through Gases*, Cambridge University Press, Cambridge, UK, 1903.
11. Tyndall, A.M., *The Mobility of Positive Ions in Gases*, Cambridge Physical Tracts, Editors Oliphant, M.L.E.; Ratcliffe, J.A., Cambridge University Press, Cambridge, UK, 1938.
12. Langevin, P., L'Ionistion des gaz, *Ann. Chim. Phys.* 1903, 28, 289–384.
13. Langevin, P., Une Formule Fondamentale de Théorie Cinétique, *Ann. Chim. Phys.* 1905, 5, 245–288.
14. Franck, J.; Pohl, R., A method for the determination of the ionic mobility in small gas volumes, *Ber. Phys. Ges.* 1908, 9, 69–74.
15. McClelland, J.A.; Kennedy, H., The large ions in the atmosphere, *Proc. R. Ir. Acad.* 1912, 30, 72–91.
16. Lattey, R.T., Effect of small traces of water vapor on the velocities of the ions produced by röntgen rays in air, *Proc. R. Soc. London A* 1911, 84, 173–181.
17. Lattey, R.T.; Tizard, H.T., The velocity of ions in dried gases, *Proc. R. Soc. London A* 1913, 86, 349–357.
18. van de Graaff, R.J., Mobility of ions in gases, *Nature* 1929, 124, 10–11.
19. Cravath, A.M., The rate of formation of negative ions by electron attachment, *Phys. Rev.* 1929, 33, 605–613.
20. Thomson, J.J., *Rays of Positive Electricity*, Green, London, 1933.
21. Bradbury, N.E.; Nielsen, R.A., Absolute values of the electron mobility in hydrogen, *Phys. Rev.* 1936, 49, 388–393.
22. Anon., Defense against the submarine, *Naval Rev.* 1970, 58(1), 9–13.
23. Authors note: The Autolycus is understood to be a detector (on antisubmarine warfare aircraft such as the *Avro Shackleton*) of airborne vapors or particulate released from diesel fumes when U-boats would surface to recharge batteries, commonly at night. This technology is not well documented and reference to the Autolycus carries cautions. Nonetheless, mention is made in references. There are two individuals with the name Autolycus in Greek mythology, one with a helmet that makes him invisible and another, an Argonaut, on the journey to find the fleece. The motivation for the name could fit either of these possibilities. *Autolycus* was also the name of a British ship sunk in 1917 by a U-boat. Diesel fume detection with Autolycus, albeit a general detector, continued in the decade after World War II; see J. Marriott, Detecting the lone submarine, *New Sci.* 1971, 50(754), 567–570.
24. Loeb, L.L., Statistical factors in spark discharge mechanisms, *Rev. Mod. Phys.* 1948, 20, 151–160.
25. Lovelock, J.E.; Wasilewska, E.M., An ionization anemometer, *J. Sci. Instrum.* 1949, 26, 367–370.
26. Lovelock, J.E., The electron capture detector—a personal odyssey, in *Electron Capture*, Editors Zlatkis, A.; Poole, C.F., Elsevier, New York, 1981, pp. 13–26.

27. Eiceman, G.A.; Stone, J.A., Ion mobility spectrometry in homeland security, *Anal. Chem.* 2004, 76(21), 390A–397A.
28. Karpas, Z., Ion mobility spectrometry: a tool in the war against terror, *Bull. Israel Chem. Soc.* 2009, 24, 26–30.
29. Eiceman, G.A.; Schmidt, H., Advances in ion mobility spectrometry of explosives, in *Aspects of Explosive Detection*, Editors Marshal, M.; Oxley, J., Elsevier, Amsterdam, 2009, pp. 171–202.
30. Mäkinen, M.; Anttalainen, O.; Sillanpää, M.E.T., Ion mobility spectrometry and its applications in detection of chemical warfare agents, *Anal. Chem.* 2010, 82(23), 9594–9600.
31. Cohen, M.J., Plasma chromatography—a new dimension for gas chromatography and mass spectrometry, Pittcon 1969, Cleveland, OH, March 1969.
32. Cohen, M.J.; Karasek, F.W., Plasma chromatography™—a new dimension for gas chromatography and mass spectrometry, *J. Chromatogr. Sci.* 1970, 8, 330–337.
33. Karasek, F.W.; Kilpatrick, W. D.; Cohen, M.J., Qualitative studies in trace constituents by plasma chromatography, *Anal. Chem.* 1971, 43, 1441–1447.
34. Karasek, F.W., Plasma chromatography of the polychlorinated biphenyls, *Anal. Chem.* 1971, 43, 1982–1986.
35. Karasek, F.W.; Cohen, M.J.; Carroll, D.I., Trace studies of alcohols in the plasma chromatograph-mass spectrometer, *J. Chromatogr. Sci.* 1971, 9, 390–392.
36. Karasek, F.W.; Tatone, O.S., Plasma chromatography of the monohalogenated benzenes, *Anal. Chem.* 1972, 44, 1758–1763.
37. Karasek, F.W.; Kane, D.M., Plasma chromatography of the n-alkyl alcohols, *J. Chromatogr. Sci.* 1972, 10, 673–677.
38. Karasek, F.W.; Tatone, O.S.; Kane, D.M., Study of electron capture behavior of substituted aromatics by plasma chromatography, *Anal. Chem.* 1973, 45, 1210–1214.
39. Shahin, M.M., Mass-spectrometric studies of corona discharges in air at atmospheric pressures, *J. Chem. Phys.* 1966, 45(7), 2600–2605.
40. Shahin, M.M., Ion-molecule interaction in the cathode region of a glow discharge, *J. Chem. Phys.* 1965, 43, 1798–1805.
41. Shahin, M.M., Use of corona discharges for the study of ion-molecule reactions, *J. Chem. Phys.* 1967, 47(11), 4392–4398.
42. Kebarle, P.; Hogg, A.M., Mass-spectrometric study of ions at near atmospheric pressures. I. The ionic polymerization of ethylene, *J. Chem. Phys.* 1965, 42(2), 668–674.
43. Good, A.; Durden, D.A.; Kebarle, P., Ion-molecule reactions in pure nitrogen and nitrogen containing traces of water at total pressures 0.5–4 torr. Kinetics of clustering reactions forming $H^+(H_2O)_n$, *J. Chem. Phys.* 1970, 52, 212–221.
43. Kebarle, P.; Searles, S.K.; Zolla, A.; Scarborough, J.; Arshadi, M., Solvation of the hydrogen ion by water molecules in the gas phase. Heats and entropies of solvation of individual reactions. $H^+(H_2O)_{n-1} + H_2O = H^+(H_2O)_n$, *J. Am. Chem. Soc.*, 1967, 89(25), 6393–6399.
44. Kebarle, P.; Chowdhury, S., Electron affinities and electron-transfer reactions, *Chem. Rev.*, 1987, 87(3), 513–534.
45. Meot-Ner (Mautner), M., The ionic hydrogen bond, *Chem. Rev.* 2005, 105(1), 213–284.
46. Meot-Ner (Mautner), M., Competitive condensation and proton-transfer reactions. Temperature and pressure effects and the detailed mechanism, *J. Am. Chem. Soc.* 1979, 101(9), 2389–2395.
47. Meot-Ner (Mautner), M.; Speller, C.V., Filling of solvent shells in cluster ions: thermochemical criteria and the effects of isomeric clusters, *J. Phys. Chem.* 1986, 90(25), 6616–6624.
48. Munson, M.S.B.; Field, F., Chemical ionization mass spectrometry. I General introduction, *J. Am. Chem. Soc.* 1966, 88, 2621–2630.

49. Grimsrud, E.P.; Kebarle, P., Gas phase ion equilibriums studies of the hydrogen ion by methanol, dimethyl ether, and water. Effect of hydrogen bonding, *J. Am. Chem. Soc.,* 1973, 95(24), 7939–7943.

50. Wentworth, W.E.; Chen, E.; Lovelock, J.E., The pulse-sampling technique for the study of electron attachment phenomena, *J. Phys. Chem.* 1966, 70, 445–458.

51. Ausloos, S.P., Editor, *Ion-Molecule Reactions in the Gas Phase, Advances in Chemistry Series No. 58,* American Chemical Society, Washington, DC, 1966.

52. Su, T.; Bowers, M.T., Ion-polar molecule collisions. The effect of molecular size on ion-polar molecule rate constants, *J. Am. Chem. Soc.* 1973, 95, 7609–7610.

53. Authors note: The period of the late 1960s to 1980 saw an explosion in ion-molecule chemistry too large to document here and included high-pressure mass spectrometry, flowing afterglow mass spectrometry, and more.

54. Mason, E.A.; Schamp, H.W., Jr., Mobility of gaseous ions in weak electric fields, *Ann. Phys. (NY)* 1958, 4, 233–270.

55. McDaniel, E.W., *Collisional Phenomena in Ionized Gases,* Wiley, New York, 1964.

56. Albritton, D.L.; Miller, T.M.; Martin, D.W.; McDaniel, E.W., Mobilities of mass-identified H_3^+ and H^+ ions in hydrogen, *Phys. Rev.* 1968, 171, 94–102.

57. Crompton, R.W.; Elford, M.T.; Gascoigne, J., Precision measurements of the Townsend energy ratio for electron swarms in highly uniform electric fields, *Austr. J. Phys.* 1965, 18, 409–436.

58. Su, T.; Bowers, M.T., Ion-polar molecule collisions: the effect of ion size on ion-polar molecule rate constants; the parameterization of the average-dipole-orientation theory, *Int. J. Mass Spectrom Ion Phys.* 1973, 12, 347–356.

59. Blyth, D.A., A vapour monitor for detection and contamination control, in *Proceedings of the Second International Symposium on Protection against Chemical Warfare Agents,* Stockholm, Sweden, June 17–19, 1983, pp. 65–69.

60. Kilpatrick, W.D., Plasma chromatography and dynamite vapor detection, *Final Report FAA-RD-71-7, Contract DOT-FA71WA-2491,* Federal Aviation Administration, Washington, DC, January 1971, AD-903108/9.

61. Wernlund, R.F.; Cohen M.J.; Kindel, R.C., The ion mobility spectrometer as an explosive or taggant vapor detector, in *Proceedings of the New Concept Symposium and Workshop on Detection and Identification of Explosives,* Reston, VA, October/November 1978, pp. 185–189.

62. Fytche, L.M.; Hupe, M.; Kovar, J.B.; Pilon, P., Ion mobility spectrometry of drugs of abuse in customs scenarios: concentration and temperature study, *J. Forensic Sci.* 1992, 37, 1550–1566.

63. Baim, M.A.; Hill, H.H., Jr., Tunable selective detection for capillary gas chromatography by ion mobility monitoring, *Anal. Chem.* 1982, 54(1), 38–43.

64. Keller, R.A.; Metro, M.M., Evaluation of the plasma chromatograph as a separator-identifier, *J. Chromatogr. Sci.* 1974, 12(11), 673–677.

65. Metro, M.M.; Keller, R.A. Plasma chromatograph as a separation-identification technique, *Sep. Sci.* 1974, 9(6), 521–539.

66. Fenn, J.B., Electrospray ionization mass spectrometry: how it all began, *J. Biomol. Technol.* 2002, 13, 101–118.

67. Gieniec, J.; Mack, L.L.; Nakamae, K.; Gupta, C.; Kumar, V.; Dole, M., Electrospray mass spectroscopy of macromolecules: application of an ion-drift spectrometer, *Biomed. Mass Spectrom.* 1984, 11(6), 259–268.

68. Smith, R.D.; Loo, J.A.; Ogorzalek, R.R.; Busman, M.; Udseth, H.R., Principles and practice of electrospray ionization-mass spectrometry for large polypeptides and proteins, *Mass Spectrom. Rev.* 1991, 10, 359–452.

69. Shumate, C.B.; Hill, H.H., Coronaspray nebulization and ionization of liquid samples for ion mobility spectrometry, *Anal. Chem.* 1989, 61, 601–606.

70. Wittmer, D.; Chen, Y.H.; Luckenbill, B.K.; Hill, H.H., Electrospray ionization–ion mobility spectrometry, *Anal. Chem.* 1994, 66, 2348–2355.
71. Bramwell, C.J.; Colgrave, M.L.; Creaser, C.S.; Dennis, R., Development and evaluation of a nano-electrospray ionization source for ion mobility spectrometry, *Analyst* 2002, 127, 1467–1470.
72. Clemmer, D.E.; Jarrold, M.F., Ion mobility measurements and their applications to clusters and biomolecules, *J. Mass Spectrom.* 1997, *32*, 577–592.
73. Stone, E.G.; Gillig, K.J.; Ruotolo, B.T.; Fuhrer, K.; Gonin, M.; Schultz, A.J.; Russell, D.H., Surface-induced dissociation on a MALDI-ion mobility-orthogonal time-of-flight mass spectrometer: sequencing peptides from an "in-solution" protein digest, *Anal. Chem.* 2001, 73, 2233–2238.
74. Limero, T.; James, J.; Reese, E.; Trowbridge, J.; Hohmann, R., The Volatile Organic Analyzer (VOA) Aboard the International Space Station, SAE Technical Paper Series 2002-01-2407, 32nd International Conference on Environmental Systems, July 2002, San Antonio, TX.
75. Eiceman, G.A.; Salazar, M.R.; Rodriguez, J.E.; Limero, T.F.; Beck, S.W.; Cross, J.H.; Young, R.; James, J.T., Ion mobility spectrometry of hydrazine, monomethylhydrazine, and ammonia in air with 5-nonanone reagent gas, *Anal. Chem.* 1993, 65, 1696–1702.
76. Bollan, H.R.; Stone, J.A.; Brokenshire, J.L.; Rodriguez, J.E.; Eiceman, G.A.; Mobility resolution and mass analysis of ions from ammonia and hydrazine complexes with ketones formed in air at ambient pressure, *J. Am. Soc. Mass Spectrom.* 2007, 18(5), 940–951.
77. Guevremont, R.; Purves, R.W.; High field asymmetric waveform ion mobility spectrometry-mass spectrometry: an investigation of leucine enkephalin ions produced by electrospray ionization, *J. Am. Soc. Mass Spectrom.* 1999, 10, 492–501.
78. Kapron, J.T.; Jemal, M.; Duncan, G.; Kolakowski, B.; Purves, R., Removal of metabolite interference during liquid chromatography/tandem mass spectrometry using high-field asymmetric waveform ion mobility spectrometry, *Rapid Commun. Mass Spectrom.* 2005, 19(4), 1979–1983.
79. Schenauer, M.R.; Leary, J.A., An ion mobility-mass spectrometry investigation of monocyte chemoattractant protein-1, *Int. J. Mass Spectrom.* 2009, 287, 70–76.
80. Morsa, D.; Gabelica, V.; De Pauw, E., Effective temperature of ions in traveling wave ion mobility spectrometry, *Anal. Chem.* 2011, 83, 5775–5782.
81. Li, H.; Giles, K.; Bendiak, B.; Kaplan, K.; Siems, W.F.; Hill, H.H., Resolving structural isomers of monosaccharide methyl glycosides using drift tube and traveling wave ion mobility mass spectrometry, *Anal. Chem.* 2012, 84, 3231–3239.
82. Tammet, H., The Aspiration Method for the Determination of Atmospheric-Ion Spectra, Israel Program for Scientific Translations, Jerusalem, 1970; also see http://ael.physic.ut.ee/tammet/am/.
83. Zeleny, J., The velocity of the ions produced in gases by Röntgen rays, *Philos. Trans. R. Soc. Lond. A* 1900, 195(262–273), 193–234.
84. Utriainen, M.; Karpanoja, E.; Paakkanen, H., Combining miniaturized ion mobility spectrometer and metal oxide gas sensor for the fast detection of toxic chemical vapors, *Sens. Actuators B* 2003, 93, 17–24.
85. Tuovinen, K.; Paakkanen, H.; Hänninen, O., Determination of soman and VX degradation products by an aspiration ion mobility spectrometry, *Anal. Chim. Acta* 2001, 440, 151–159.
86. Fernandez de la Mora, J.; de Juan, L.; Eichler, T.; Rosell, J., Differential mobility analysis of molecular ions and nanometer particles, *Trend Anal. Chem.* 1998, 17(6), 328–339.
87. Hogan, C.J., Jr.; Fernández de la Mora, J., Ion mobility measurements of non-denatured 12–150 kDa proteins and protein multimers by tandem differential mobility analysis-mass spectrometry (DMA-MS), *J. Am. Soc. Mass Spectrom.* 2011, 22, 158–172.

88. Miller, R.A.; Eiceman, G.A.; Nazarov, E.G., A micromachined high-field asymmetric waveform-ion mobility spectrometer (FA-IMS), *Sens. Act. B Chem.* 2000, 67, 300–306.

89. Miller, R.A.; Nazarov, E.G.; Eiceman, G.A.; King, T.A., A MEMS radio-frequency ion mobility spectrometer for chemical agent detection, *Sens. Act. A. Phys.* 2001, 91, 301–312.

90. Eiceman, G.A., Characterization of Draper Labs MEMS FAIM Spectrometer 1999, Annual Report on Grant No. AS99-0307 from Draper Laboratory University IR&D Grant, Office of Grants and Contracts, New Mexico State University, Las Cruces, NM, Dec. 1999.

91. Wu, W.J.; Blethen, G.; Griffin, M.T.; Harden, S.; Ince, B.; McHugh, V.; Rauch, P.J., Differential mobility spectrometry (DMS) for the detection of explosive vapors, Poster Number 2450-1, PITTCON2012, Orlando, FL, March 15, 2012.

92. Eiceman, G.A.; Nazarov, E.G.; Miller, R.A.; Krylov, E.V.; Zapata, A., Micromachined planar field asymmetric ion mobility spectrometer as gas chromatographic detectors, *Analyst* 2002, 127, 4, 466–471.

93. Limero, T.; Reese, E.; Cheng, P., Optimization of the microanalyzer to detect trace organic compounds in a complex mixture, 17th Annual Conference on Ion Mobility Spectrometry, Ottawa, Canada, July 20–25, 2008.

94. Eiceman, G.A.; Krylov, E.V.; Krylova, N.S.; Nazarov, E.G.; Miller, R.A., Separation of ions from explosives in differential mobility spectrometry by vapor-modified drift gas, *Anal. Chem.* 2004, 76(17), 4937–4944.

95. Levin, D.S.; Miller, R.A.; Nazarov, E.G.; Vouros, P., Rapid separation and quantitative analysis of peptides using a new nanoelectrospray—differential mobility spectrometer–mass spectrometer system, *Anal. Chem.* 2006, 78(15), 5443–5452.

96. Brown, L.J.; Toutoungi, D.E.; Devenport, N.A.; Reynolds, J.C.; Kaur-Atwal, G.; Boyle, P.; Creaser, C.S., Miniaturized ultra high field asymmetric waveform ion mobility spectrometry combined with mass spectrometry for peptide analysis, *Anal. Chem.* 2010, 82(23), 9827–9834.

97. Walter, C.; Karpas, Z.; Lorber, A., New technology for diagnosis of bacterial vaginosis, *Eur. J. Obstet. Gynecol. Reprod. Biol.* 2003, 111, 83–87.

98. Eiceman, G.A., Workshop on Ion Mobility Spectrometry, Final Report, Task Control No. 92057, Delivery Order No. 0127, CRDEC SSP92-04, to Battelle Columbus Division, Research Triangle Park, NC, New Mexico State University, Las Cruces, NM Sept. 1992.

3 Sample Introduction Methods

3.1 INTRODUCTION

The efficient transfer of an analyte from its original condition to the ionization region of an ion mobility spectrometer (IMS) is the topic of this chapter. Successful detection and identification of an analyte by IMS depend on many steps but none more important than those by which a sample is introduced into an instrument. IMS instruments are used for the detection and identification of analytes found in air, water, biological fluids and tissues, industrial solvents and on surfaces. Because ion mobility spectrometry is such a universal analytical instrument, sample introduction methods are diverse and depend on the type of sample analyzed. Atmospheric pressure operation makes IMS suitable for interfacing with several sample introduction systems as a detector as well as a selective filter for mass spectrometric techniques.

Because IMS is a method that separates gas phase ions through collisions with a buffer gas, all analytes must be transported from the sample matrix and converted to a gas phase ion before ion mobility separation and detection can be performed. Thus, the type of introduction method largely depends on the physical characteristics of the analyte. The remainder of this chapter is divided into four sections based on the characteristics of the sample: vapor, semivolatile, aqueous, and solid. While these categories are somewhat arbitrary with significant sample overlap, it is useful to think of volatile samples as those compounds that exist or partially exist as vapors under ambient temperature and pressure; semivolatile samples as those compounds that can be volatilized but have vapor pressures too low to detect by IMS under ambient temperature and pressure; aqueous samples as those compounds that are not volatile but can be dissolved in water; and solid samples as compounds not in a solution. Table 3.1 lists a number of example analytes according to the categories discussed in this chapter.

3.2 VAPOR SAMPLES

Initially, IMS was developed as an analytical method for the detection of vapors. An early analytical application was described by Cohen and Karasek[3] and is shown in Figure 3.1.

Here, approximately 2 parts per billion by volume of dimethyl sulfoxide (DMSO) in dry air were introduced continually and directly to the ionization region of the first atmospheric pressure IMS. The abscissa of the ion mobility spectrum is the arrival time of each ion species in milliseconds. This arrival time is inversely proportional to the mobility of the ion species represented by the individual peaks in the spectrum.

TABLE 3.1
Sample Categories Suitable for Ion Mobility Spectrometry

Sample Types	Vapor	Semivolatile	Aqueous	Solid Phase
Environmental	VOCs	Pesticides	Anions and cations	
Security and safety	CWAs			
TICs	Explosives			
Forensics		Explosives	Metabolites	
Clinical	Breath	Pharmaceuticals	Metabolites	Proteins and peptides
Law enforcement		Drugs of abuse		

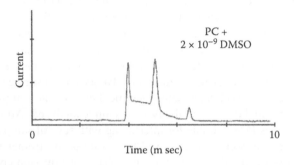

FIGURE 3.1 Early ion mobility spectrum at atmospheric pressure. Positive ions of dimethyl sulfoxide (DMSO) at about 2×10^{-9} mole parts in dry air at 760 torr and 150°C. (From Cohen and Karasek, Plasma chromatography—a new dimension for gas chromatography and mass spectrometry, *J. Chromatogr. Sci.* 1970, 8, 330–337.)

Thus, in Figure 3.1, the peak with an arrival time of approximately 4 ms has the highest mobility of the three peaks produced from the air sample of DMSO. For the peaks shown in Figure 3.1, the one with an arrival time of about 4 ms was identified as the hydronium reactant ions $(H_2O)H^+$ and $(H_2O)_2H^+$ in rapid equilibrium; the peaks at about 5 and 6.5 ms were the product ions of DMSO and were identified as the protonated species $(DMSO)H^+$ and the proton-bound dimer $(DMSO)_2H^+$, respectively. Spectral interpretation is discussed in more detail in this chapter. The high background between the reactant ions and the product ions is called *bridging* or *ion stripping* and is presumably due to decomposition reactions of the product ion with water in the drift region of the spectrometer to produce the reactant ion as shown in Reaction 3.1. More is discussed about ion–molecule reactions occurring in IMS in this chapter.

$$(DMSO)H^+ + H_2O \rightarrow DMSO + (H_2O)H^+ \qquad (3.1)$$

3.2.1 DIRECT INJECTION

The simplest method for introducing a sample into an IMS is by flowing a carrier gas containing the neutral vapor of the analyte, or analytes, directly into the ionization region of the IMS. The first mixture separated and detected by IMS was a mixture of

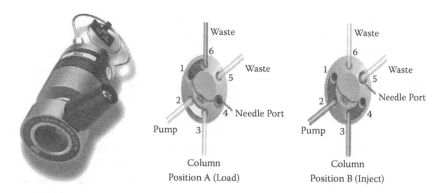

FIGURE 3.2 A schematic diagram of a six-port valve for direct IMS injection. The picture shows a Rheodyne 7725-type six-port valve with an external sample loop. Position A is the "load" position in which the loop is filled, and position B is the "inject" position in which the sample in the loop is directed into a chromatographic column or directly in an IMS. (From Rheodyne Web site.)

DMSO, malathion, and triethyl phosphite in air.[4] In these first investigations of IMS as a separation and detection method for organic mixtures, the air containing these compounds was directly introduced into the ionization region of an IMS using a radioactive ^{63}Ni source. Ion molecule reactions of each neutral analyte with the reactant ions produced stable characteristic product ions from each of these compounds that were subsequently separated and detected in the IMS. This approach is also used for continuous monitoring (as opposed to discrete samples, described in the following).

Today, direct injection of vapor samples into an IMS is best accomplished with the use of a six-port valve (see Figure 3.2). This can be used for "sample-and-detect" operations. Sample-and-detect operations, unlike continuous monitoring, can be used for discrete measurement. Systems for continuous monitoring and measurement are described in this chapter.

The valve is operated in the following manner: Each port is connected to a transfer tube. Port 1 is connected directly to the sample loop. Sample loops can be external or internal to the valve. In Figure 3.2, the picture of a Rheodyne 7725 valve shows an external sampling loop that is 20 μL in volume. The volume of an external sampling loop may be varied from microliter to milliliter volumes. The handle on the valve is toggled from the "load" to the "inject" position. In the load position, the carrier gas comes from the pump or buffer gas tank through port 2 and through a chromatographic column into the IMS or directly into the IMS. The loop is filled (loaded) with the sample by introduction through port 4 into the loop, then through port 1 and out to waste at port 6. When the valve is toggled to the inject position, the carrier gas enters the valve through port 2, travels to the loop through port 1, and sweeps the sample out of the loop through port 4 and into the chromatographic column and IMS through port 3.

3.2.2 MEMBRANES

Direct injection (for continuous monitoring) of vapor samples into the IMS can create problems with ionization if the sample contains large amounts of water (moisture).

This is particularly a problem when the IMS is used for the detection of vapors in real samples where the humidity and other contaminants in the air can vary in concentration in the sample, interfering with the analytical response of the target analyte. To prevent unwanted contaminants from entering the ionization region of the IMS, a membrane can be placed between the sampling section and the ionization section of the IMS. The practice of using membranes for selective introduction of samples to analytical instruments is at least 50 years old.[5] They were initially used to separate the sampling from the ionization regions of mass spectrometers. Both direct sampling and sampling from gas chromatographs (GCs)[6] have benefited from the use of membranes for selective sample introduction. If the membrane is chosen correctly, target analytes such as organic compounds will pass through the membrane while interfering species such as water will be retarded by the membrane. Membranes also prevent particles contained in the sample from entering the IMS. In theory, membranes can be selected for discrimination of a wide variety of contaminants; in practice, however, small polar molecules are excluded by silicone membranes, while volatile and semivolatile organic molecules preferentially pass through the membrane into the ionization region of the IMS. Applications using membrane inlets include quality control of pharmaceuticals,[7] environmental analysis,[7,8] online process control,[9,10] and the detection of flavors and fragrances.[11,12]

Sometimes, membrane inlets are operated at the same temperature as the IMS cell, and sometimes they are operated at elevated temperatures. More recently, advantages of controlling the temperature of the membrane separate from that of the cell have been reported.[13] The primary advantage of temperature-controlled membranes for sample introduction to an IMS is the ability to concentrate the sample and then heat the membrane to provide low-resolution separations as the analytes diffuse through the membrane. One application of this active membrane technology is the detection of benzene in water by photoionization differential mobility spectrometry.[14] Drawbacks of using membranes are reduced sensitivity, increased response times, and longer clearance times (i.e., memory effects).

3.2.3 Active Inlets and "Sniffing"

One potential application of IMS technology is to replace expensive, high-maintenance, and singly selective canines with low-cost, durable, and tunable-selective instruments for the detection of chemical agents, explosives, toxic industrial chemicals (TICs), drugs, and other trace vapors. Several sampling approaches have been utilized, mostly by the military for the rapid detection of chemical agents. In general, sniffing is accomplished by drawing gas samples past a membrane with a small air-sampling pump. Detection relies on the ability of the analyte to dissolve into the membrane and then diffuse through the membrane into the ionization region of the spectrometer. These inlets can be "active" in that the temperature of the membrane can be controlled as described previously to cool and concentrate the sample and then heated to inject the sample into the ionizer.

An alternative method for "sniffing" a gas sample into an IMS was developed by Smiths Detection for the lightweight chemical detector (LCD). In this approach, the gas sample is drawn through a plenum with a small fan. A pinhole in the plenum serves as the entrance to the IMS ionizer. To inject a gas sample into the IMS, the pressure is briefly decreased inside the IMS by a vibrating speaker diaphragm, and a pulse of vapors is sucked into the IMS through the pinhole from the sample plenum. The vibration of the speaker diaphragm can be set to a specific frequency for rapid and continuous "sampling." A single IMS spectrum can be obtained within 20 ms, so the sampling frequency can be as high as 50 Hz. High sensitivity and fast response times required for some applications involving continuous monitoring of highly toxic chemical vapors at very low concentrations can be achieved by directly sampling ambient air.

3.3   SEMIVOLATILE SAMPLES

As stated, the distinction between semivolatile and volatile samples is somewhat arbitrary. For this chapter, semivolatile samples are defined as those compounds with vapor pressures too low to be directly detected by IMS. Because most ionization sources used with IMS require neutral vapor samples to be submitted to the ionizing radiation, semivolatiles must be converted to the vapor state before detection. Volatilization is typically accomplished by increasing the temperature of the introduction platform through thermal desorption or heated chromatographic inlets and in some cases by changing the chemical form of the compound through simple chemical treatment of the sample (e.g., using an alkaline solution to convert acidic amines into volatile free base). Chemical conversion and heat can be combined to enhance vaporization.

3.3.1   PRECONCENTRATION

For volatiles and semivolatiles with vapor pressures too low to be detected by direct IMS introduction methods, preconcentration techniques are often employed. The most common approach is to pack a glass tube or stainless steel column with an adsorbing material such as Tenax® or Carbosieve®. Other methods, such as microfabricated adsorbents, have also been proposed, but these microfabricated concentrators are relatively new, with limited field exposure. Nevertheless, for handheld and mini-IMS instruments, these preconcentration devices appear promising, offering rapid concentration and desorption with low power.[15]

For a typical preconcentration operation, a vapor sample is pulled through an adsorption material at a flow rate much larger than is possible with direct injection into an IMS. For example, the flow rates used in a standard preconcentrator are on the order of 160 L/s. With a miniaturized preconcentrator, these flows are approximately 2 L/s. For preconcentrators used for explosive detection in airports, the filters that are used are known as metal felt. Metal felt consists of a high-density mesh of metal filaments that provides minimal restriction to the sample airflow but efficiently adsorbs organic vapors in the air sample. After the adsorption step, the metal felt is heated to 200°C, volatilizing the adsorbed species into a clean carrier gas, which is directed into the IMS. The carrier gas flow is very low compared with the sampling gas flow. Thus, the net result of the adsorption/desorption operation is to increase the concentration of the analyte by a factor of 10 or more. A comprehensive review of the rapidly developing area of microconcentration devices can be found in the work of Rodriguez and Vidal.[16]

It should be mentioned that these sampling devices can be used when directly interfaced with the IMS, as shown previously, or for sample collection in the field and subsequent analysis in the laboratory. Regardless of the type of adsorbent, all preconcentrators have several desirable features in common: They should have high flow rates in the adsorption step so that a large quantity of sample can be rapidly concentrated. They should be stable to thermal heating so that the adsorbed analyte can be released to the IMS, and they should have the ability to adsorb target analytes selectively while letting interfering species pass through the collection device. The gas sample is passed over the adsorbent and concentrated onto the surface of the material. The adsorbent material is then transferred to the laboratory and thermally desorbed, converting the adsorbed analyte into a vapor, which is then swept by an inert carrier gas into the ionization region of an IMS. The advantage of microfabricated preconcentration devices is that they can be directly attached to an IMS and rapidly cycled to produce rapid injections of concentrated samples into the IMS.

While most preconcentration methods focus on the collection and concentration of analytes from vapor samples, the purge-and-trap method is used for collection and concentration of volatile and semivolatile analytes from liquid (usually aqueous) samples.

Figure 3.3a shows a schematic of a purge-and-trap system for the collection of volatiles contained in an aqueous sample. Clean air is bubbled through the water sample to purge the water of the volatiles and then carry the volatiles into the IMS. If the concentrations of the analytes are too low for direct detection, a trap (preconcentrator) device is inserted between the purged water sample and the IMS. Figure 3.3b is a schematic of an exponential dilution system used to calibrate IMS instruments for

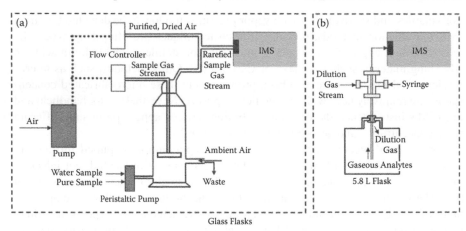

FIGURE 3.3   (a) Glass flask for extracting volatile organic compounds from water samples. (From Borsdorf, H.; Rammler, A.; Schulze, D.; Boadu, K.O.; Feist, B.; Weiss, H., Rapid on-site determination of chlorobenzene in water samples using ion mobility spectrometry, *Anal. Chim. Acta* 2001, 440, 63–70.) (b) Exponential dilution system. (From Sielemann et al., Detection of alcohols using UV-ion mobility spectrometers, *Anal. Chim. Acta* 2001, 431, 293–301. With permission.)

quantitative analysis. Here, a calibrant gas or volatile liquid is injected with an initial concentration of $C(0)$ into a container for which the volume is precisely known (V in milliliters). As a carrier gas is introduced into the container, the gaseous analytes at a flow rate of F mL/min are swept out into the IMS, and their concentration decreases exponentially with time (t in minutes), so that after t minutes it is $C(t)$ (Equation 3.2).

$$C(t) = C(0) * \text{Exp}\left(-F * t/V\right) \tag{3.2}$$

Equation 3.2 holds for analytes that are not absorbed on the surface of the container and for perfect mixing of the analyte. The actual behavior of analytes, especially polar molecules, would show a slower decrease in signal intensity than that predicted by true exponential dilution.

3.3.2 THERMAL DESORPTION

The most common approach for semivolatile detection is through the thermal desorption of samples from the solid state to produce neutral vapors, which are then introduced into the IMS as discussed. The first thermal desorption method was simply a platinum wire onto which 1 or 2 µL of the liquid sample, or dissolved sample solution, was placed and the solvent evaporated, leaving the semivolatile components adsorbed on the wire's surface.[17] This wire was then inserted into the heated injection port of an IMS, where the sample was thermally desorbed into the carrier gas and the vapors transferred to the ionization region of the spectrometer.

The main driving force for the development of sampling methods for semivolatile compounds has been the need for rapid and sensitive explosive detectors. The principal purpose for these explosive detectors is to detect bombs that are being smuggled onto planes, trains, and buses or into crowed public areas, such as athletic events, shopping centers, and amusement parks. Explosive materials are "sticky" compounds and will adsorb onto many surfaces. Traces of explosive material will be adsorbed onto shoes from walking across a floor contaminated with explosives. Through handling, accidental contact, or deposition of aerosols, explosive materials can be transferred to a variety of surfaces. Thus, the typical approach for explosive detection is to wipe suspected surfaces such as shoes, briefcases, computers, and hands with special cloth materials treated in a manner that enables the efficient collection of small quantities of semivolatile particles of the explosive and deliver them for thermal desorption into an IMS.

The primary problems in explosive detection are (1) small sample sizes, (2) the fact that semivolatile compounds condense on cooler surfaces of the inlet and transfer lines, and (3) the requirement of rapid detection.[18] Thus, the challenge for detection is the rapid, quantitative vaporization and transfer of trace amounts of explosive material from the site of collection to the ionization region of the IMS. Figure 3.4 shows a typical IMS used for the detection of explosives and other semivolatile compounds, such as drugs of abuse and pesticides.[19]

While IMS instruments are described elsewhere in this text, it is worth discussing in detail the sample introduction part of Figure 3.4 as it relates to thermal desorption

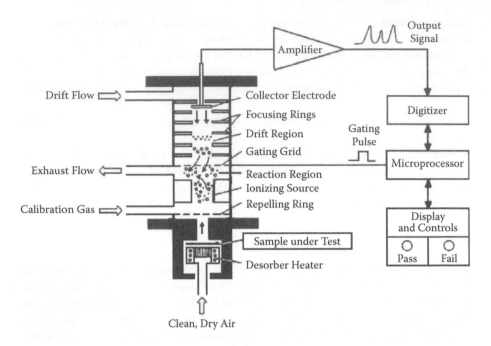

FIGURE 3.4 Block diagram of an ion mobility spectrometer used for thermal desorption introduction of samples. (From Fetterolf and Clark, Detection of trace explosive evidence by ion mobility spectrometry, *J. Forensic Sci.* 1993, 38(1), 28–39. With permission.)

and the detection of semivolatile chemicals. The sample introduction method of the IMS shown in Figure 3.4 consists of the following: An entrance port for *clean, dry air* serves as the carrier gas to transfer the sample from the thermal desorber to the ionization region of the spectrometer; a *desorber heater* supplies the heat to the sample that has been captured on the collector material described and inserted onto the desorber heater block, where the semivolatile material is vaporized into the clean, dry air carrier gas; a *repelling ring* applies the voltage to the IMS, setting an electrical field through which the ions migrate in the time-of-flight (TOF) spectrometer; and a *calibration gas* inlet uses IMS dopants to obtain selective ionization, and IMS standards used to calibrate the instrument can be introduced. For explosive detection, an ionization dopant would typically be methylene chloride, which undergoes electron capture dissociation in the ionization region of the spectrometer to form the chloride reactant ion that is used to ionize explosive vapors selectively through attachment or clustering (see Chapter 12).

The desorber heater and the drift tube with the drift gas can be operated at a variety of temperatures. In general, the desorber temperature is 210°C, but some explosives will decompose at these temperatures and thus must be desorbed at lower temperatures. Programmed temperature desorption appears to be a promising approach for obtaining IMS spectra of a broad range of explosives as well as for simple and quick graduated thermal desorption to differentiate between compounds of different volatility.

3.3.3 SOLID PHASE MICROEXTRACTION

Thermal desorption from a solid phase microextraction (SPME) fiber has shown considerable potential for selectively introducing semivolatile chemicals into an IMS.[20] The SPME approach is a simple design patterned after the early platinum wire introduction thermal desorption system described. With SPME, semivolatile compounds are extracted by either absorption or adsorption onto a nonvolatile polymeric coating or solid sorbent phase that has been coated onto a small fiber. Normally, the adsorption fiber is housed in the needle of a syringe to permit puncture of a sample bottle septum and to protect the fiber from contamination during transfer of the fiber from the sample to the IMS instrument. After the analytes are adsorbed onto the SPME fiber, the fiber is retracted into the needle and then injected in a normal syringe technique such that the fiber is extended into the heated region of the IMS and the analytes are desorbed from the fiber into the clean carrier gas of the IMS.

SPME affords a means of enhancing selectivity and specificity as well as increasing sensitivity by preconcentration of vapors or dissolved compounds. SPME can also serve to transport samples that were collected in the field to the laboratory, where the analysis may take place.

Other extraction and preconcentration methods patterned after the SPME approach include stir-bar sorptive extractors in which sorptive materials are coated onto a stir bar, and as the bar is stirred in solution, selected organics or other analytes adsorb onto the surface. As with the previous methods, these stir bars are heated to release their adsorbed analytes into the IMS.

3.3.4 GAS CHROMATOGRAPHY

When a mixture of compounds is introduced directly into the ionization region of an IMS, charge competition occurs among analytes. In positive mode, if the proton affinity of one compound A is greater than another compound B in the sample, the compound with the greater proton affinity will preferentially acquire the charge. Thus, under equilibrium conditions, compound A may be completely ionized before compound B can acquire a charge and be protonated. For example, if the reactant ion H_3O^+ reacts with compound B to form BH^+ and if BH^+ collides with compound A having a greater proton affinity than compound B, then $A + BH^+ \rightarrow AH^+ + B$ will occur, and only AH^+ will be detected even though compound B was present in the mixture. Gas phase ionization sources are not always operated under equilibrium conditions because the analyte is often present in trace amounts, and the probability of an $A + BH^+$ type collision is low. Thus, both AH^+ and BH^+ may be observed from a single sample, although true quantification is difficult due to the lack of information on the ion suppression process. The use of chromatographic injection methods reduces the effects of charge competition in IMS (and in mass spectrometry), enabling quantitative analysis.

IMS matches well as a detector for GC and can be interfaced directly to the exit of a gas chromatograph as a selective detector. Because ion mobility spectra are obtained at a frequency of 20 to 50 Hz, many IMS spectra can be obtained for

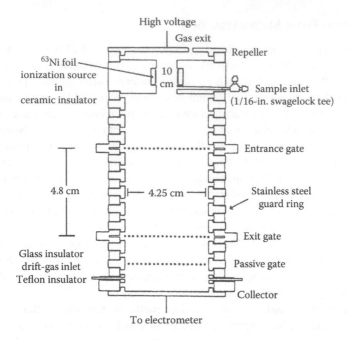

High voltage

Gas exit

Repeller

^{63}Ni foil ionization source in ceramic insulator

10 cm

Sample inlet (1/16-in. swagelock tee)

Entrance gate

4.8 cm

4.25 cm

Stainless steel guard ring

Exit gate

Glass insulator drift-gas inlet
Teflon insulator

Passive gate

Collector

To electrometer

FIGURE 3.5 Schematic of the first sealed ion mobility spectrometer. This was used as a chromatographic detector for high-resolution capillary gas chromatography demonstrating the ability of IMS for high-throughput analysis. (From Baim and Hill, Tunable selective detection for capillary gas chromatography by ion mobility monitoring, *Anal. Chem.* 1982, 54(1), 38–43. With permission.)

each chromatographic peak. The first example of GC as an introduction method for IMS was demonstrated by the GC separation of freons with ion mobility spectra for each separated peak as they eluted from a GC packed column into the IMS.[21] High-resolution separation of a complex mixture of gasoline using a capillary chromatographic column interfaced directly to a unique IMS design demonstrated that IMS could be used for high-throughput detection.[22]

Figure 3.5 shows a schematic of the first capillary gas chromatograph (CGC)-IMS in which the drift region was completely enclosed and the buffer gas was introduced at the detector end of the spectrometer, exiting through the ionization region.

The effluent from the CGC was introduced orthogonally to the buffer gas flow between the ion gate and the ^{63}Ni foil and then swept with the full velocity of the buffer gas back through the ionization region. This reverse flow, which is essentially unidirectional through the ionization region of the spectrometer, enables more efficient ionization of neutral analytes and reduces the effects of charge competition on the sample. Thus, extremely low detection limits can be achieved even in mixtures. Sensitivity can be improved further by the axial introduction of the GC effluent to the IMS tube.[23] It has also been suggested that sensitivity can be improved by modifying the axial approach by including a sleeve with a makeup flow to create turbulence and efficient mixing of the analyte vapors with the reactant ions.[24]

CGC can provide high-resolution separations of complex mixtures before ion mobility interrogation of individual components of the mixture. IMS after CGC offers a number of modes of detection. First, when only the positive reactant ion peak is monitored, the IMS behaves similar to that of a flame ionization detector in that it responds nonselectively to most compounds introduced into the ionization region, and compound identification is based solely on the retention time of the GC. For water-based ion chemistry, reactant ions are primarily hydronium ions $(H_2O)_n H^+$ that react with neutral analytes eluting from the CGC column to produce MH^+ ions that become the product ions characteristic of the analyte. The decrease of the hydronium ion signal intensity in the IMS depicts a detection event and follows the elution concentration profile of the chromatographic peak. If, however, the mobility of the product ion (MH^+) is continuously monitored, then the IMS becomes a selective detector for a specific compound of interest. By continuously monitoring negative reactant ions, the IMS behaves similar to an electron capture detector (ECD), and when negative product ions are monitored, the IMS becomes a mobility-specific detector for electron-capturing compounds. Alas, due to the 1% or so duty cycle of the IMS and ion losses in the drift section, the sensitivity of the ECD is two or three orders of magnitude higher. When an IMS is operated in a manner in which only specific mobilities are monitored as described, it is referred to as an ion mobility detector (IMD). The IMD provides two-dimensional (2D) detection and identification.

Figure 3.6 is a diagram of a CGC that is interfaced to an IMS connected to a TOF mass spectrometer (TOFMS).

The speed of IMS is perfect for interfacing to both CGC and mass spectrometry. CGC separations of complex mixtures normally take minutes. Each peak in a high-resolution chromatogram may take a few seconds. IMS spectra normally require 10–100 ms for full spectra, while mass spectrometric spectra require only 10–100 μs for full TOF mass spectra. Thus, for each analyte it is possible to obtain three-dimensional information based on retention time in the CGC, ion mobility, and mass. Figure 3.7 is a 2D CGC-IMS spectrum of lavender oil.

Today, IMS is often coupled to multicapillary columns to better match the flow rates of several milliliters per minute while maintaining good chromatographic resolving powers. Figure 3.8 shows a commercially available multicapillary column, and a schematic representation of the device is shown in Figure 3.9.

The application of GC for the introduction of complex samples into field asymmetric IMS (FAIMS) and differential mobility spectrometry (DMS) instruments is also used extensively. Fast capillary chromatography in which relatively simple mixtures can be separated in less than a second provides a rapid separation-and-introduction method for DMS.[25] One specific advantage of FAIMS (or DMS) as a chromatographic detector is that both positive and negative ions can be monitored simultaneously from the GC effluent.[26] Figure 3.10 provides a schematic of a typical capillary GC/DMS instrument in which SPME is used to inject semivolatile compounds into the capillary column with DMS detection.

In this configuration, thermal desorption from a SPME fiber introduces the vapors of a mixture of the adsorbed volatile and semivolatile compounds into a helium

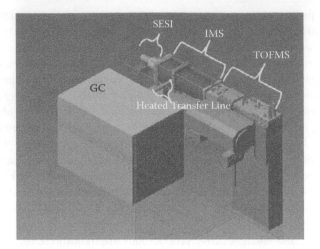

FIGURE 3.6 Diagram of a CGC-SESI-IM-TOFMS (time-of-flight mass spectrometer). The CGC was connected to an IM-TOFMS via a heated transfer line, enabling the introduction of gaseous analytes into the reaction region of the IMS. These CGC-separated analytes were subsequently ionized by secondary electrospray ionization. The analyte ions were then pulsed into the IMS drift region for mobility separation. The ions were further separated and detected according to their mass-to-charge ratios by the TOFMS.[48] (From Crawford et al., The novel use of gas chromatography-ion mobility-time of flight mass spectrometry with secondary electrospray ionization for complex mixture analysis, *Int. J. Ion Mobil. Spectrom.* 2010, 14, 23–30. With permission.)

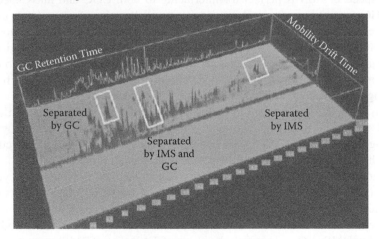

FIGURE 3.7 Graphical representation of CGC retention time (in minutes), mobility drift time (in milliseconds), and total ion intensity (in arbitrary units) of lavender oil. The highlighted peaks (in white boxes) show separation of analyte peaks by GC only, IMS only, and both CGC and IMS; this dual CGC and IMS technique enables a greater degree of separation for a complex mixture than with either technique alone.[48] (From Crawford et al., The novel use of gas chromatography-ion mobility-time of flight mass spectrometry with secondary electrospray ionization for complex mixture analysis, *Int. J. Ion Mobil. Spectrom.* 2010, 14, 23–30. With permission.)

FIGURE 3.8 Cross section of a multicapillary column.[49] (From Sielemann et al., *Int. J. Ion Mobil. Spectrom.* 1999, 1, 15–21.)

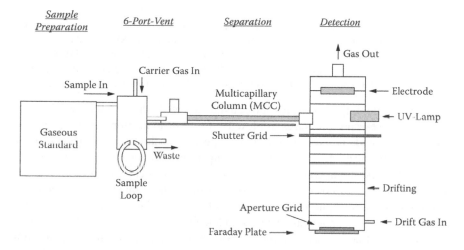

FIGURE 3.9 Schematic presentation of multicolumn GC-IMS.[49] (From Sielemann et al., *Int. J. Ion Mobil. Spectrom.* 1999, 1, 15–21. With permission.)

carrier gas stream that carries the mixture to the chromatographic column for separation. After GC separation, the 250 mL/min of DMS carrier gas flow of nitrogen is mixed with the 1 mL/min flow of helium from the GC column to produce the flow through the DMS. At the entrance to the DMS, the sample is ionized by either a radioactive source or an ultraviolet (UV) ionization source.

3.3.5 Supercritical Fluid Chromatography

In addition to gases, supercritical fluids can be used to carry samples into an IMS after chromatography or directly after extraction. Supercritical fluids have two primary advantages for sample introduction. First, unlike gases, supercritical fluids

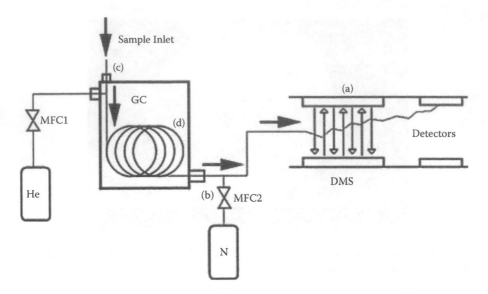

FIGURE 3.10 Experimental setup of GC–DMS system. Chemical analysis is performed using gas chromatography differential mobility spectrometry (GC/DMS). Several user-defined parameters were selected factorial experiments: (a) the RF voltage of the DMS sensor, (b) nitrogen carrier gas flow rate through the DMS, (c) solid phase microextraction (SPME) filter type, and (d) GC cooling profile. (From Molina et al., *Anal. Chim. Acta* 368(2), 2008.)

provide some solubility for samples and can thus solvate high molecular weight compounds that have vapor pressures too low for GC. For ionization detectors such as IMS, carbon dioxide is the most common supercritical fluid used. Carbon dioxide becomes supercritical when the operating temperature is above its critical temperature of 31.3°C and the operating pressure is above its critical pressure of 73.9 bar. The disadvantage of carbon dioxide as a supercritical fluid is that it is nonpolar and is best used for solvating nonpolar compounds. One rule of thumb is that if an analyte is soluble in hexane, then it would be a candidate for supercritical fluid chromatography (SFC) and extractions using carbon dioxide as the supercritical fluid. Polar fluids such as water have also been proposed and investigated as supercritical fluids, but operating conditions are too extreme for practical application. For example, water has a critical temperature of 374.4°C, well above that of carbon dioxide, and a critical pressure of 229.8 bar, five times higher than carbon dioxide.

The second advantage of CO_2 as a supercritical fluid for introducing samples into an IMS is that CO_2 is considered a "green" technology. Although CO_2 is released to the atmosphere after SFC or supercritical fluid extraction (SFE), it was originally purified from the atmosphere and thus is carbon neutral. Unlike liquid chromatography (LC) or solvent extraction, for which the solvents used in these technologies have to be contained and appropriately disposed, CO_2 is simply vented to the atmosphere as a gas after chromatography or extraction. Finally, CO_2 serves as a unique buffer gas for IMS. Due to its polarizability, CO_2 provides unique ion mobility separation patterns relative to those more nonpolarizable buffer gases of He or N_2 that have low polarizability values (see Chapter 11).

IMS has been compared with the UV detector after separation with SFC of benzoates and esters, demonstrating the ability of IMS for the detection of compounds that do not have sensitive chromophores for UV detection.[27] Often, polymers do not have sufficient UV-visible (Vis) absorbance for detection after LC or SFC, while SFC-IMS can be used for both efficient separation and detection of a variety of polymeric materials.[28] IMS detection of a variety of drugs, such as various steroids, opiates, and benzodiazepines, after SFC separation demonstrated the ease and utility of acquiring ion mobility spectra at ambient pressure.[29] While SFC has only captured a small portion of the separation market, the potential of IMS for detecting compounds that cannot be easily seen with a standard UV-Vis approach has led to its use as a stand-alone detector for LC.

3.4 AQUEOUS SAMPLES

Most compounds that dissolve in nonpolar solvents are nonpolar and may have significant vapor pressures. Thus, they can be introduced into an IMSs by the direct injection of the solvent or by deposition of the sample solution on a surface where the solvent is evaporated and the analyte is then thermally desorbed into the IMS. While this approach was acceptable for semivolatile compounds dissolved in nonpolar solvents, it still excludes important biological, environmental, and industrial samples dissolved in water. Water is the ubiquitous solvent on Earth. Thus, our living systems have evolved around the use of water to transport chemicals through our environment and within organisms. Most compounds important to chemistry or life on Earth have polar moieties that reduce or eliminate vapor pressure while increasing water solubility. Thus, application of IMS to aqueous samples significantly expands its utility as an analytical tool.

3.4.1 ELECTROSPRAY

The primary process by which polar and ionic analytes are introduced into an IMSs is electrospray ionization (ESI). The first attempt to interface ESI to IMS was by Dole in 1972; he obtained the first ion mobility spectrum for an organic compound with ESI.[30,31] In these early investigations, liquid flow rates were so large and heat transfer from the carrier gas was so inefficient that the solvent was not evaporated from the analyte ions, diminishing drift time separation.

Figure 3.11 shows an early design of an electrospray IMS in which a counterflow of heated buffer gas was used to evaporate the solvent from the analyte ions such that ions were desolvated from solvent molecules and introduced into the drift region of an ambient pressure IMS.

Figure 3.12 is the first successful ion mobility spectrum of an analyte ion greater than 1000 amu. An electrospray source was fitted into a drift tube IMS in which heated nitrogen buffer gas was flowing counter to the electric field. The electrospray solvent was evaporated from the electrosprayed droplets so that the bare, desolvated ion was injected into the drift region of the spectrometer, and ion separation could occur. In Figure 3.12, the product ion of erythromycin estolate (MW = 1,056 Da) was separated from the background ions produced from the electrospray process.

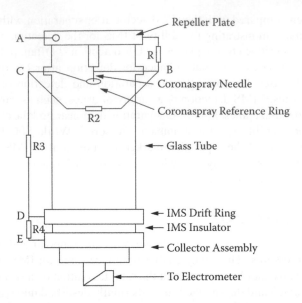

FIGURE 3.11 Electrospray prototype: corona ring, 1-cm diameter; R1, R2, R3, and R4 represent resistors at values of 1.25, 2.5, 6.2, 6.2, and 50 kΩ, respectively. Conditions to initiate spray were (A) 4,000 V, (B) 3,500 V, (C) 2,500 V, (D) 20 V, and (E) ground. Typical total current was 7 nA. (Taken from Shumate, An electrospray nebulization/ionization interface for liquid introduction into an ion mobility spectrometer, thesis, Washington State University, Pullman, 1989.)

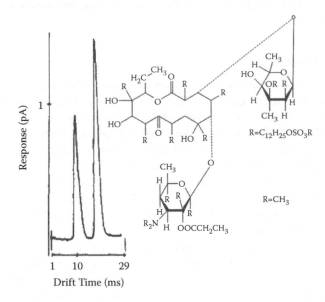

FIGURE 3.12 First electrospray ion mobility spectrum: erythromycin estolate (MW = 1,056 Da). (From Shumate and Hill, Coronaspray nebulization and ionization of liquid samples for ion mobility spectrometry, *Anal. Chem.* 1989, 61(6), 601-606. With permission.)

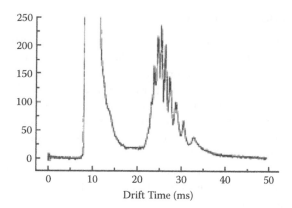

FIGURE 3.13 ESI of cytochrome c demonstrating the first example of multiply charged ions, 11–18 positive charges/ion, separated by IMS.[51] (From Wittmer, Workshop on Ion Mobility Spectrometry, Mescalero, NM, June 1992.)

Other early electrospray ion mobility spectra include those of peptides, showing a sharp peak for peptide ions that were clearly resolved from the reactant (or background) ions generated from the electrospray solvent. Mixtures of biological compounds such as amines can also easily be electrosprayed into an IMS and separated by their respective mobilities. One of the key advantages of introducing samples into an IMS by ESI is that large biomolecules can be introduced into an IMS. For example, Figure 3.13 demonstrates the first example of a protein being electrosprayed into an IMS. In this case, the protein is cytochrome c, and a number of separated charge states are clearly visible in the spectrum. The large peak around 10 ms is the solvent background, but the sharp peaks on the right are due to pure cytochrome c. Each peak represents cytochrome c with a different charge state. In this IMS spectrum, the protein has 11 protons on the peak at about 34 ms, and each successive peak located at a shorter drift time is the same protein with an additional proton. Thus, the next faster peak has 12 protons, the next 13 protons, and so on. From this spectrum, charge states from $(M + 11H)^{11+}$ to $(M + 18H)^{18+}$ could be observed for cytochrome c. These charge states were initially confirmed by comparison with the electrospray mass spectrum of cytochrome c and later by coupling the IMS directly to a mass spectrometer.

Inorganic and organic ions can also be introduced into an IMS through the electrospray process, opening up the possibility for the rapid separation of anions and cations in the gas phase rather than the slow processes that are involved using ion chromatography or capillary electrophoresis. In an early application, the anions chloride, nitrite, formate, nitrate, and acetate were all separated by IMS after ESI introduction.

In summary, by using a heated reverse drift flow to both evaporate and prevent solvent vapors from entering the drift region of the IMS, ESI has become a common ionization source for IMS.[32–35] Today, the ESI source serves as the favorite direct introduction method of aqueous samples into an IMS for stand-alone separation or for rapid preseparation before mass spectrometry. It is also the preferred interface between LC and IMS.

3.4.2 LIQUID CHROMATOGRAPHY

One early method for interfacing LC to IMS instruments consisted of a rotating metallic tape. The sample was deposited on the tape at the exit of the chromatographic column, transferred to the entrance of the IMS, thermally desorbed into the IMS, and then recycled through a cleaner, which served to thermally clean the tape, and then the tape was cycled back to the chromatographic exit for another sample. By keeping the rotation of the tape constant, an ion mobility tracing of an LC separation could be obtained. In general, only compounds with some volatility could be separated and detected by LC–IMS.

As with mass spectrometry, the advent of ESI enabled the effluent from a LC to be directly introduced into an IMS.[36] Today, ESI–IMS is used as a stand-alone detector for LC and as an effective interface between LC and mass spectrometry, separating such complex biological mixtures as carbohydrates, protein digests, and metabolomes.

One unique development in relation to electrospray is paperspray,[37] a soft ionization method in which the sample is ionized into the IMS in an uncomplicated manner without the need for a pump or capillary. The advantage is that an analyte can be ionized directly from filter paper containing the sample or after separation by paper chromatography. Examples of the use of paperspray-IMS include chlorpromazine, reserpine, and 2,6 Di-t-butylpyridine.

3.5 SOLID SAMPLES

Perhaps the most common method for the introduction of solid samples into an IMS is through thermal desorption as described for semivolatile compounds. When semivolatile compounds are present in solid samples (e.g., drug tablets or explosives adsorbed onto surfaces or inert particles), heating them to produce vapors and transferring those vapors via a carrier gas into an IMS is an efficient method of sample introduction. In this section, however, we use the term *solid samples* to refer to those target molecules that are not volatile (or have extremely low volatility), such as high molecular weight biological and industrial polymers, ionic species, or highly polar compounds.

Nonvolatile solid samples can be bulk crystalline samples or target analytes adsorbed onto a surface. One approach to generate gas phase ions from solid samples is to insert a probe containing the sample directly into the ionization region of the IMS. More convenient methods, however, have recently been developed in which the ionizing process is brought directly to the surface to be analyzed. In these cases, the ionization source also serves as the method for sample introduction into the IMS as the sample is already introduced in ionic form. Three common sources for surface ionization of solid samples are direct analysis in real time (DART), desorption electrospray ionization (DESI), and matrix-assisted laser desorption ionization (MALDI). Here, the sample introduction method and the ionization method overlap. They differ significantly in ionization mechanism: The DART source uses energetic neutrals, DESI uses ions, and MALDI uses light to generate ions for analysis from solid samples on surfaces. A more detailed discussion of each of these ion sources follows (see also Chapter 4).

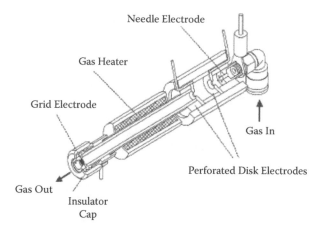

Needle Electrode

Gas Heater

Grid Electrode

Gas In

Perforated Disk Electrodes

Gas Out

Insulator
Cap

FIGURE 3.14 Cutaway view of a DART (direct analysis in real time) source. (From Cody et al., Versatile new ion source for the analysis of materials in open air under ambient conditions, *Anal. Chem.* 2005, 77, 2297–2302. With permission.)

3.5.1 DIRECT ANALYSIS IN REAL TIME

A novel ionization source, DART has been developed with the goal of replacing radioactive sources in IMS systems such as the CAM (chemical agent monitor).[38] The DART ionization source, shown in Figure 3.14, consists of a discharge chamber containing a cathode and anode; this produces plasmas containing ions, electrons, and excited-state species.

Metastable helium atoms or nitrogen molecules are extracted from the plasma and directed to a sample surface, where Penning ionization makes product ions that can be analyzed by mass spectrometry or IMS. DART sources have been used to ionize a number of classes of compounds as solids or adsorbed directly onto a variety of surfaces.

3.5.2 DESORPTION ELECTROSPRAY IONIZATION

A somewhat similar but mechanistically different ion source from DART is DESI.[39] DESI ionization is accomplished by directing electrosprayed charged droplets and ions of the solvent onto a surface of the sample. Figure 3.15 provides a schematic of the DESI process.

The ion source is a typical ESI source using water/methanol/acid as the electrospray solvent to produce a charged spray of ions and droplets from the electrospray needle. A sheath flow of nitrogen is used to assist the spray and direct the ions to the surface for analysis. DESI is particularly useful for the detection of small molecules such as drugs but can also ionize larger compounds such as peptides and proteins. One feature of DESI is the ability to deliver chemical reagents to a surface for selective chemical ionization. For example, the addition of an enzyme substrate to the spray solution can generate the enzyme-substrate complex ion for analysis when sprayed onto a surface containing the enzyme.[39] DESI detection of compounds from plant stems and directly from living tissue surfaces has been demonstrated.[39]

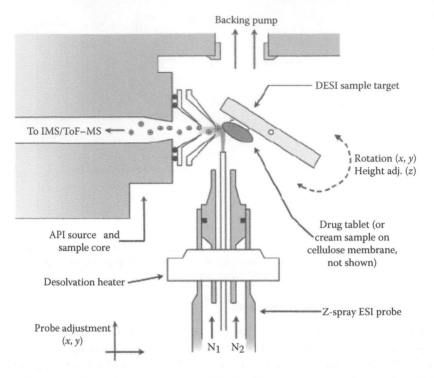

FIGURE 3.15   Schematic of a typical DESI experiment. (From Weston et al. Direct analysis of pharmaceutical drug formulations using ion mobility spectrometry/quadrupole-time-of-flight mass spectrometry combined with desorption electrospray ionization *Anal. Chem.* 2005, 77(23), 7572–7580.

3.5.3   Matrix-Assisted Laser Desorption Ionization

The use of light to ionize solid samples requires a laser source or a special photo-emission source. Laser desorption ionization (LDI) has long been employed for the ionization of samples on surfaces, and although the addition of a specific matrix to assist LDI has made sample preparation more difficult than that required for DART and DESI, it has provided an important tool for both mass spectrometry and IMS. MALDI is particularly useful for the sensitive analysis of large molecules such as peptides and proteins. Although ambient pressure MALDI-IMS has been reported,[40] most MALDI sources used for introducing gas phase ions of large non-volatile molecules into an IMS have required reduced pressures and are thus more complex than DART or DESI. Nevertheless, the data that can be obtained from a MALDI-IMS-mass spectrometric experiment is quite impressive. Figure 3.16 shows a schematic of the first MALDI-IMS instrument.[41] It is a high-pressure MALDI source (1–10 torr) coupled to a low-pressure IMS (1–10 torr) followed by a small TOFMS with a resolution up to 200. (Note: A value of 1–10 torr is high pressure for MALDI and low pressure for IMS.)

The IMS had a resolving power of about 25 in this design. One primary advantage of MALDI over DART and DESI is that the last two ion sources are continuous, and

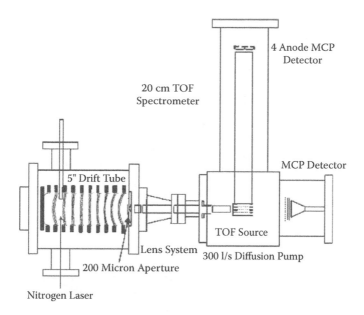

FIGURE 3.16 Schematic diagram of a high-pressure MALD/IM-o-TOF apparatus. (From Gillig et al., Coupling high-pressure MALDI with ion mobility/orthogonal time-of flight mass spectrometry, *Anal. Chem.* 2000, 7217, 3965–3971. With permission.)

the ions produced must be gated or accumulated in an ion trap[42] before introduction into the drift region of an IMS. MALDI, on the other hand, is pulsed such that the duty cycle of a MALDI source can be matched to that of the IMS spectrum. With the apparatus shown in Figure 3.16, the sample is introduced directly into the ion mobility drift cell with a direct insertion probe. The sample is placed on the end of the probe and irradiated with a 337-nm, 20-Hz pulse frequency nitrogen laser at a grazing angle of about 30° from the axis of the drift cell. The pulse duration of the laser is about 4 ns, producing a sharp ion pulse for separation in the ion mobility cell without the requisite ion gate or shutter. In more recent designs of MALDI-IMS-o-TOF instruments, both IMS and TOF resolving powers have been improved. In addition, a multisampling wheel has been incorporated into the instrument so that several samples can be deposited.[43] Application of the sample is in the traditional dried droplet manner for the preparation of MALDI samples.

Figure 3.17 shows an example of an ion mobility–mass spectrum for a sample of peptides, lipids, and nucleotides demonstrating one of the strengths of combining IMS with mass spectrometry, the formation of mobility–mass correlated "trend lines".

In Figure 3.17 the mobility information is on the y-axis while the mass information is on the x-axis. Because ion mobility measures the ion's collision cross section Ω to charge eZ ratio and the TOF measures an ion's mass m to charge eZ ratio, the 2D mobility-mass spectrum provides a measure of the ion's Ω/m ratio. Larger Ω/m ratios indicate that an ion is larger per unit mass than an ion with a smaller Ω/m ratio. That is, ions with larger Ω/m ratios are less dense than those with smaller Ω/m ratios. For example, protein tertiary structures can be determined to be folded (denser)

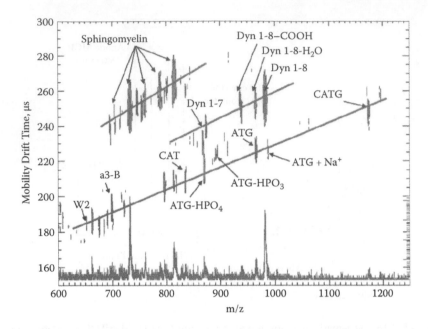

FIGURE 3.17 IM–MS 2D plot of a short oligonucleotide, a peptide, and a lipid. Note the various trend lines along which these compounds and their fragments are aligned. (From Woods et al., Lipid/peptide/nucleotide separation with MALDI-ion mobility-TOF MS, *Anal. Chem.* 2004, 76(8), 2187–2195. With permission.)

or unfolded (less dense) depending on their Ω/m ratios. Compound classes can be categorized according to their ion densities or Ω/m ratios. In Figure 3.17, lipids form a mobility–mass trend line with the largest Ω/m ratio, while nucleotides are the most dense ions with the smallest Ω/m trend line. Peptides and proteins have ion densities that fall in between those of lipids and nucleotides.

Mobility–mass trend lines were first demonstrated to be specific for classes of compounds by Karasek, Kim, and Rokushika in 1978 in a study of reduced mobility as a function of compound class.[44] In this study, they demonstrated that Ω/m trend lines (they reported them as K_o/m) were larger in the following order: primary alkylamines > secondary alkylamines > n-alkanes > tertiary alkylamines > benzene derivatives. These samples were introduced into the IMS as vapors. Thus, the use of Ω/m for class identification appears to be useful for all types of sample introduction.

3.5.4 Laser Ablation

In laser ablation, a solid sample is irradiated with a laser pulse that ablates the point of laser-solid contact to produce a plume of ions and neutrals in the vapor space just above the point of laser-solid contact with the sample surface. If this plume is swept into an ionization source or if reactant ions are electrically focused into the ablated sample plume, product ions are formed. These ions can be electrically focused into an IMS for ion mobility analysis. Direct laser ablation followed by ionization from

O_2^- ions formed from electrons during the ablation process has been demonstrated.[45] Both corona ionization[46] and ESI[47] have been used to assist the laser ablation process in creating analyte ions for IMS.

3.6 SUMMARY

IMS is a qualitative and quantitative separator–detector with applications to a broad array of samples. For gases and thermally desorbed analytes in the vapor state, ionization sources such as the standard ^{63}Ni radioactive ionization source or the promising secondary electrospray ionization (SESI) source offer sensitive, stable, and well-characterized ionization processes (see Chapter 4 for a comprehensive discussion of ionization sources). Vapor or gas samples, such as chemical warfare agents and TICs can be introduced through semipermeable membranes or by discrete sampling with a six-port valve. Semivolatile compounds are commonly introduced as vapors after thermal desorption from a sample collector. When gaseous, volatile, and semivolatile compounds are introduced to an IMS from a GC or SFC, 2D separations are achieved. In some methods, sample introduction is part of the ionization process, as is the case with ESI, DESI, and MALDI. Aqueous samples, for example, are commonly introduced into an IMS with electrospray nebulization and ionization, while solid samples are introduced using DART, DESI, or MALDI sources. These combinations of sample delivery and ionization methods make IMS a highly versatile analytical instrument.

REFERENCES

1. Borsdorf, H.; Rammler, A.; Schulze, D.; Boadu, K.O.; Feist, B.; Weiss, H., Rapid on-site determination of chlorobenzene in water samples using ion mobility spectrometry, *Anal. Chim. Acta* 2001, 440, 63–70.
2. Sielemann, S.; Baumbach, J.I.; Schmidt, H.; Pilzecker, P., Detection of alcohols using UV-ion mobility spectrometers, *Anal. Chim. Acta* 2001, 431, 293–301.
3. Cohen, M.J.; Karasek, F.W., Plasma chromatography—a new dimension for gas chromatography and mass spectrometry, *J. Chromatogr. Sci.* 1970, 8, 330–337.
4. Karasek, F.W., *Research/Development* 1970, March, 34–37.
5. Hoch, G.; Kok, B., A mass spectrometer inlet system for sampling gases dissolved in liquid phases, *Arch. Biochem. Biophys.* 1963, 101, 160.
6. Llewelynn, P.; Littlejohn, D. 1969.
7. Creaser, C.S.; Stygall, J.W., *Anal. Proc.* 1995, 32, 7.
8. Harland, B.J.; Nicholson, P.J., Continuous measurement of volatile organic chemicals in natural waters, *Sci. Total Environ.* 1993, 135, 37–54.
9. Johnson, R.C.; Srinivasan, N.; Cooks, R.G.; Schell, D., Membrane introduction mass spectrometry in pilot plant: on-line monitoring of fermentation broths, *Rapid Commun. Mass Spectrom.* 1997, 11, 363.
10. Bohatka, S., Process monitoring in fermentors and living plants by membrane inlet mass spectrometry, *Rapid Commun. Mass Spectrom.* 1997, 11, 656–661.
11. Hansen, K.F.; Degn, H., *Biotechnol. Tech.* 1996, 10, 485.
12. Wong, P.; Srinivasan, N.; Kasthurikrishnan, N.; Cooks, R.G.; Pincock, J.A.; Grossert, J.S., On-line monitoring of the phytolysis of benzyl acetate and 3,5-dimethoxybenzyl acetate by MIMS, *J. Org. Chem.* 1996, 61, 6627.

13. Rezgui, N.D.; Kanu, A.B.; Waters, K.E.; Grant, B.M.B.; Reader, A.J.; Thomas, C.L.P., Separation and preconcentration phenomena in internally heated poly(dimethylsilicone) capillaries: preliminary modelling and demonstration studies, *Analyst* 2005, 130, 755–762.

14. Kanu, A.B.; Thomas, C.L.P., The presumptive detection of benzene in water in the presence of phenol with an active membrane-UV photo-ionization differential mobility spectrometer, *Analyst* 2006, 131, 990–999.

15. Hannum, D.W.; Parmeter, J.E.; Linker, K.L.; Rhykerd, J., C.L.; Varley, N.R., Miniaturized explosives preconcentrator for use in man-portable field detection system, *Sandia Rep.* 1999, SAND99-2000C.

16. Rodriguez, J.; Vidal, S.L., Sampling Methods for Ion Mobility Spectrometers: Sampling, Preconcentration and Ionization (D300.2), *European Commission Report* 2009, Seventh Framework Programme (2007–2013) (Project No. 217925 Localisation of Threat Substances in Urban Society (LOTUS)), 1–43.

17. Karasek, F.W.; Kim, S.H.; Hill, H.H., Mass identified mobility spectra of p-nitrophenol and reactant ions in plasma chromatography, *Anal. Chem.* 1976, 48, (8), 1133–1137.

18. Conrad, F.; Kenna, B.T.; Hannuir, D.W., in *An Update on Vapor Detection of Explosives*, Nuclear Materials Management, 31st Annual Meeting, Los Angeles, 1990, Institute of Nuclear Materials, Northbrook, IL, 1990, p. 902.

19. Fetterolf, D.D.; Clark, T.D., Detection of trace explosive evidence by ion mobility spectrometry, *J. Forensic Sci.* 1993, 38(1), 28–39.

20. Perr, J.M.; Furton, K.G.; Almirall, J.R., Solid phase microextraction ion mobility spectrometer interface for explosive and taggant detection, *J. Sep. Sci.* 2005, 28, 177–183.

21. Cram, S.P.; Chesler, S.N., Coupling of high speed plasma chromatography with gas chromatography, *J. Chromatogr. Sci.* 1973, 11(August), 391–401.

22. Baim, M.A.; Hill, H.H.J., Tunable selective detection for capillary gas chromatography by ion mobility monitoring, *Anal. Chem.* 1982, 54(1), 38–43.

23. St. Louis, R.H.; Siems, W.F.; Hill, H.H.J., Detection limits of an ion mobility detector after capillary gas chromatography, *J. Microcol. Sep.* 1990, 2, 138–145.

24. Eiceman, G.A.; Karpas, Z., *Ion Mobility Spectrometry*, CRC Press, Boca Raton, FL, 1994.

25. Eiceman, G.A.; Gardea-Torresday, J.; Overton, E.; Carney, K.; Dorman, F., Gas chromatography, *Anal. Chem.* 2004, 76, 3387–3394.

26. Cagan, A.; Schmidt, H.; Rodriguez, J.E.; Eiceman, G.A., Fast gas chromatography-differential mobility spectrometry of explosives from TATP to Tetryl without gas atmosphere modifiers, *Int. J. Ion Mobil. Spectrom.* 2010.

27. Rokushika, S.; Hatano, H.; Hill, H.H., Jr., Ion mobility spectrometry after supercritical fluid chromatography, *Anal. Chem.* 1987, 59(1), 8–12.

28. Eatherton, R.L.; Morrissay, M.A.; Siems, W.F.; Hill, J.H.H., Ion mobility detection after supercritical fluid chromatography, *J. High Res. Chromatogr. Chromatogr. Commun.* 1986, 9(March 2986), 154–160.

29. Eatherton, R.L.; Morrissey, M.A.; Hill, H.H., Comparison of ion mobility constants of selected drugs after capillary gas chromatography and capillary supercritical fluid chromatography, *Anal. Chem.* 1988, 60(20), 2240–2243.

30. Gieniec, J.; H.L. Cox, J.; Teer, D.; Dole, M., in *20th ASMS Conference on Mass Spectrometry and Allied Topics*, Dallas, TX, 1972, pp. 276–280.

31. Dole, M.; Gupta, C.V.; Mack, L.L.; Nakamae, K., *K. Polym. Prepr.* 1977, 18(2), 188.

32. Shumate, C.; Hill, H.H., Jr., Electrospray ion mobility spectrometry after liquid chromatography, presented at 42nd Northwest Regional Meeting of the American Chemical Society, Bellingham, WA, June, 1987.

33. Shumate, C.B., An electrospray nebulization/ionization interface for liquid introduction into an ion mobility spectrometer, thesis, Washington State University, Pullman, 1989.

34. Shumate, C.B.; Hill, H.H., Jr., Coronaspray nebulization and ionization of liquid samples for ion mobility spectrometry, *Anal. Chem.* 1989, 61(6), 601–606.

35. Shumate, C.B.; Hill, H.H., Jr., Electrospray ion mobility spectrometry: its potential as a liquid stream process sensor, in *Pollution Prevention and Process Analytical Chemistry*, Editors Breen, J.; Dellarco, M., ACS Books, Washington, DC, 1992, pp. 192–205.

36. Hill, H.H., Jr.; Siems, W.F.; Eatherton, R.L.; St. Lewis, R.H.; Morrissay, M.A.; Shumate, C.B.; McMinn, D.G., Gas, supercritical fluid, and liquid chromatographic detection of trace organics by ion mobility spectrometry, in *Instrumentation for Trace Organic Monitoring*, Editors Clement, R.E.; Siu, K.W.M.; Hill, J.H.H., Lewis, Chelsea, MI, 1991, pp. 49–64.

37. Sukumar, H.; Stone, J.A.; Nishiyama, T.; Yuan, C.; Eiceman, G.A., Paper spray ionization with ion mobility spectrometry at ambient pressure, *Int. J. Ion Mobil. Spectrom.* 2011, 14, 51–59.

38. Cody, R.B.; Laramee, J.A.; Durst, H.D., Versatile new ion source for the analysis of materials in open air under ambient conditions, *Anal. Chem.* 2005, 77, 2297–2302.

39. Takats, Z.; Wiseman, J.M.; Gologan, B.; Cooks, R.G., Mass spectrometry sampling under ambient conditions with desorption electrospray ionization, *Science (Washington, DC)* 2004, 306, 471–473.

40. Steiner, W.E.; Clowers, B.H.; English, W.A.; Hill, H.H., Jr., Atmospheric pressure matrix-assisted laser desorption/ionization with analysis by ion mobility time-of-flight mass spectrometry, *Rapid Commun. Mass Spectrom.* 2004, 18(8), 882–888.

41. Gillig, K.J.; Ruotolo, B.; Stone, E.G.; Russell, D.H.; Fuhrer, K.; Gonin, M.; Schultz, A.J., Coupling high-pressure MALDI with ion mobility/orthogonal time-of flight mass spectrometry, *Anal. Chem.* 2000, 72(17), 3965–3971.

42. Henderson, S.C.; Valentine, S.J.; Counterman, A.E.; Clemmer, D.E., ESI/ion trap/ion mobility/time-of-flight mass spectrometry for rapid and sensitive analysis of biomolecular mixtures, *Anal. Chem.* 1999, 71, 291.

43. Woods, A.S.; Ugarov, M.; Egan, T.; Koomen, J.; Gillig, K.J.; Fuhrer, K.; Gonin, M.; Schultz, J.A., Lipid/peptide/nucleotide separation with MALDI-ion mobility-TOF MS, *Anal. Chem.* 2004, 76(8), 2187–2195.

44. Karasek, F.W.; Kim, S.H.; Rokushika, S., Plasma chromatography of alkyl amines, *Anal. Chem.* 1978, 50, 2013–2016.

45. Eiceman, G.A.; Young, D.; Schmidt, H., Rodriguez, J.; Baumbach, J.I.; Vautz, W.; Lake, D.A.; Johnston, M.V., Ion mobility spectrometry of gas-phase ions from laser ablation of solids in air at ambient pressure, *Appl. Spectrosc.* 2007, 61, 1076–1083.

46. Steiner, W.E.; Clowers, B.H.; English, W.A.; Hill, H.H., Jr., Atmospheric pressure matrix-assisted laser desorption/ionization with analysis by ion mobility time-of-flight mass spectrometry, *Rapid Commun. Mass Spectrom.* 2004, 18(8), 882–888.

47. Harris, G.A.; Graf, S.; Knochenmuss, R.; Fernandez, F.M., Coupling laser ablation/ desorption electrospray ionization to atmospheric pressure drift tube ion mobility spectrometry for the screening of antimalarial drug quality, *Analyst* 2012, 137, 3039–4044.

48. Crawford, C.L.; Graf, S.; Gonin, M.; Fuhrer, K.; Zhang, X.; and Hill, H.H., Jr., The novel use of gas chromatography-ion mobility-time of flight mass spectrometry with secondary electrospray ionization for complex mixture analysis, *Int. J. Ion Mobil. Spectrom.* 2010, 14, 23–30.

49. Sielemann, S., Baumback, J.I.; Pilzecker, P.; Walendzik, G., *Int. J. Ion Mobil. Spectrom.* 1999, 1, 15–21.

50. Davis et al., *IEEE Sensors* 2010.

51. Wittmer, Workshop on Ion Mobility Spectrometry, Mescalero, NM, June 1992.

4 Ion Sources

4.1 INTRODUCTION

The formation of gas phase ions must precede the processes of ion separation and detection by mobility measurements. Ionization can occur after the sample vapors enter the ionization or reaction region of the drift tube or in the same step as sample introduction, depending on the type of ion source used. Here, we present the ion sources that are suitable for atmospheric pressure ionization that have been used for ion mobility spectrometry (IMS) studies. Ionization in analytical IMS commonly occurs in air at ambient pressure; consequently, the methods or reactions used to produce ions must operate with the levels of moisture and oxygen found in ambient air. In this chapter, the more popular ion sources are discussed in some detail, while others that can be classified as experimental are viewed only briefly. The commonly deployed ion sources include radioactive sources, corona discharge (CD) sources, photodischarge lamps, and lasers as well as variations of electrospray ion sources. Ion sources that are used mainly for research and experimental purposes include radio-frequency (RF) sources, flames, surface ionization sources, as well as some proposed new approaches, like glow discharge and helium plasma. For each of these ion sources, we present the principle of operation, bring examples of application in IMS, and summarize its advantages and limitations. During the past decade, a few trends were evident, with increased interest in nonradioactive sources and ion sources suited for liquid and solid samples. This last trend is especially evident with increasing interest in semivolatile and nonvolatile molecules, particularly macromolecules for biological and environmental applications. Some of the features of ionization sources that are in use in commercial and research instruments are summarized in Table 4.1. A comprehensive survey of ionization sources for IMS can be found in a recent report by Rodriguez and Lopez-Vidal.[1]

4.2 RADIOACTIVITY: NICKEL, AMERICIUM, AND TRITIUM

Radioactive sources are favored for use in IMS analyzers because they provide stable and reliable operation, with ionization chemistry that is well suited for most current applications of IMS. Furthermore, radioactive foils do not require an external power supply and have no moving parts or maintenance requirements. At present, the most widely used and best understood of all ion sources for IMS is still the long-favored radioactive ^{63}Ni source, which is also widely used in electron capture detectors (ECDs) for gas chromatography (GC). The preferred radioactive source is 10 mCi ($3.7*10^8$ Bq) of ^{63}Ni coated as a thin layer on a metal strip, generally nickel or gold.[2] The maximum energy of the electrons emitted from the ^{63}Ni source

TABLE 4.1

Summary of Ionization Techniques Used in Ion Mobility Spectrometry

Ion Source	Type of Chemicals	Maintenance	Cost	Comments
Radioactive	Universal	Low	Medium/low	Licensing required
Corona discharge	Universal	High	Medium	Electrode replacement required
Photoionization UV, laser	Selective	Medium	Medium	Low efficiency
Surface ionization	Selective (N, P, As, S)	High	Medium	Complex
Electrospray	Liquid samples	Medium	Medium	Long clearance time
DESI, DART	Solid samples	Medium	Medium	
SESI	Solid, liquid, and vapor	Medium	Medium	Research stage
MALDI	Macromolecules	High	High	Biological mainly
Flame	Selective	Medium	Low	Structural information lost
Plasma	Universal	Medium	Medium	Research stage
Glow discharge	Universal	Medium	Medium	Research stage
Alkali cation	Selective	Medium	Medium	Research stage

is 67 keV, with an average energy near 17 keV. Almost all the energy of this source is dissipated in air at an ambient pressure within 10 to 15 mm, from the surface of the metal, establishing guidelines for the optimum diameter when the source is used as a cylinder, a common geometry in IMS analyzers.[3] The electrons emitted from the ^{63}Ni produce ions and secondary electrons (see Chapter1), and this process is repeated until the secondary electrons are no longer energetic enough to ionize the gas molecules of the supporting atmosphere. The formation of an ion pair requires about 35 eV, so each beta particle emitted from the source could ideally produce on average about 250 ions if we assume that 50% of the electrons are directed into the foil.[4,5] Negative ions may be produced also through electron attachment processes, which, in most cases, proceed efficiently when the electrons are at thermal energy, thus further increasing the ion yield.

Radioactive isotopes other than ^{63}Ni have been used with IMS drift tubes, including a beta-emitting tritium source and an alpha-emitting ^{241}Am isotope similar to the source used in household smoke detectors.[6] Alpha particles emitted from ^{241}Am are highly energetic, with energies above 5.4 MeV, and have a short effective range in air, so that ionization is efficient in small-volume sources.[7] Tritium poses less radiation hazard than ^{63}Ni sources and has been used as an ionization source in a study of environmental monitoring of part-per-billion levels of toxic compounds in ambient air.[8]

Despite the attractiveness of radioactive ion sources in IMS, use of ^{63}Ni, Am, or T sources is discouraged today due mainly to regulatory issues as well as financial, organizational, and technical reasons. Radioactive sources require special permits and licensing procedures, and general licenses for alpha-emitting sources with appropriate activity are particularly difficult to obtain. Once in service, radioactive-based drift tubes require periodic checks for leakage of radioactivity and general laboratory hygiene. The burdens of the cost of assays and keeping records and the legal implication of workplace hygiene may discourage prospective and current users.

Another concern with radioactive sources is the care needed to prevent chemically reactive conditions in the IMS drift tube; for example, elevated drift tube temperatures in the air atmosphere and acid gases may eventually cause oxidation of the metal foil and the formation of nickel oxides or salts. These salts or oxides are mechanically unstable and may be released into the ambient environment if the drift tube is vented without a particulate filter. A paper filter fitted on the drift tube effluent will trap radioactive particulate matter. Finally, proper disposal of drift tubes equipped with radioactive sources requires permits and can be expensive.

4.3 CORONA DISCHARGES

The ion chemistry and electrical stability of a point-to-plane corona discharge (CD) ion source was described using a tandem mass spectrometer equipped with an atmospheric pressure source.[9] This source was a continuous current-regulated discharge with a direct current (DC) power supply and demonstrated the regions of stability with potential and distance between the needle and plane. A schematic overview of the mechanism of formation of ions in a CD is shown in Figure 4.1.

CD sources with various designs and electronics control have also been developed or explored with IMS,[10-13] and CDs are available in some production instruments.[14] To form a CD, a sharp needle or thin wire is placed 2 to 8 mm from a metal plate or discharge electrode with a voltage difference of 1 to 3 kV between the needle and plate. An electric discharge develops in the gap between the needle or wire and the opposing conductor, and the ions formed in the gap can closely resemble those found in the ^{63}Ni ion source. These ions are then available for subsequent ion-molecule reactions with the sample.

CD ion sources have been used in pure nitrogen for formation of ions with negative polarity[12] in which a large number of electrons were produced and used for ionization. However, electronegative substances like oxygen or ozone can quench the discharge, and to overcome this, a drift tube was designed and evaluated for detection of halogenated methane and some nitro compounds. Positive and negative ion formation was shown to be consistent with the Townsend formula (I/V is a linear function of V), and the total ion current obtained from the corona ionization source was about 10 times greater than that of the ^{63}Ni source. Although continuous DC corona discharges are possible with IMS,[11,12] a pulsed corona may be attractive due to the reduced power needed to operate the source. However, a pulsed source introduces an element of time or time-resolved chemistry that begins with the start of the discharge. Changes in ion composition occur as the ions move away from the center of the discharge.[13] A complication of this chemistry can be found in the negative polarity; nitrogen oxides and ozone formed in the corona may interfere with gas phase ion–molecule chemistry and degrade the response of an IMS analyzer to certain chemicals. Two different solutions to this problem have been proposed: increasing the distance between the electrodes,[9] and reversing the airflow past

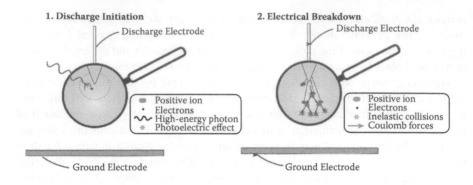

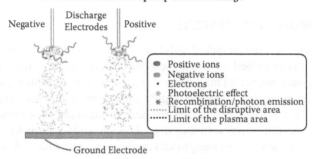

FIGURE 4.1 A schematic of the mechanism of corona discharge: (1) A neutral atom or molecule of the medium in a high potential gradient is ionized to create a positive ion and a free electron. (2) The electric field separates the charged particles and imparts them with kinetic energy. (3) As a result, further electron/positive ion pairs may be created by collision with neutral atoms. (4) The energy of these plasma processes is converted into further initial electron dissociations to seed further avalanches. (5) An ion species created in this series of avalanches is attracted to the uncurved electrode, completing the circuit and sustaining the current flow. (From Wikipedia; Ernest Galbrun, author. Translated from original French. http://en.wikipedia.org/wiki/File:Corona_discharge_upkeep.svg; http://en.wikipedia.org/wiki/File:Corona_discharge_upkeep.svg; http://en.wikipedia.org/wiki/File:Corona_discharge_upkeep.svg)

the corona needle.[10] CD sources require maintenance because electrodes undergo erosion and require replacement. Nonetheless, several applications of CD-based drift tubes have been described, including the determination of chlorobenzene in water samples[15] and the monitoring of acetone in air.[16] An IMS drift tube with a CD ion source has been used to characterize biogenic amines in saliva samples.[17] In that case, a stainless steel syringe needle served as the electrode.

Another approach to minimize these effects is to use very short (150-ns rise time and 500-ns pulse width) 12-kV pulses to generate a CD in air.[18] These short pulses may even make it possible to remove the ion gate and thus simplify the drift tube design. Changing the geometry of the CD source, which is constructed of two small-diameter (0.16- and 4-mm) cylinder electrodes, has been proposed as an ambient negative CD source.[19]

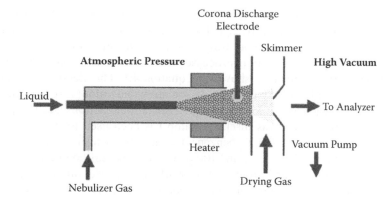

FIGURE 4.2 Formation of ions by corona discharge from a nebulized microspray. (From Wikipedia. http://en.wikipedia.org/wiki/File:Apci.gif)

CD ionization may also be used to form ions in a microspray of droplets produced by nebulization of a liquid sample (Figure 4.2). Although this has not yet been applied to IMS measurements, it does have potential for such.

In summary, CD ion sources have been deemed valuable because there is no radioactivity, the ion currents are relatively high, design and assembly are simple, and certain applications (e.g., the direct analysis of liquid samples) may be best made using a CD source. The disadvantages include the need for an external high-voltage power supply, especially a current regulated power supply, corrosion and erosion of components, maintenance requirements of the discharge, and the formation of corrosive chemical vapors such as NOx and ozone. Stability may be degraded by corrosion of the needle; this is cumulative and is governed by use.

4.4 PHOTOIONIZATION: DISCHARGE LAMPS AND LASERS

Photodischarge lamps[20–27] and lasers[28–37] may be used as a means of ionizing neutral molecules in air at ambient pressure. The lasers may also serve as a means of vaporizing solid samples, such as adsorbed films and solids before ionization.[30,34] Photodischarge lamps emit photons from the electrical excitation of gases filled in the lamp,[38] and lamps that are commercially available can provide energies of 9.5, 10.2, 10.6, and 11.7 eV.[39] The formation of positive ions with photons has been described as direct ionization through the reaction given in Equation 4.1:

$$h\nu + M \rightarrow M^{+\cdot} + e^- \tag{4.1}$$

where $h\nu$ is the photon energy, and M is the neutral molecule. The energy needed for this, the ionization potential, with organic compounds is generally between 7 and 10 eV, and $M^{+\cdot}$ ions are routinely observed with aromatic hydrocarbons.[30]

The exact mechanism of ion formation in air at ambient pressure should be considered as incompletely understood for compounds of high proton affinity or compounds for which protons can be released. For example, product ions for ketones with a discharge lamp include MH^+ species rather than M^+. This may suggest that intermediate reactive species play a role in ambient pressure ionization, although the reactions are not fully described. Negative ions are not formed directly through photoionization processes but arise in the course of chemical reactions with the electron shown in Equation 4.1. The electron may attach directly to a molecule or may undergo dissociative ionization; also, the electron may react with oxygen and proceed through association reactions (see Chapter 1).

A 10.6-eV low-pressure gas discharge lamp has been used for the continuous detection of alcohols[26] and BTEX (benzene, toluene, ethylbenzene, and xylene)[27] in the concentration range between 1 and 100 ppmv (parts per million by volume), and in other studies aliphatic and aromatic hydrocarbons were measured by ultraviolet (UV) IMS. The main advantage of photodischarge sources is that some selectivity in response may be ensured by the choice of an appropriate ionization energy or wavelength. The disadvantages of photodischarge lamps are the requirement for an external power supply, the cost of the lamps, and the need to replace them periodically due to the finite lifetimes of such lamps. It should be noted that the ionization efficiency strongly depends on the cross section for photoabsorption, so that certain types of compounds (like those containing double bonds) may be ionized selectively, while others will have very low yields of ions. In one case, a drift tube without an ion shutter was equipped with a pulsed xenon lamp photoionization light source and characterized for parameters and performance with negative ion detection.[40] The performance of this design was found to be comparable to that of conventional IMS drift tubes. In a recent publication, a vacuum-UV Kr lamp was used as a bipolar ionization source for production of positive and negative ions [assumed to be hydrated ozone $O_3^-(H_2O)_n$], and several compounds such as SO_2, CO_2, and H_2S were detected in the negative mode.[41]

Lasers could be used, in principle, to provide any photon energy in the ultraviolet-to-infrared range of wavelengths with dye lasers. However, the common Nd-YAG (yttrium-aluminum-garnet) laser conveniently offers four wavelengths: a fundamental at 1,064 nm and harmonics at 532, 355, and 266 nm. Laser-based, single-photon ionization is possible only for photons in the UV part of the spectrum, and lasers that operate in this energy range are mainly excimer lasers. Little may be said about gas phase ionization by lasers with IMS because only a handful of studies have been described.[28,29,42]

4.5   ELECTROSPRAY IONIZATION AND ITS DERIVATIVES

Ionization of fine droplets of a liquid sample, called electrospray ionization (ESI), may occur when an aerosol is formed between a needle tip under a potential of several thousand volts relative to a grid or plate (Figure 4.3).[43] The sample entering the drift tube is already ionized, so an additional ionization source inside the drift tube is not required. The mechanism of ion formation was described in detail by Kebarle and Tang.[44] The liquid sample, usually consisting of the analyte dissolved in a solution of water with a volatile organic solvent, is dispersed as a fine aerosol of charged droplets by electrospray. The solvent evaporates, and the radius of the charged droplet decreases so that coulomb fission takes place. Two mechanisms have

FIGURE 4.3 A photograph of an electrospray ionization (ESI) source near the pinhole entrance of a mass spectrometer. (From Wikipedia. http://en.wikipedia.org/wiki/File:NanoESIFT.jpg)

been proposed for ion formation in ESI: the ion evaporation model[45] and the charged radius model.[46]

The first reference to ESI-IMS was in a talk by Hill in 1987[47] and was later described in peer-reviewed papers from his group.[48–51] ESI has become attractive as a method for IMS determination of biomolecules[52–57] and for the analysis of samples of environmental interest.[58–61]

An example of the value of an ESI source can be seen in the determination of chemical warfare agents (CWAs) and their degradation products in water through analysis by ESI/IMS/MS (mass spectrometry).[58] Water samples were injected through an ESI system into a high-resolution IMS drift tube, in which separation according to mobility occurred. A mass spectrum of each peak in the mobility spectrum was obtained using a time-of-flight mass spectrometer (TOF-MS). Traces of CWAs and their degradation products were rapidly detected and identified using this approach. ESI may also be used for detection of inorganic ions such as uranyl acetate in water.[60] One noteworthy difference between ESI and the other ion sources used in IMS is the composition of the ions: While other ion sources produce predominantly, and usually solely, monovalent (singly charged) ions, ESI of proteins and other macromolecules often leads to formation of multiply charged ions that are generally not only multiply protonated but also sodiated.[62]

The advantages of ESI are that liquid samples may be introduced directly into the IMS, molecular information is retained due to the soft ionization processes, and in several cases multiply charged ions may be formed. The main limitation with ESI for practical analytical IMS is that relatively long rinsing times are needed between samples because memory effects in the fluid delivery system can be quite significant.

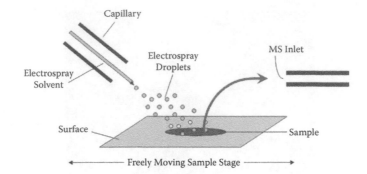

FIGURE 4.4 A schematic of desorption electrospray ionization (DESI). The electrosprayed droplets from an ESI source impinge on the surface of the sample, and the ions desorbed from the surface are analyzed by MS or IMS. (From Wikipedia. http://en.wikipedia.org/wiki/File:DESI_ion_source.jpg)

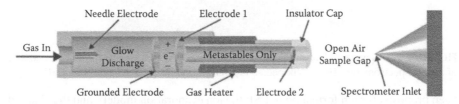

FIGURE 4.5 Schematic representation of direct analysis in real time (DART); ions from the sample are produced from metastable, excited species. (From Wikipedia. http://en.wikipedia.org/wiki/File:DART_ion_source_schematic.gif)

Several innovative ion sources that are based on ESI were developed, including secondary electrospray ionization (SESI),[63,64] desorption electrospray ionization (DESI)[65] (Figure 4.4), and nanoelectrospray ionization (nESI).[66] These were described in some detail in Chapter 3 in the context of sample introduction techniques and are only briefly discussed here.

While standard ESI is used for analysis of liquid samples, DESI and SESI have been applied to direct sampling of surfaces, and determination of surface contamination, by bombarding the surface with a jet of ions from an ESI source and introducing the ions that are desorbed from the surface into an IMS or MS for analysis. nESI deploys an ion funnel for axial focusing, accumulation, and generation of short duration (10–30 μs) ion pulses for use with the IM–MS.[67] Another technique for sampling solid surfaces is direct analysis in real time (DART), shown in Figure 4.5, developed by Cody et al.[68] The main difference between DART and DESI or SESI is that the former uses excited metastable species that impinge on the sample surface, and the ions that are emitted are introduced into the IMS or MS for analysis.

4.6 MATRIX-ASSISTED LASER DESORPTION IONIZATION

A variation of laser-based ionization that merits separate discussion is matrix-assisted laser desorption ionization (MALDI).[69] The sample is placed on a crystal matrix, so most of the laser energy is absorbed initially by the matrix. Part of this energy is transferred to the analyte molecules, which are ionized and desorbed with little or no fragmentation. Thus, a solid sample may be directly desorbed, vaporized, and ionized inside an IMS drift tube when an intense laser pulse is directed at the sample, as described in Chapter 3. MALDI has gained popularity in combination with MS for the direct ionization of solid samples, particularly for large, nonvolatile macromolecules and as a way of introducing metal ions into the gas phase and is being applied also in various IMS-MS studies.[70–72]

The rapid vaporization of macromolecules from biological samples prevents, or minimizes, their dissociation and fragmentation and thus provides a way to obtain molecular and structural information. Also, MALDI provides a way of producing different types of cationized molecular ions with Li^+, Na^+, Cu^+, and Ag^+, in addition to the protonated species usually formed in atmospheric pressure ionization processes.[70] The most serious drawbacks of using MALDI are the price of a suitable laser and the complexity and limited applicability to certain types of solid samples. Although early studies were made with MALDI-MS with the ion source region under vacuum, the demonstration of atmospheric pressure MALDI with MS encouraged the development of MALDI-IMS.

Some examples of studies with MALDI and IMS include the use of MALDI to generate sodiated parent ions of a number of oligosaccharides without fragmentation[70] and cationized forms of bradykinin. Cross sections of the ions were obtained and compared with predictions made by molecular mechanics or molecular dynamics calculations.[71] Comparison of MALDI/IM/TOF-MS with MALDI/IM/MS analysis yielded results similar to that by nanoelectrospray MS.[71] Dipeptides and biogenic amines were analyzed with a system that consisted of an atmospheric pressure MALDI source, an IMS, and an orthogonal TOF-MS, and sensitivity was improved when a localized CD source was added.[72]

4.7 SURFACE IONIZATION SOURCES

An ion source based on surface ionization principles was developed primarily by Rasulev's group in Uzbekistan and was demonstrated on a mobility spectrometer.[73–76] The source consists of an emitter made from a single crystal of molybdenum doped with iridium (or another platinum group metal) and heated to 300–500°C. Certain types of molecules, mainly nitrogen bases, will undergo electron transfer on collision or contact with the heated reactive surface, resulting in the formation of a positive ion. A platinum film, fabricated by a microelectronics micromachining (microelectromechanical system, MEMS) process, can also be used as a surface ionization source for trimethylamine.[77,78] An ion thermoemitter was developed as an ionization source for nitrogen-, sulfur-, phosphorus-, and arsenic-based organic compounds.[79]

Ionization of tertiary amines was shown to be favored over that of secondary amines, which were more efficiently ionized than primary amines. Compared with a conventional ^{63}Ni source, the surface ionization source was reported to have a relatively large dynamic range, a selective response to certain types of compounds, and none of the regulatory problems associated with radioactive ionization sources.

Amines, tobacco alkaloids, and triazine herbicides were all shown to exhibit picogram limits of detection with a dynamic range of five orders of magnitude.[73] The success of a surface ionization source is dependent on the preparation of stable emitters with a simple design.[74] One of the problems with this source is the poisoning of the surface after exposure to certain compounds. The surface can be regenerated but only under special conditions in a controlled atmosphere. Another complication in the use of this source for several substances is that the response is highly dependent on the structure of the analyte, and in general it is mainly suitable for group V (N, P, As) and VI (S) compounds.[79]

4.8   FLAMES

Flames were among the earliest sources of ions at ambient pressure, and the mobility of ions in flames was measured in 1978 to explore the properties of a flame ionization detector widely used in GC. Studies with IMS demonstrated that ions in the flame as residual current were hydrated protons, and ions such as protonated monomers of a compound were not observed when a chemical was introduced into the flame.[80] The possibility of a flame source, however, was suggested in the work of Atar et al., who described the use of ions and electrons produced in a hydrogen or hydrocarbon flame to ionize sample molecules, in a fashion similar to a flame ionization detector.[81] Several possibilities for utilizing this idea were proposed. Among these was a pulsed flame with considerably reduced fuel consumption. This ionization method may increase sensitivity to certain types of compounds and extend the range of IMS applications to include chemicals used in the microelectronics industry, such as hydrides and fluorides. The hydrides, which have relatively low proton affinities,[82] do not readily form stable positive ions in the IMS drift tube but may be converted in the flame into the corresponding oxides, readily forming negative ions. The advantages of flame ionization sources are the possibility of producing high ion currents, capability to detect additional types of compounds, and elimination of a radioactive or photoionization source. The main disadvantages are adding considerable complexity to the drift tube and losing some of the specificity of the detector because molecular identity is lost when oxides are formed. Although promising, this source has not yet been incorporated into an IMS drift tube.

4.9   PLASMA-BASED ION SOURCES

A pulsed RF discharge was proposed as a means of extending the life of the needle, improving the reliability of the source, and reducing the interference from unwanted oxides.[83] The idea of using plasma ionization as an ion source for atmospheric pressure IMS has recently been demonstrated.[84–86] In one case, the source consisted of a high-RF voltage applied across two electrodes that were separated by a thin film of dielectric material, and the plasma formed in air contained both positive and negative ions.[84] In a different approach, a miniaturized helium plasma was used as an ion source for the IMS.[85,86] In this case, the helium ions coming out of the plasma jet were used to ionize air molecules, and these became the reactant ions, in a similar fashion to the conventional IMS sources.[86]

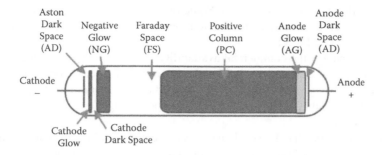

FIGURE 4.6 Schematic of a glow discharge tube that can serve as an ion source. (From Wikipedia. http://en.wikipedia.org/wiki/File:Electric_glow_discharge_schematic.png)

4.10 GLOW DISCHARGE ION SOURCE

A schematic of a glow discharge tube is shown in Figure 4.6. Ions that are formed in an atmospheric pressure glow discharge can be used as an ion source for IMS.[87–90] In one case, this ion source was used to characterize perfume odors.[87] In another study, the sensitivity toward m-xylene was enhanced compared to the [63]Ni and CD sources, and the dynamic range was extended to three orders of magnitude.[90]

4.11 OTHER ION SOURCES

A patent describing a thermionic emitter for generating positive ions that incorporated a mixture of beta-alumina and inert material such as charcoal positioned on a filament for heating the mixture has been issued.[91] Two decades later, an alkali cation emitter, based on intercalated alkali ions in a graphite matrix, was proposed.[92] When heated on a red hot filament, this source emits ions from alkali salts, which subsequently can be used to form product ions.

A truly innovative soft ionization source, based on a nanometer-thick membrane, was developed.[93] The gas sample passes through a porous membrane that is coated on both sides with a metallic conductor film. A low voltage (10 V) produces a large electric field (>10^7 V cm^{-1}) that causes soft and efficient ionization of molecules passing through the membrane. Despite its apparent advantages, this ionization method has not found its way to commercial devices.

4.12 SUMMARY

A variety of methods has been developed to produce ions under atmospheric pressure conditions that can be introduced into an IMS. Some of these methods are universal and can be employed for several applications, while others are limited to specific types of compounds (like surface ionization) or to certain types of samples (gaseous, liquid, or solid). Some require expensive and large systems (like MALDI), while others are particularly suitable for handheld devices (like the radioactive ion sources). Some are useful for studies of macromolecules, mainly of biological interest (like ESI and MALDI), while others can only be used to detect volatile compounds that form gases or vapors. Some of these ion sources are external to the drift tube and serve for sample introduction such as all ESI-based techniques including SESI, DESI and DART.

Another aspect is the commercial availability of the different ion sources. Some sources can, at present, be considered only as exotic and suitable only for research purposes and have yet to be tested in the field and in the laboratory, while others are mature techniques that are well understood, with advantages and limitations that are well recognized.

Finally, the Appendix shows schematic representations of the principle of operation of some of the ionization processes presented.[94] Among these are the theory of ESI, atmospheric pressure chemical ionization (APCI), and atmospheric pressure photoionization (APPI). Also shown are schematics of the ionizers and typical experimental conditions for the APCI and APPI sources as well as that of an ESI-APCI mixed source. It should be noted that these schematics are for sources that are interfaced with a mass spectrometer but are similar to IMS interfaces.

APPENDIX

Extracted from E. Naegele, Agilent Technologies, Waldbronn, Germany.

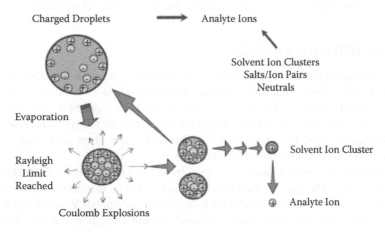

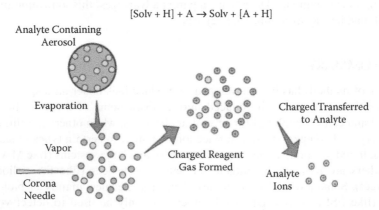

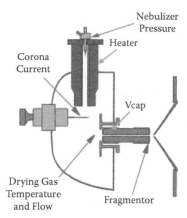

HPLC Flow Rate > 500 µL/min
Nebulizer pressure
• 60 psig
Drying Gas Temperature
• Start with 350°C
Drying Gas Flow
• 4 L/min
Vaporizer Temperature
• Optimize with
  flow injection analysis (FIA)
Vcap
• Optimize with FIA (2000–6000)
• Start with 2500 V
Corona Current
• Optimize with FIA
• Start with 25 µA (neg) or 4 µA (pos)

APPI

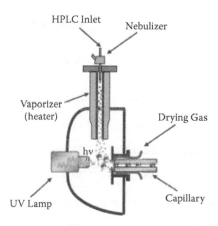

HPLC Flow Rate > 500 µL/min
Nebulizer Pressure
• 35 psig
Drying Gas Temperature
• Start with 275°C
Drying Gas Flow
• 11 L/min
Vaporizer Temperature
• Optimize with FIA
Vcap
• Optimize with FIA (2000–6000)
• Start with 2500 V

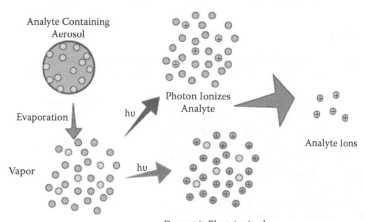

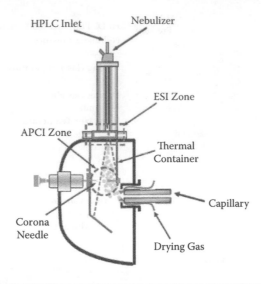

Parameters	ESI	APCI	Mixed Mode
Capillary Voltage (Vcap)			
Single Ion Polarity	2000 V	2000 V	2000 V
Polarity Switching	1000 V	1000 V	1000 V
Charging Electrode	2000 V	2000 V	2000 V
Corona Current	0 µA	4 µA	2 µA
Drying Gas Flow	5 L/min	5 L/min	5 L/min
Drying Gas Temperature	300°C	300°C	300°C
Nebulizer Pressure	60 psig	30 to 60 psig	40 to 60 psig
Vaporizer Temperature	150°C	250°C	200°C

REFERENCES

1. J. Rodriguez, J. and S. Lopez-Vidal, S. D300.2 Sampling methods for ion mobility spectrometers: Sampling, preconcentration & ionization, Project No. 217925, LOTUS TR-09-007, 2009.
2. Simmonds, P.G.; Fenimore, D.C.; Pettitt, B.C.; Lovelock, J.E.; Zlatkis, A., Design of a nickel-63 electron absorption detector and analytical significance of high temperature operation, *Anal. Chem.* 1967, 39, 1428–1433.
3. Siegel, M.W., Rate equations for prediction and optimization of chemical ionizer sensitivity, *Int. J. Mass Spectrom. Ion Phys.* 1983, 46, 325–328.
4. Thomson, J., Ionizing efficiency of electronic impacts in air, *Proc. R. Soc. Edinburgh* 1931, 51, 127–141.
5. Jesse, W.P., Absolute energy to produce an ion pair in various gases by ß-particles from sulfur-35, *Phys. Rev.* 1958, 109, 2002–2004.
6. Yun, C.-M.; Otani, Y.; Emi, H., Development of unipolar ion generator—separation of ions in axial direction of flow, *Aerosol Sci. Technol.* 1997, 26, 389–397.

7. Paakanen, H., About the applications of IMCELLTM MGD-1 detector, *Int. J. Ion Mobil. Spectrom.* 2001, 4, 136–139.

8. Leonhardt, J.W., New detectors in environmental monitoring using tritium sources, *J. Radioanal. Nucl. Chem.* 1996, 206 (2, International Conference on Isotopes, Proceedings, 1995, Pt. 4), 333–339.

9. Eiceman, G.A.; Kremer, J.H.; Snyder, A.P.; Tofferi, J.K., Quantitative assessment of a corona discharge ion source in atmospheric pressure ionization-mass spectrometry for ambient air monitoring, *Int. J. Environ. Anal. Chem.* 1988, 33, 161–183.

10. Bell, A.J.; Ross, S.K., Reverse flow continuous corona discharge ionization, *Int. J. Ion Mobil. Spectrom.* 2002, 5, 95–99.

11. Tabrizchi, M.; Khayamian, T.; Taj, N., Design and optimization of a corona discharge ionization source for ion mobility spectrometry, *Rev. Sci. Instrum.* 2000, 7, 2321–2328.

12. Tabrizchi, M.; Abedi, A., A novel electron source for negative-ion mobility spectrometry, *Int. J. Mass Spectrom.* 2002, 218, 75–85.

13. Hill, C.A.; Thomas, C.L.P., A pulsed corona discharge switchable high resolution ion mobility spectrometer-mass spectrometer, *Analyst* 2003, 128, 55–60.

14. Taylor, S.J.; Turner, R.B.; Arnold, P.D., Corona-discharge ionization source for ion mobility spectrometer, PCT Int. Appl. 1993, 32 pp. CODEN: PIXXD2 WO 9311554 A1 19930610.

15. Borsdorf, H.; Rammler, A.; Schulze, D.; Boadu, K.O.; Feist, B.; Weiss, H., Rapid on-site determination of chlorobenzene in water samples using ion mobility spectrometry, *Anal. Chim. Acta* 2001, 440, 63–70.

16. Khayamian, T.; Tabrizchi, M.; Taj, N., Direct determination of ultra-trace amounts of acetone by corona-discharge ion mobility spectrometry, *Fresenius' J. Anal. Chem.* 2001, 370, 1114–1116.

17. Barnard, G.; Atweh, E.; Cohen, G.; Golan, M.; Karpas, Z., Clearance of biogenic amines from saliva following the consumption of tuna in water and in oil, *Int. J. Ion Mobil. Spectrom.* 2011, 14, 207–211.

18. An, Y.; Aliaga-Rossel, R.; Choi, P.; Gilles, J.-P., Development of a short pulsed corona discharge ionization source for ion mobility spectrometry, *Rev. Sci. Instrum.* 2005, 76, 085105/1-085105/6.

19. Tang, F.; Wang, X.; Liu, K.; Zhang, L., Ambient negative corona discharge ion source with small line-cylinder electrodes, *Guangxue Jingmi Gongcheng* 2009, 17, 1953–1957.

20. Baim, M.A.; Eatherton, R.L.; Hill, H.H., Jr., Ion mobility detector for gas chromatography with a direct photoionization source, *Anal. Chem.* 1983, 55, 1761–1766.

21. Leasure, C.S.; Fleischer, M.E.; Anderson, G.K.; Eiceman, G.A., Photoionization in air with ion mobility spectrometry using a hydrogen discharge lamp, *Anal. Chem.* 1986, 58, 2142–2147.

22. Sielemann, S.; Baumbach, J.I.; Schmidt, H., IMS with non radioactive ionization sources suitable to detect chemical warfare agent simulation substances, *Int. J. Ion Mobil. Spectrom.* 2002, 5, 143–148.

23. Borsdorf, H.; Nazarov, E.G.; Eiceman, G.A., Atmospheric pressure chemical ionization studies of non-polar isomeric hydrocarbons using ion mobility spectrometry and mass spectrometry with different ionization techniques, *J. Am. Soc. Mass Spectrom.* 2002, 13, 1078–1087.

24. Xie, Z.; Sielemann, S.; Schmidt, H.; Li, F.; Baumbach, J.I., Determination of acetone, 2-butanone, diethyl ketone and BTX using HSCC-UV-IMS, *Anal. Bioanal. Chem.* 2002, 372, 606–610.

25. Miller, R.A.; Nazarov, E.G.; Eiceman, G.A.; King, A.T., A MEMS radio-frequency ion mobility spectrometer for chemical vapor detection, *Sensor. Actuat. A: Phys.* 2001, A91, 301–312.

26. Sielemann, S.; Baumbach, J.I.; Schmidt, H.; Pilzecker, P., Detection of alcohols using UV-ion mobility spectrometers, *Anal. Chim. Acta* 2001, 43, 293–301.

27. Sielemann, S.; Baumbach, J.I.; Schmidt, H.; Pilzecker, P., Quantitative-analysis of benzene, toluene, and m-xylene with the use of a UV-ion mobility spectrometer, *Field Anal. Chem. Tech.* 2000, 4, 157–169.
28. Lubman, D.M.; Kronick, M.N., Plasma chromatography with laser-produced ions, *Anal. Chem.* 1982, 54, 1546–1551.
29. Lubman, D.M.; Kronick, M.N., Resonance-enhanced two-photon ionization spectroscopy in plasma chromatography, *Anal. Chem.* 1983, 55, 1486–1492.
30. Young, D.; Douglas, K.M.; Eiceman, G.A.; Lake, D.A.; Johnston, M.V., Laser desorption-ionization of polycyclic aromatic hydrocarbons from glass surfaces with ion mobility spectrometry analysis, *Anal. Chim. Acta* 2002, 453, 231–243.
31. Illenseer, C.; Lohmannsroben, H.-G., Investigation of ion–molecule collisions with laser-based ion mobility spectrometry, *Phys. Chem. Chem. Phys.* 2001, 3, 2388–2393.
32. Gormally, J.; Phillips, J., The performance of an ion mobility spectrometer for use with laser ionization, *Int. J. Mass Spectrom. Ion Proc.* 1991, 107, 441–451.
33. Eiceman, G.A.; Anderson, G.K.; Danen, W.C.; Ferris, M.J.; Tiee, J.J., Laser desorption and ionization of solid polycyclic aromatic hydrocarbons in air with analysis by ion mobility spectrometry, *Anal. Lett.* 1988, 21, 539–552.
34. Huang, S.D.; Kolaitis, L.; Lubman, D.M., Detection of explosives using laser desorption in ion mobility spectrometry/mass spectrometry, *Appl. Spectrosc.* 1987, 41, 1371–1376.
35. Laiko, V.V.; Baldwin, M.A.; Burlingame, A.L., Atmospheric pressure matrix assisted laser desorption/ionization mass spectrometry, *Anal. Chem.* 2000, 72, 652–657.
36. Bramwell, C.J.; Creaser, C.S.; Reynolds, J.C.; Dennis, R., Atmospheric pressure matrix-assisted laser desorption/ionization combined with ion mobility spectrometry, *Int. J. Ion Mobil. Spectrom.* 2002, 5, 87–90.
37. Steiner, W.E.; Clowers, B.H.; English, W.A.; Hill, H.H., Jr., Atmospheric pressure matrix-assisted laser desorption/ionization with analysis by ion mobility time of-flight mass spectrometry, *Rapid Commun. Mass Spectrom.* 2004, 18, 882–888.
38. Driscoll, J.N., Evaluation of a new photoionization detector for organic compounds, *J. Chromatogr.* 1977, 134, 49–55.
39. Accessed August 31, 2012 from http://www.hnu.com/index.php?view=Portable&cmd= Home
40. Eiceman, G.A.; Vandiver, V.J.; Leasure, C.S.; Anderson, G.K.; Tiee, J.J.; Danen, W.C., Effects of laser beam parameters in laser-ion mobility spectrometry, *Anal. Chem.* 1986, 58, 1690–1695.
41. Begley, P.; Corbin, R.; Foulger, B.E.; Simmonds, P.G., Photo-emissive ionization source for ion mobility detectors, *J. Chromatogr.* 1991, 588, 239–249.
42. Chen, C.; Dong, C.; Du, Y.; Cheng, S.; Han, F.; Li, L.; Wang, W.; Hou, K.; Li, H., Bipolar ionization source for ion mobility spectrometry based on vacuum ultraviolet radiation induced photoemission and photoionization, *Anal. Chem.* 2010, 82, 4151–4157.
43. Fenn, J.F., Electrospray wings for molecular elephants (Nobel lecture), *Angew. Chem. Ind. ED.* 2003, 42, 3871–3894.
44. Tanaka, K., The origin of macromolecule ionization by laser irradiation (Nobel lecture), *Angew. Chem. Ind. ED.* 2003, 42, 3861–3870.
45. Kebarle, P.; Tang, L., From ions in solution to ions in the gas phase, *Anal. Chem.* 1993, 65, 972A–986A.
46. Iribarne, J.V.; Thomson, B.A., On the evaporation of small ions from charged droplets, *J. Chem. Phys.* 1976, 64, 2287–2294.
47. Dole, M.; Mack, L.L.; Hines, R.L.; Mobley, R.C.; Ferguson, L.D.; Alice, M.B., Molecular beams of macro ions, *J. Chem. Phys.* 1976, 49, 2240–2249.
48. Hill, H.H., Jr., Electrospray ion mobility spectrometry, Society of Western Analytical Professors (SWAP), Utah State University, January 29–February 1, 1987.

49. Shumate, C.B.; Hill, H.H., Jr., Coronaspray nebulization and ionization of liquid samples for ion mobility spectrometry, *Anal. Chem.*, 1989, 61, 601–606.
50. Hill, H.H., Jr.; Eatherton, R.L., Ion mobility spectrometry after chromatography: accomplishments, goals, challenges, *J. Res. Nat. Bur. Std.*, 93, 1988, 425–426.
51. St. Louis, R.H.; Hill, H.H., Jr., Ion mobility spectrometry in analytical chemistry, *CRC Crit. Rev. Anal. Chem.* 1990, 21, 321–355.
52. McMinn, D.G.; Kinzer, J.A.; Shumate, C.B.; Siems, W.F.; Hill, H.H., Jr., Ion mobility detection following liquid chromatographic separation, *J. Microcol. Sep.* 1990, 2, 188–192.
53. Shumate, C., Electrospray ion mobility spectrometry, *Trends Anal. Chem.* 1994, 13, 104–109.
54. Wittmer, D.; Luckenbill, B.K.; Hill, H.H.; Chen, Y.H., Electrospray-ionization ion mobility spectrometry, *Anal. Chem.* 1994, 66, 2348–2355.
55. Srebalus, C.A.; Li, J.W.; Marshall, W.S.; Clemmer, D.E., Gas-phase separations of electrosprayed peptide libraries, *Anal. Chem.* 1999, 71, 3918–3927.
56. Henderson, S.C.; Valentine, S.J.; Counterman, A.E.; Clemmer, D.E., ESI/ion trap/ion mobility/time-of-flight mass-spectrometry for rapid and sensitive analysis of biomolecular mixtures, *Anal. Chem.* 1999, 71, 291–301.
57. Hoaglund, C.S.; Valentine, S.J.; Sporleder, C.R.; Reilly, J.P.; Clemmer, D.E., 3-Dimensional ion mobility TOFMS analysis of electrosprayed biomolecules, *Anal. Chem.* 1998, 70, 2236–2242.
58. Hudgins, R.R.; Jarrold, M.F., Conformations of unsolvated glycine-based peptides, *J. Phys. Chem. B* 2000, 104, 2154–2158.
59. Steiner, W.E.; Clowers, B.H.; Matz, L.M.; Siems, W.F.; Hill, H.H., Jr., Rapid screening of aqueous chemical warfare agent degradation products: ambient pressure ion mobility mass spectrometry, *Anal. Chem.* 2002, 74, 4343–4352.
60. Wu, C.; Siems, W.F.; Asbury, G.R.; Hill, H.H., Electrospray-ionization high-resolution ion mobility spectrometry–mass-spectrometry, *Anal. Chem.* 1998, 70, 4929–4938.
61. Dion, H.M.; Ackerman, L.K.; Hill, H.H., Jr., Detection of inorganic ions from water by electrospray ionization-ion mobility spectrometry, *Talanta* 2002, 57, 1161–1171.
62. Gidden, J.; Bowers, M.T.; Jackson, A.T.; Scrivens, J.H., Gas-phase conformations of cationized poly(styrene) oligomers, *J. Am. Soc. Mass Spectrom.* 2002, 13, 499–505.
63. Lee, S.; Wyttenbach, T.; Bowers, M.T., Gas-phase structures of sodiated oligosaccharides by ion mobility ion chromatography methods, *Int. J. Mass Spectrom.* 1997, 167, 605–614.
64. Chen, Y.H.; Hill, H.H., Jr.; Wittmer, D.P., Analytical merit of electrospray ion mobility spectrometry as a chromatographic detector, *J. Microcol. Sep.* 1994, 6, 515–524.
65. Wu, C.; Siems, W.F.; Hill, H.H., Jr., Secondary electrospray ionization ion mobility spectrometry - mass spectrometry of illicit drugs, *Anal. Chem.* 2000, 72, 396–403.
66. Takáts, Z.; Wiseman, J.M.; Gologan, B.; Cooks, R.G., Mass spectrometry sampling under ambient conditions with desorption electrospray ionization, *Science* 2004, 306(5695), 471–473.
67. Colgrave, M., Nanoelectrospray ion mobility spectrometry and ion trap mass spectrometry studies of the non-covalent complexes of amino acids and peptides with polyethers, *Int. J. Mass Spectrom.* 2003, 229, 209–216.
68. Sundarapandian, S.; May, J.C.; McLean, J.A., Dual source ion mobility-mass spectrometer for direct comparison of electrospray ionization and MALDI collision cross section measurements, *Anal. Chem.* 2010, 82, 3247–3254.
69. Cody, R.B.; Laramée, J.A.; Durst, H.D., Versatile new ion source for the analysis of materials in open air under ambient conditions, *Anal. Chem.* 2005, 77, 2297–2302.
70. Koomen, J.M.; Ruotolo, B.T.; Gillig, K.J.; Mclean, J.A.; Russell, D.H.; Kang, M.J.; Dunbar, K.R.; Fuhrer, K.; Gonin, M.; Schultz, J.A., Oligonucleotide analysis with MALDI-ion-mobility-TOFMS, *Anal. Bioanal. Chem.* 2002, 373, 612–617.

71. Woods, A.S.; Koomen, J.M.; Ruotolo, B.T.; Gillig, K.J.; Russel, D.H.; Fuhrer, K.; Gonin, M.; Egan, T.F.; Schultz, J.A., A study of peptide-peptide using MALDI-ion mobility o-TOF and ESI mass-spectrometry, *J. Am. Soc. Mass Spectrom.* 2002, 13, 166–169.

72. Steiner, W.E.; Clowers, B.H.; English, W.A.; Hill, H.H., Atmospheric pressure matrix-assisted laser desorption/ionization with analysis by ion mobility time-of-flight mass spectrometry, *Rapid Commun. Mass Spectrom.* 2004, 18, 882–888.

73. Rasulev, U.K.; Khasanov, U.; Palitcin, V.V., Surface-ionization methods and devices of indication and identification of nitrogen-containing base molecules, *J. Chromatogr. A* 2000, 896, 3–18.

74. Rasulev, U.K.; Iskhakova, S.S.; Khasanov, U.; Mikhailin, A.V., Atmospheric pressure surface ionization indicator of narcotic, *Int. J. Ion Mobil. Spectrom.* 2001, 4, 121–125.

75. Rasulev, U.Kh.; Nazarov, E.G.; Palitsin, V.V., Surface ionization gas-analysis devices with separation of ions by mobility, *Fourth Intern. Workshop Ion Mobility Spectrometry*, Cambridge, UK, 1995.

76. Wu, C.; Hill, H.H.; Rasulev, U.K.; Nazarov, E.G., Surface-ionization ion mobility spectrometry, *Anal. Chem.* 1999, 71, 273–278.

77. He, X.L.; Guo, H.Y.; Li, J.P.; Jia, J.; Gao, X.G., A micro electro mechanical system surface ionization source for ion mobility spectrometer, *Sensor Lett.* 2008, 6, 970–973.

78. Guo, H.-Y.; He, X.-L.; Jia, J.; Gao, X.-G.; Li, J.-P., An ion mobility spectrometer with a micro-hotplate surface ionization source [in Chinese], *Fenxi Huaxue* 2008, 36, 1597–1600.

79. Kapustin, V.I.; Nagornov, K.O.; Chekulaev, A.L., New physical methods of organic compound identification using a surface ionization drift spectrometer, *Tech. Phys.* 2009, 54, 712–718.

80. Bolton, H.C.; Grant, J.; McWilliam, I.G.; Nicholson, A.J.C.; Swingler, D.L., Ionization in flames. II. Mass-spectrometric and mobility analyses for the flame ionization detector, *Proc. R. Soc. London, A: Math. Phys. Eng. Sci.* 1978, 360, 265–277.

81. Atar, E.; Cheskis, S.; Amirav, A., Pulsed flame—a novel concept for molecular detection, *Anal. Chem.* 1991, 63, 2061–2064.

82. Lias, S.G.; Liebman, J.F.; Levin, R.D., Evaluated gas phase basicities and proton affinities of molecules; heats of formation of protonated molecules, *J. Phys. Chem. Ref. Data* 1984, 13, 695–808.

83. Marr, A.J.; Cairns, S.N.; Groves, D.M.; Langford, M.L., Development and preliminary evaluation of a radio-frequency discharge ionization source for use in ion mobility spectrometry, *Int. J. Ion Mobil. Spectrom.* 2001, 4, 126–128.

84. Waltman, M.J.; Dwivedi, P.; Hill, H.H.; Blanchard, W.C.; Ewing, R.G., Characterization of a distributed plasma ionization source (DPIS) for ion mobility spectrometry and mass spectrometry, *Talanta* 2008, 77, 249–255.

85. Vautz, W.; Michels, A.; Franzke, J., Micro-plasma: a novel ionisation source for ion mobility spectrometry, *Anal. Bioanal. Chem.* 2008, 391, 2609–2615.

86. Michels, A.; Tombrink, S.; Vautz, W.; Miclea, M.; Franzke, J., Spectroscopic characterization of a microplasma used as ionization source for ion mobility spectrometry, *Spectrochim. Acta B Atomic Spectrosc.* 2007, 62B, 1208–1215.

87. Zhao, Q.; Soyk, M.W.; Schieffer, G.M.; Fuhrer, K.; Gonin, M.M.; Houk, R.S.; Badman, E.R., An ion trap-ion mobility-time of flight mass spectrometer with three ion sources for ion/ion reactions, *J. Am. Soc. Mass Spectrom.* 2009, 20, 1549–1561.

88. Heng, L., Glowing discharge ion source of ion mobility spectrometry [in Chinese], *Liaoning Shifan Daxue Xuebao, Ziran Kexueban* 2008, 31, 485–486.

89. Dong, C.; Wang, L.; Wang, W.; Hou, K.; Li, H., Dopant-enhanced atmospheric pressure glow discharge ionization source for ion mobility spectrometry, *Rev. Sci. Instrum.* 2008, 79(10, Pt. 1), 104101/1–104101/7.

90. Dong, C.; Wang, W.; Li, H., Atmospheric pressure air direct current glow discharge ionization source for ion mobility spectrometry, *Anal. Chem.* 2008, 80, 3925–3930.
91. Spangler, G.E.; Carrico, J.P.; Campbell, D.N., Thermionic ionization source, U.S. Patent Number 4,928,033, May 22, 1990.
92. Tabrizchi, M.; Hosseini, Z.S., An alkali ion source based on graphite intercalation compounds for ion mobility spectrometry, *Measurement Sci. Technol.* 2008, 19, 075603/1–075603/6.
93. Hartley, T.F.; Kanik, I., A nanoscale soft-ionization membrane: a novel ionizer for ion mobility spectrometers for space applications, *Proc. SPIE* 2002, 4936, 43. http://dx.doi.org/10.1117/12.484271.
94. Naegele, E. Making your LC method compatible with mass spectrometry, technical overview. Agilent Technologies, Inc. Waldbronn, Germany. Accessed August 31, 2012 from www.chem.agilent.com/Library/technicaloverviews/Public/5990--7413EN.pdf

5 Ion Injection and Pulsed Sources

5.1 INTRODUCTION

A measurement of the mobility for an ion swarm is initiated when ions from a source or reaction region are introduced into an electric field continuously, as in field asymmetric ion mobility spectrometry (FAIMS), differential mobility spectrometry (DMS), differential mobility analysis (DMA), or aspiration IMS (aIMS) or periodically, as in traditional configurations of ion mobility spectrometry (IMS) or in traveling wave methods. In conventional drift tubes, peak separation in a mobility spectrum is defined by differences in drift times and peak widths whose minimum value is established by pulse width of ion injection with the ion shutter. Resolving power in continuous ion flow methods is defined by these same terms, although minimum peak width is established by the dimensions of the inlet aperture for such an analyzer. The methods and practices of ion introduction regardless of ion mobility method are essential in establishing the best possible performance; consequently, this should have received intense scrutiny and development of IMS. Strangely, interest in ion shutters and in the dynamics of ion flow at ambient pressure has grown only in the past decade, resulting in new descriptions on the limits of shutter performance, alternative methods of ion injection, and improved analytical results. These portend further improvements in resolving power, precision of measurements, and drift tube designs.

Imperfect as shutters and apertures of today may be viewed, a balance in perspective is needed: Existing technology provides enough resolving power for measurements today; however, improved resolving power may increase the precision and accuracy of measurements. The earliest wire grid sets for ion shutters in mobility methods were described in the late 1920s, and a valuable description of the origins and development of ion shutters can be found in a 1938 monograph;[1] Arthur Mannering Tyndall wrote:

> Following upon a method devised by the writer and Grindley and adopted later by Bradbury, a second method was developed by Powell, Starr and the writer which has been extensively used in the Bristol laboratory. ... The actual priority of publication must go the van de Graaff who developed an essentially similar method independently in the Oxford Electrical laboratory; he did not however apply it to any important cases.

Indeed, Robert J. van de Graff not only described[2] ion shutters known today as Bradbury–Nielson (BN) shutters[3] but also was the first to publish a mobility spectrum recognizable today in a standard format of detector response versus drift time (Chapter 2, Figure 2.1). Regardless of implementation, wire grids are today the standard technology for ion injection in IMS, and the design, function, performance, and limitations of ion shutters are described in the first part of this chapter, Other methods of introducing ions into drift tubes, including continuous methods such as needed with FAIMS, DMS, and aIMS, are described further in the chapter.

5.2   OPERATION AND STRUCTURES OF ION SHUTTERS

Ion shutters found ordinarily in IMS drift tubes are made of wire grids, or etched metal analogs, that span the internal cross section of a drift tube and are typically located between a reaction region and the drift region. Differences in voltages between adjacent wires (Figure 5.1a) are used to establish an electric field over the cross section of the tube, and ions moving in the drift tube with drift fields of 200 to 400 V/cm encounter the shutter fields of $\geq$ 600 V/cm and are drawn to wire surfaces. Collisions of ions on the wires of the shutter result in neutralization of the ions, and neutral products are swept from the drift tube with gas flow, either through the ion source with a unidirectional flow design or elsewhere depending on flow patterns.

A mobility spectrum is initiated when the electric field in the grid sets is eliminated or weak enough for ions to drift through the grids under the influence of the external field and enter the drift region. In this condition, the potentials of all wires in the shutter are identical and should be located in the voltage gradient of the drift tube commensurate with its physical position between neighboring drift rings. While some losses in ion flux could occur from distortion of field contours between the shutter and neighboring drift rings, significant losses of flux occur simply through collisions of ions on wires. Such losses can be as large as 30%, although this will depend on the thickness of wires and quality of the grid. After some interval, typically 100 to 400 μs, the electric field between wires in the grid sets is restored to the maximum value, and ions are again blocked from entering the drift region. The waveform used to apply voltages to wires in the ion shutter can be described as a boxcar as shown in Figure 5.1b, and the rising and falling edges occur in a few microseconds or less. This pulse of ions is repeated after an interval of 10 to 30 ms or more, depending on the ion residence time in the drift tube. Several arrangements of voltage have been employed for controlling the field, as shown in Figure 5.1B. In one, the "downfield" grid is elevated in potential to stop ion flow into the drift region, and potential on this grid is pulsed low to match the other grid for an injection. In another design, wires are placed in potential symmetrically off the gradient potential and pulsed to this common potential. There is as yet no clear fundamental advantage in performance between these two methods of controlling electric fields in ion shutters.

Although ion shutters are useful and effective in IMS at ambient and subambient pressures, difficulties range from practical details, including the manufacture of grids that tolerate swings in temperature and are mechanically robust, to fundamental facets, including mobility biases in sampling. More critically, conventional ion shutters produce a very low-duty cycle for sampling ions in the ionization region, as low as about 1% for shutter pulses of 200 μs over 20-ms intervals. Much of the innovation in grid performance or operation during the past decade was concerned with this poor duty cycle and consequences for measurements. Also, variations exist in the

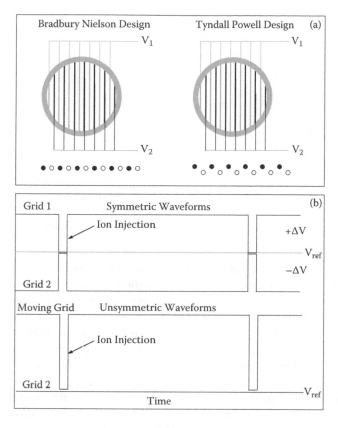

FIGURE 5.1 Designs of ion shutters with Bradbury–Nielson on left and Tyndall–Powell on right with end view and side view (a). Two waveform plans (b); a waveform is applied to each grid. A voltage difference on adjacent wires creates an electric field, and ions are drawn to wires and collide. Resultant neutrals are swept with drift gas from the analyzer. In each plan, wires are brought to a common potential, which is referenced to the voltage divider of the drift tube.

selection of wires, grid designs, and manufacture, while the basic concept of pulsing the gate using electric fields with grids is found in all conventional drift tubes.

5.2.1 Bradbury Neilson Design

In the BN design of an ion shutter,[3] wires are placed in a parallel pattern with close separations in a single plane as shown in Figure 5.1a. Wires are supported on a non-conducting frame, and adjacent wires are isolated mechanically and electrically. Although this style of shutter is understood as the design with best performance, a main limitation is difficulty of manufacture, particularly for high temperatures. For example, the snapping or sagging of wires from thermal expansion or contraction of the frame or wires during heating and cooling cycles often experienced with IMS analyzers can degrade performance when wires are not parallel or contain bends. Distortions of electric field in such changed grid sets will affect performance of the ion shutter if electric

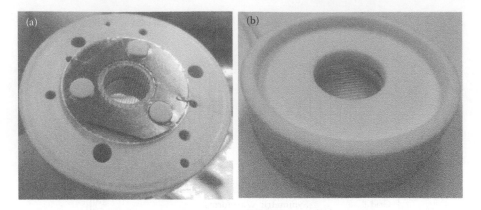

FIGURE 5.2 Photographs of (a) Tyndall–Powell ion shutter and (b) Bradbury–Neilson shutter.

fields are not even throughout the cross section. In an extreme condition, an ion shutter will fail when wires are snapped or touch, shorting shutter electronics.

In solution to the challenge of thermal expanstion,[4] the frame was made from two ceramic half circles, which were aligned and separated by springs suitable for use at high temperatures. The wires were held on the ceramic halves at fixed tautness since changes in thermal expansion were compensated by the spring tension. In another method employed at New Mexico State University,[5] wires were placed under tension in a frame and cycled repeatedly through extremes of temperature before wires were attached to the frame. Wires were bonded to the frame using high-temperature epoxies; one such BN shutter is shown in Figure 5.2b. Other BN shutters have been made using insulating frames of circuit boards, though such shutters cannot be used at elevated temperatures.

5.2.2 THE TYNDALL POWELL DESIGN

Another design of wire-based grids is an ion shutter described by Tyndall and Powell[6] (TP); this shutter consists of a set of grids with parallel wires, and the grids are placed in separate planes (Figure 5.1a), offset by distances of 0.01 to 1 mm. The distance of separation of the two grids is established by the thickness of an insulator between the grids. When assembled and aligned, the wires of the grids appear parallel and in an interdigitated pattern, as with the BN design, when viewed from the direction of flow of an ion swarm. The function and electrical control of the TP design can be identical to a BN shutter, and the main advantage of this design is the convenience of manufacture. In the TP design, shutters may be crafted in three parts: two geometrically correct and independent wire grids and an insulator. These can be assembled conveniently and economically to form a functional ion shutter as seen in Figure 5.2a.

In the TP design of Figure 5.2a, individual grids are etched metal with parallel "wires" of 0.05 mm diameter at distances of 0.5 mm center to center. The wires, with rectangular cross section, span a circular cross section with an internal diameter of 13 mm on a disk with a 30 mm outer diameter. The two grids were separated by

0.3 mm with a thin mica sheet, allowing use at high temperatures, and the grids were positioned so the wire of the combined grids made an interdigitated arrangement, with wire-to-wire distances of 0.25 mm, center to center when viewed line of sight.[7] This shutter design has been extensively modeled and tested empirically (Section 5.3).

In a rugged and simple variation on a TP design, the wire grids are replaced with mesh grids as shown in Figure 5.3. An ion shutter is made from two mesh grids, which are separated with a thin Teflon ring, overlapping the circumference of the grids, and with potentials as used in BN or TP designs. The grids are aligned at 45°, although an attractive feature of this shutter is that precision of grid alignment, surprisingly, is not particularly important, and the shutter functions whether the grids are overlapped or phased. This is not characteristic of parallel wire configurations with the TP design, for which performance is sensitive to misalignment, although this has not been understood, modeled, or even well described.

5.2.3 MICROMACHINED ION SHUTTERS

In principle, methods of microfabrication should allow the mass production of ion shutters at low cost and ultimate convenience. Unfortunately, technology for micromachining silicon establishes some limits on wire size and grid structures. In one effort intended for ion mobility instruments,[8] shutters were fabricated using wafers of silicon-on-insulator with thicknesses of 6 to 30 μm for the upper silicon layer. Standard anisotropic wet etching was used with an automatic etch stop in the buried oxide layer. Finally, the conductive wires were made by reactive ion etching (RIE). Design parameters were limited to voltage differences of 5 to 17.1 V; height or thickness of wires of 6 to 30 μm; distance between wires of 10 to 81 μm; and width between wires of 20 to 100 μm. Although no actual data were obtained with these micromachined ion shutters grids, modeling of electric fields was described as promising over the range of combinations of these parameters.

Micromachined shutters of comparable dimensions and structures were described in detail and used in a time-of-flight mass spectrometer (MS).[9] The wires or "electrodes" were spaced at distances ranging from 25 to 100 μm and with a thickness of 20 μm. Other parameters for this shutter were 400 μm total thickness; 20-μm electrode depth; 100-μm electrode pitch; 128-pF capacitance; 80% transmission; and 25 mm² total area of the grid with a 5 by 5 mm shape. The transmission efficiency was comparable to wire shutters, and performance was good, with rise and fall times of 0.2 μs.

Nearly the same ion shutters were incorporated in prototype microscale IMS drift tubes, which were 25 mm long. Ion injections could be reduced to a few microseconds since ion travel through the shutter grids was microns.[10] The microshutter showed behavior comparable to larger wire-based constructions, except for faster injections as expected with small sizes, and behavior otherwise was regular as shown in Figure 5.4. Disadvantages were the small cross section through which ions could flow and the mechanically fragile wires, which were easily broken by touch. Once installed, these ion shutters were rugged and unaffected by drift gas flow rates. While microfabrication methods could be useful for small drift tubes and interesting with future innovations, applications to large drift tubes presently seem neither cost effective nor compelling.

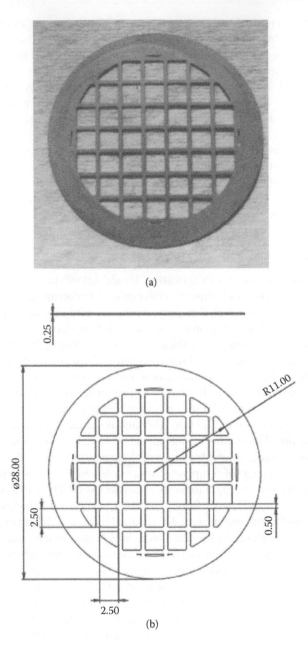

(a)

(b)

FIGURE 5.3 Grid for ion shutter. Two grids of this design have been employed on drift tubes built for 3QB in Israel. (Courtesy Z. Karpas.)

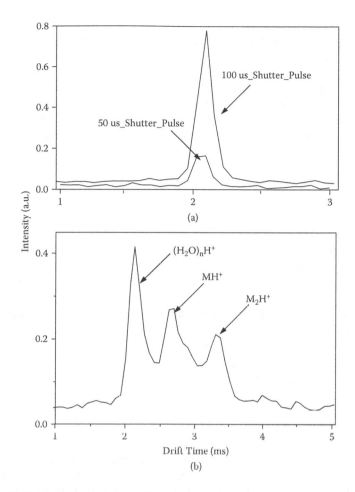

FIGURE 5.4 Mobility spectra obtained using microfabricated ion shutters in a small DMS IMS instrument for which the length of the drift region was 10 mm, showing (a) reactant ions with two pulse widths on the ion shutter and (b) mobility spectrum of 4-heptanone. (Unpublished data adapted from Eiceman et al., Characterization of positive and negative ions simultaneously through measures of K and ΔK by tandem DMS–IMS, 14th International Symposium on Ion Mobility Spectrometry, 2005.)

5.3 MODELS AND MODES OF OPERATION

5.3.1 SHUTTER BEHAVIOR AND MODELS

The width of the waveform pulse applied to wires in an ion shutter is the key parameter in establishing peak widths in a mobility spectrum since the minimum possible width of an ion swarm, before normal diffusion and other influences in the drift region, is set by the shutter pulse time. The minimum time of a pulse, even in a well-designed and well-built instrument, is fixed by the time for ions, under mobility control of the superimposed electric field of the drift tube, to pass from one side of a shutter grid set (reaction region) to the other side (drift region). Injection pulses of about 100 μs are commonly the lowest possible injection times for contemporary drift tubes operated at ambient pressure, with mobility values of 0.5 to 2.5 cm^2/Vs, although this can be reduced with subambient pressure drift tubes, in which drift velocities are increased. At present, there is no standard or common value for shutter pulse widths in IMS applications or research, and the injection pulse width is chosen based on desired resolving power and signal-to-noise (S/N) level. The influence of variables, besides pulse width, in an ion shutter for IMS was described in the mid-1980s,[11] and a thorough quantitative description was provided only in 2005.[12] This significantly clarified the effects on instrument response by control of parameters in an ion shutter and supplemented understandings of pulse width as summarized next.

5.3.2 Ion Shutter Performance and Peak Shape

5.3.2.1 Peak Widths in Spectra and Shutter Pulse Width

The performance of an ion shutter is controlled, and limited, by the drift velocities of ions in electric fields, and this governs ultimate performance of a shutter. In a simple calculation, an ion swarm with velocity of 5 m/s, or 5 mm/ms, will require an injection pulse of 0.2 ms to allow ions 1 to 2 mm distant from the shutter to reach the shutter itself, if infinitely thin. Since shutters have finite width, a minimum time of 0.1 ms is also needed for ions to pass through a 0.5-mm thick shutter. At shutter pulse times below this, ion intensities drop rapidly to zero. The effect is not only geometric, however, and electric fields of the drift tube and the wire grids create an "effective width" for a shutter that is variable and larger than simple geometric widths. There are consequences of increased pulse widths in a unipolar environment found inside the drift region, and in 1975 these were quantitatively described;[13] the rectangular pulse on a BN design ion shutter was modeled to two error functions, which reduced to a Gaussian peak shape as shown in Figure 5.5. In addition to shutter pulse width, normal diffusion and ion–ion repulsions were shown as essential terms to completely describe the peak shape of a chloride ion in purified gas atmosphere at ambient pressure. As the shutter pulse time exceeded 200 μs with a 10-mCi ^{63}Ni foil as the ion source, ion–ion repulsion became significant, and peak broadening could be discerned in mobility spectra at ion densities of 5.26×10^7 ions/cm^3 or greater. This is seen in Figure 5.5; a correction for ion–ion repulsion is first visible for shutter pulse times of 200 μs or longer for a chloride ion with mobility coefficient of 2.99 cm^2/Vs. This becomes clearly apparent at injection pulses of 500 μs. At pulse widths of only 50 μs, peak distortion was negligible and could be attributed to electronics of the drift tube. This first work demonstrated that peak shape in a mobility spectrometer with a clean gas atmosphere could be described principally by the initial pulse width and corrected for diffusion and ion–ion repulsions at elevated levels of ion loading into the drift region. There was no accounting in this early report for ion motion through the shutter or for synergistic effects between electric fields of the shutter and in the drift tube.

5.3.2.2 Effects of Drift and Shutter Fields in Performance

Static and dynamic models for ion shutters were developed as a joint effort of computational models and experimental measurements for which electric fields were

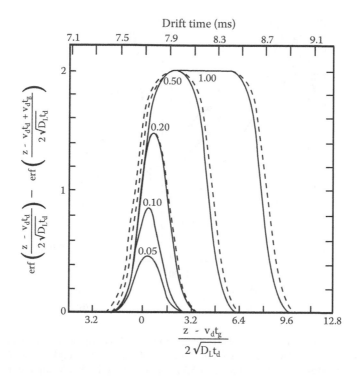

FIGURE 5.5 Mobility spectra for chloride anion as predicted (solid lines) as a function of gate width setting.[13] The corrections for electrostatic repulsion to unite theory and experiment are shown by broken lines. (From Spangler and Collins, Peak shape analysis and plate theory for plasma chromatography, *Anal. Chem.* 1975. With permission.)

considered between wires in the shutter and for the shutter inside the electric field of the drift tube.[12] When potentials between wires are low, significant penetration by ions through the shutter occurs during the time when the field between the wires should block ion flow. This is seen as an elevated baseline in any mobility spectrum; ion peaks will have small intensity on a large offset in detector current. The peak width will still be governed largely by the shutter pulse waveform and will not disclose ion penetration in the shutter. As the potential difference between the wire set is increased, ion flow between wires will be stopped proportionally, and the spectrum will show a peak of comparable absolute intensity as before, yet now on a lower absolute baseline value. That is, the DC component of the spectrum arising from continuous ion leakage will be reduced and at some field strength between wires E_s will be eliminated. As E_s is increased, the percentage of ions stopped by the shutter is improved, approaching a maximum, total stopping of ions; at this point, a maximum is seen for peak intensity (peak apex minus baseline) and peak area. When the field strength in the shutter is increased further, well past the point of maximum ion stopping, peak intensity and area are decreased. This is attributed to a decrease in ion populations near the shutter due to losses from ion annihilation by collision on wires aided by high electric fields extending from the wires, both in the direction of the ion source and in the direction of the drift region. The combination of these two effects is

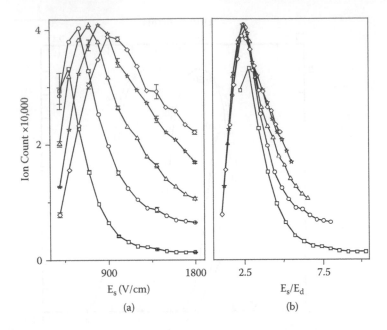

FIGURE 5.6 Plots of ion count for (a) reactant ions as a function of electric field within the ion shutter E_s at ambient temperature for the following E_d values: rectangles, 175 V/cm; circles, 225 V/cm; triangles, 275 V/cm; stars, 325 V/cm; and diamonds, 375 V/cm; and (b) when E_s was normalized to corresponding drift fields E_d. (From Tadjimukhamedov et al., A study of the performance of an ion shutter for drift tubes in atmospheric pressure ion mobility spectrometry: computer models and experimental findings, *Rev. Sci. Inst.* 2009. With permission.)

a maximum value for E_s, for which ion penetration is low and ions are brought through the shutter rapidly, without ion depletion before or after injection (Figure 5.6a).

The drift field E_d that surrounds the ion shutter is not passive and controls ion penetration through the shutter and the field in the shutter wires for optimum performance. The influence of the externally imposed field is seen as a displacement of absolute voltage needed to balance ion leakage and ion transmission through the shutter and results in a shift of the field for optimal performance as shown in Figure 5.6b. The maximum in peak height at various magnitudes of E_d arose from the same phenomenon, a balance of ion penetration versus ion losses, and this was demonstrated when the plots were normalized and the maxima under all conditions of drift tube and shutter showed a ratio for $E_s:E_d$ of 2.5:1 on average. Changes in electric fields of drift tubes should be accompanied by resetting the electric fields in the shutter for maximum performance of the shutter. While these studies were made using a TP design, nearly comparable results from modeling were observed with a BN design.[14] A detailed examination of the BN shutters also disclosed the importance of wire diameter on performance and imperfections of the formation of an ion cloud or swarm even with the best possible design for wire-based shutters. The influence of mobility on the very act of swarm injection into the drift region contains distortions and limits the overall performance with regard to sensitivity and resolving power of an IMS analyzer.

5.3.2.3   Repulsive Interactions with Shutter-Produced Ion Swarms

In an IMS, coulombic repulsions can be anticipated from the unipolar condition with gas phase ions at reasonably high density. This has for decades been considered negligible, when drift tube are operated carefully, based on understanding provided in early measurements and modeling described previously in this chapter.[13] Occasionally, ion–ion repulsion can be observed with broadened peak widths owing to an intense ion source, such as laser ionization or ablation; however, effects have been ignored or marginalized when ion shutters are used to control ion densities in the drift region. In a detailed work,[15] findings were presented for which differences in drift times for ion swarms were attributed to the effects of repulsions in ordinary or standard IMS measurements. Possibly relevant in this observation was the precision of reporting drift times in 1-μs intervals (e.g., a drift time of 6.016 ms), while investigators routinely reported 10-μs intervals (i.e., 6.02 ms). Differences of 10 to 40 μs were observed when ion swarms had relatively close drift times. These effects may have been overlooked for the past 40 years of modern analytical IMS, perhaps owing to the imprecision of time measurements.

5.3.3   The Use of Two Ion Shutters in a Drift Tube

The earliest mobility spectrometers were operated using an identical pair of ion shutters of the TP design, or of mesh, with ion shutters at the front and end of a drift region to define the distance of drift for the ion swarms. Each grid set was identically provided a sinusoidal wave of voltage establishing an oscillating electric field, in phase for both ion shutters.[1] On one portion of the wave, where the electric field in the shutter was inverted with respect to drift field, ion entry into the drift region was blocked. When the voltage from the waveform created a field in the direction of the drift field, in Tyndall's terms,[1] a "forward bias of the field in the grids existed," ions were injected through the first shutter. Ions passed through the second shutter under identical field constraints and only when the time of swarm drift matched a subsequent waveform "opening" the second shutter for ions to move to the detector. Mobility of ions was not determined using a time of drift measured directly but one measured indirectly through control of frequency of the sinusoidal waveform. High frequencies could lead to overtones, so a mobility measurement for a given ion of certain mobility was comprised of several peaks with appearances at f, $f/2$, $f/3$, and so on. The frequency could be converted to drift time and then mobility. This concept of two-shutter operation reappeared nearly half a century later with Fourier transform (FT) methods to increase duty cycle and improve S/N ratios (Section 5.3.4.1).

While useful in certain research studies, this type of control of ion shutters and drift tubes was impractical for more complex chemical measurements and analytical applications. Nonetheless, a nearly identical strategy was employed with the Franklin GNO model Beta VI, the inaugural commercial IMS instrument in 1970 (Figure 2.2a); ion shutters of the BN design were operated with rectangular waveforms instead of the sinusoidal waveforms of Tyndall. A dual-shutter approach was used largely due to relatively slow data acquisition capability of that era, and mobility spectra were generated using a boxcar integrator controlling the two ion shutters and X-Y plotter to record the spectrum. In a boxcar integrator, the ion shutter at the beginning of the drift region was pulsed continuously at some duty cycle and frequency, nominally 1% and 10 Hz, respectively. The second ion shutter, at the end of the drift region, was operated at the same duty cycle and frequency as the first shutter but with a delay from the first shutter pulse; this delay was incremented on each waveform by some small value. An ion swarm arriving at the second shutter with a given delay (i.e., drift time) passed through the second ion shutter, a Faraday plate detector. A plot of ion current versus delay yielded a mobility spectrum in which delay and drift time were identical. Practically, a mobility spectrum of acceptable S/N and peak width could be obtained in 2 to 5 min. Apart from the additional cost and loss in ion flux with two ion shutters, rapid changes of sample composition in the ion source could not be monitored comprehensively. When a substance of certain mobility was anticipated, the second shutter could be fixed at the appropriate delay and response monitored continuously, that is, at the frequency waveform, usually 10 to 30 Hz.

While dual shutters were a common feature of early IMS drift tubes, there may seem to be little value for this approach with high-speed digital processing of signals; however, a dual ion shutter design can be helpful when drift tubes are interfaced with MSs, such as relatively slow quadrupole MSs. In such ion mobility–mass spectrometry (IM–MS) combinations, a second ion shutter is useful in isolating ions from a region of the mobility spectrum to pass into an MS. A mass spectrum is obtained then for only the mobility selected ions. Poor ion transmission from IMS to MS in the ambient air to vacuum interfaces with pinhole inlets can be compensated by integrating the response in the MS for a single mobility region when slow quadrupole MSs are employed. The concept of dual shutters in the IM–MS has been revisited.[16] The benefit of a dual shutter in such an IMS–MS configuration with poor efficiency of ion sampling is the possibility of signal averaging over a long period of time. An alternative approach to an IM–MS with slow quadrupole MSs is synchronization of data systems of the IMS and the MS and selecting only one ion for mass detection. This was employed in the PCP, Incorporated, line of IM–MS instruments and was termed mass-resolved mobility spectra.

Another example today for which dual ion shutters are still useful is the development of a kinetic IMS[16] comprised of a reaction region and two drift regions with an ion shutter between each region. The first ion shutter is used to sample ions in the reaction region, and a second ion shutter is used with a specific delay from the first shutter to isolate a particular ion swarm in the first drift region. Ions passing through the second ion shutter enter a second drift region as an isolated swarm. In this second drift region, the ion swarm undergoes thermal decomposition or dissociation during ion passage, leading to the formation of mobility spectra with distortions of the baseline or peak shape. The original ion and the ion produced throughout the drift region allow rates of reactions to be determined at ambient pressure where ions are fully thermalized. This technique has been developed[17,18] to explore the lifetimes of ions as part of an effort to explore their formation and the appearance of mobility spectra.

5.3.4 ALTERNATE MODES OF ION INJECTION

Regardless of the disadvantages of crafting ion shutters or losses in ion transmission from ion collisions on wires, a fundamental objection to such technology is the signal intensity, which is limited by duty cycle. Typically, ions are sampled from the reaction region for only 100 to 300 µs every 20 to 30 ms; thus, only 0.5–1% from the ion source or reaction are injected into the drift region. This is hugely wasteful of ion flux, which limits the total signal intensity and disadvantages the effort to improve the S/N. A shutter pulse of 500 to 1,000 µs would admit more ions into the drift region, although spectral resolving power would be degraded. Thus, comparatively narrow shutter widths are necessary to preserve the moderate spectral resolution achievable with IMS analyzers at ambient pressure. Although efforts at signal processing can be made to reduce the noise level and boost S/N values,[19] alternatives to ion shutter control as described have been sought. That is, narrow peak shape in mobility spectra retained with the duty cycle is increased, and S/N is improved. This has been approached by the deployment of FT and Hadamard transform (HT) techniques, so-called multiplex methods, in which the simple boxcar waveform used in conventional ion shutter operation is replaced with complex control of the shutter.

5.3.4.1 Multiplex Control of Ion Shutters to Improve Duty Cycle of Ion Injection through Shutters

In 1985, Knorr et.al., introduced the FT-IMS to improve the duty cycle of IMS instruments, resulting in a 25% duty cycle and improvement of 3 in S/N.[20] In their FT-IMS method, two ion shutters were employed in a manner similar to that described by Tyndall nearly a half-century earlier.[1] One ion shutter was used to inject ions into the drift region, and a second ion shutter was located at the end of the drift region and directly in front of the detector. A square waveform (i.e., 50% duty cycle), was applied synchronously to both shutters. Ion drift velocities are comparatively slow, and since the delay between the two ion shutters was nil, ions pass the second gate not on the same waveform pulse as used in ion injection, but rather on subsequent pulses. Ions reach the detector, passing through the second shutter when the drift time for an ion swarm and the waveform pulse on the second gate coincide. Thus, waveform frequency and drift times must be coincident, and a range of mobilities can be measured with a sweep of frequency from a few hertz to 5 kHz. As described by Knorr et al.: "With a gate modulation frequency υ, the only ions that reach the detector with full intensity have transit times $0, 1/\upsilon, 2/\upsilon, 3/\upsilon$. ... The only ions which do not reach the detector at all have transit times $1/2\upsilon, 3/2\upsilon, 5/2\upsilon$."[20] The resultant signal is a mixture of frequency-derived waveforms, and an inverse, fast FT can be performed to recover the time domain mobility spectrum. While the signal to noise ratio should be enhanced with increased duty cycle from the shutter and a 5× increase in S/N was expected, an increase of only 3× was observed. Eighteen seconds were needed to obtain a spectrum.

Part of the FT-IMS spectrum, from 0 to 1 ms, showed a decline in intensity, and this was attributed to a "gate depletion effect," which they described as "a lower concentration of ions just outside the entrance gate due to the high electric field of the gate.

At low square wave frequencies, sampling extends deep into the reaction region, but at high frequencies the depleted region supplies a larger proportion of the sample. This interpretation was supported by experiments in which the gate voltage was varied."[20]

Tarver[21] simplified the FT–IMS instrument by replacing one of the ion shutters, that at the end of the drift region, with a "virtual shutter" in which the function of the second shutter was controlled with electronics. The synchronously pulsed virtual shutter beats against the ion signal to generate the interferogram instead of beating against the streaming ions in the drift tube, avoiding ion losses caused by ion collision on wires of an actual exit shutter. The expected transmission rate was reduced by postmeasurement signal analysis, yet a 7× gain in S/N was obtained versus conventional single ion shutter measurements. This was better than three obtained by Knorr with two wire grid ion shutters.

The long-sought increase in ion transmission was reported using HT methods; a full 50% duty cycle was reported by two teams independently.[22,23] Unlike FT–IMS, for which a sweep of frequency was needed and elevated current applies only to ions of a specific drift time at any given moment, ion injection with the ion shutter occurs using a pseudorandom binary pattern. Signal generated using Hadamard-based encoding could be deconvoluted to a mobility spectrum, and S/N was enhanced by 10 times that of a signal-averaged spectra method with the same drift tube. Spectral artifacts were generated in HT–IMS methods, and the limitation of ion depletion near the ion shutters was not addressed. Artifact-free spectra were obtained with a flexible digital multiplexing method, with user-generated arbitrary binary sequences of variables allowing the duty cycles to be adjusted from 0.5% to 50%. At reduced frequency, the effects of ion depletion were reported to be lessened, and "the cumulative effect of imperfect gating events are mitigated in comparison to standard Hadamard multiplexing."[24] Compared to standard HT–IMS methods, this method exhibited false peaks typically present in Hadamard multiplexing techniques and did not exhibit the same gains in S/N as HT–IMS. As of yet, these methods have not been incorporated into any commercial IMS analyzer and are available generally even to IMS researchers.

5.3.4.2   Inverse IMS

In inverse IMS,[25] ions are entered continuously though an ion shutter and into the drift tube. An "injection" is made when fields within the ion shutter are pulsed to temporarily stop ion flow into the drift region. The absence of ion flux is measured on a drift time scale disclosing mobility coefficients. More than this, resolving power is reported to increase for their drift tube by 63%, from 44 to 72. The height of the best-resolved peak in the inverse mode was 10 times higher than the narrowest peak in the normal mode, and the reason for improvement in the inverse mode was attributed to coulomb repulsion. The following was reported[25]:

> In the case of normal operation, a swarm of ions is traveling, and the ions diffuse out of the swarm during their travel, which results in broadening. In addition, the ions inside the swarm push each other away, causing further broadening. In fact, diffusion and repulsion both contribute to the broadening. However, in the case of inverse operation, the width of the traveling dip tends to decrease because of the repulsion between the ions outside the dip.

Inverse IMS was tested using two compounds with closely spaced ion peaks in a mobility spectrum and was resolved at baseline, unlike conventional IMS, for which a partially resolved double peak was observed. In the normal mode, the space between the two close peaks is thought to be filled

> due to diffusion and repulsion inside the two adjacent ion packets. However, in the inverse mode, in which two neighboring dips are traveling, a layer of ions is trapped between the two dips. These ions are under a sandwich force from the ion clouds on both sides. As a result, the trapped swarm of ions gets narrower, thus increasing the baseline resolution for the two neighboring dips.

Their findings are consistent with this interpretation.

A theoretical treatment of inverse IMS was provided by Spangler and was based on the effects of space charge on peak broadening in an ambient pressure drift tube.[26] He concluded that peak broadening in the absence of repulsions was indeed controlled by the coefficient of mutual diffusion, and as ion density is increased, mutual repulsion between ions, independent of diffusion, causes peak broadening in conventional IMS and peak narrowing in inverse IMS. The mutual repulsion produces additional edge velocities, causing the moving ion swarm to expand and the analog depleted zone to contract. The theory reveals that inverse IMS is effective only when ion densities are large, necessitating bright ion sources, and resolving power may be significantly dependent on concentration, hence product ion densities, of substances in a sample.

5.3.4.3 Mechanical Ion Injection

A remarkable adaptation of technology borrowed from another analytical method was demonstrated for an electrospray ionization (ESI) IMS instrument; ion injection to a drift tube was achieved through modulation with a mechanical chopper as found in atomic absorption spectrometry.[27] The chopper was a disk with a small hole that would align with the source and drift tube and would operate as an ion injector. The disk had a second window that was used with optical sensors to synchronize ion injection and drift time, and ion injections were made at pulse rates of 5 to 200 Hz with pulse widths of 200 to 500 μs.

The drift tube dimensions or characteristics were as follows: length of 45.0 cm, a voltage divider with 3.34-MΩ resistors, electric field of about 400 V/cm, and 78 drift rings (0.12 cm thick, 4.90 cm outside diameter, 2.55 cm inside diameter). The front flange of the drift tube was placed at ground potential, and the detector, and housing of the drift tube, was floated to −20.0 kV. Thus, the capillary of the ESI source was operated only at +5.0 kV with +500 V applied to the chopper wheel. The distance between the ESI source and inlet window of the chopper was 2 mm, and that between the inlet window and inlet flange of the drift tube was 5 mm.

An attractive feature of this chopper-based method was the exclusion of ESI spray from the drift tube except for the brief period of ion injection. Aerosols were blocked from entering the drift tube most of the time, protecting the drift tube from the large burden of mass flux; this made this design inherently clean, unlike other ESI IMS designs, for which a large demand exists for a drift tube to accept an aerosol-rich flow, which must be desolvated and swept from the drift tube. Injection pulse widths were adjusted by reducing the area of the sample inlet window with metal tape. Baseline separations of selected benzodiazepines, antidepressants, and antibiotics were obtained, and resolving power was about 70.

5.4 ION INJECTION TO DRIFT REGIONS WITHOUT WIRE-BASED ION SHUTTERS

The presence of wires in the inside cross section of a drift tube has disadvantages of mechanical construction and associated failure rates. The costs and complexity of manufacturing IMS drift tubes could be significantly improved if wire grid or mesh designs for ion shutters could be eliminated. This is one central motivation for developing gridless methods for ion injection. When an ion source is periodic or discontinuous and ions are formed in periods short enough to approximate the injection time of an ion shutter, shutters may be discarded or considered supplemental. This source-mediated ion injection can occur with certain corona discharges (CDs) or photoionization sources, such as pulsed lasers and matrix-assisted laser desorption and ionization.

5.4.1 DRIFT FIELD-SWITCHING METHODS

5.4.1.1 Field Switching Inside Source Region

In 1992, a patent was filed by Anthony Jenkins for a drift tube design in which ions were "trapped" and then passed to the drift region.[28] In this design, a sample vapor is passed through an ion source, and a grid, located near the reaction region in the cross section of the drift tube, is used to establish a field-free volume inside the source or reaction region by holding this grid at the same potential as the surrounding surfaces inside the reaction region. Another grid near this first grid is fixed at a potential suitable to establish an electric field in the drift region. When an injection of ions is made, the entire reaction region and the first grid are rapidly reset to voltages suitable for ions to flow from the source and reaction volume toward this first grid, the second grid, and finally into the drift region. The intention of this design was to accumulate ions over 20 ms and then "compress the swarm into a pulse of 0.2 ms." This is said to increase ion density and signal current by a factor of 100. This design was commercialized with the Itemiser instruments of Iontrack Instruments, later GE Security and now Morpho Detection.

5.4.1.2 Field Switching Inside Drift Tube

Blanchard demonstrated ion injection by switching fields in drift rings in the absence of wire grids.[29,30] Unlike the method of Jenkins, the voltage gradient or electric field in the drift tube was wholly conventional, except in the volume of the "ion shutter," where a drift ring was set at a potential well below that needed to establish a linear voltage gradient (Figure 5.7a). In the drift tube volume between this and neighboring drift rings, ions flow rapidly from the adjacent upfield drift ring and are blocked from moving further by the downfield drift ring. The original term used for this method was "ion well," and indeed ion flow is predictable, entering the interring volume

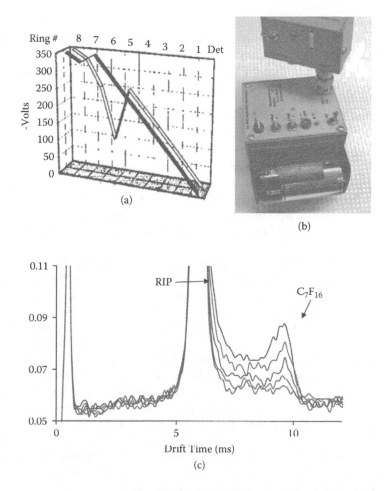

(a)

(b)

(c)

FIGURE 5.7 Ion injection without wire grids. In this concept (a), potentials on rings are placed low to accumulate ions and then pulsed high to inject ions into the drift region. (b) A small demonstrator instrument and (c) a mobility spectrum in negative polarity for perfluoroheptane are shown. This instrument was designed as an inexpensive replacement for a smoke alarm. (From Blanchard et al. *Int. J. Ion Mobility Spectrom.* 2002, 5(3), 15–18; Blanchard ion detecting apparatus and methods, patent number 6924479, filed January 24, 2003, issued August 2, 2005.)

of ions without passing into the remainder of the drift tube. To inject ions, the ring potentials are reset by elevating two rings to push ions in the "well" volume into the drift region. In addition, an upfield ring is set low to block ion passage into the well volume. Blanchard noted correctly that ions enter the volume between drift rings based on mobility, and that the "accumulation of ions is dynamic." Indeed, ion flow is dynamic, and ions, once inside the well volume, move outward toward the inner walls of the drift tube. Ions reaching the wall are annihilated and removed from a measurement. Times of about 1 to 25 ms were used to "accumulate" ions flowing from the source (here ^{241}Am) in an air atmosphere at ambient pressure. Demonstrator instruments (Figure 5.7b) were fabricated around this concept and produced peaks with discernible mobilities and peak shapes (Figure 5.7c). This resolving power was suitable for use as an advanced smoke alarm, the original intention for this device. While this concept was simple for design and construction of drift tubes, the highly dynamic environment of "holding" ions suggested a need for timing and field control not possible with the technology of the demonstrator instruments.

5.4.1.3   Ion Trap for Injecting Ions into a Drift Tube

Although an ESI source provides a continuous flow of ions, the amount of this ion flux used in a drift tube is limited by the duty cycle of the ion shutter. One method to improve S/N with drift tubes operated at subambient pressure is to place a quadrupole ion trap between source and drift tube. Ions from the source could be accumulated in the trap and then injected into the IMS drift tube.[31] Improvements in S/N by factors of 10 to 30 were reported, and the experimental duty cycle was nearly 100%. Detection limits of 1.3 pmol for the oligosaccharide maltotetraose were obtained in 10 s or less.

In operation of the trap injector, ions enter the trap continuously through a 0.32-cm diameter orifice and experience a 1.1-MHz electric field, applied to the ring electrode and variable from 0 to 5 kV depending on the m/z ratios of the ions being trapped. When the trap is biased within a few volts (≤ 30 V) of that at the exit of the ESI source, ions are retained inside the trap. Ions are ejected from the trap when the radio frequency (RF) is switched off and a 0.6-μs negative pulse is applied to the entrance electrode. Neither pulse voltages from 40 to 400 V nor delays of 0.2 to about 4 μs between RF off and pulse on affected the ion signal. Control of the bias voltage needed to trap ions was critical; otherwise, ions either were not ejected during the injection pulse or were not efficiently trapped, leaking continuously from the trap. The pressure inside the trap was varied from 5×10^{-4} to 8×10^{-4} torr by changing the drift tube pressure, with no noticeable effect on the performance of the ion trap. This injection method was designed for use with drift tubes at subambient pressure in ESI IM–MS experiments and will likely remain associated with such designs with no obvious translation to drift tubes at ambient pressure.

5.4.2   Source-Mediated Injection

Certain ion sources are not continuous, as with ESI and ^{63}Ni, and instead are intermittent, requiring electrical pulses, such as with CDs or optical pulsing like that with multiphoton ionization to make ions from a sample. Ion injection with such sources can be coincident with ion formation, and ion injection may be arranged without ion shutters. In this instance, an ion shutter may be seen as optional or unnecessary. While a pulsed ion source could be attractive for reduced costs of drift tube manufacture and simplicity of design, pulsed sources can introduce time-dependent chemistry or ion intensity into a mobility measurement, and neither of these is easily controlled or desirable.

5.4.2.1   Pulsed Corona Discharge

In a pulsed CD, the formation of ions is initiated with a fast increase in the electric field, commonly near a point-to-plane or point-to-surface discharge. The common direct current voltage discharge moves with increasing voltage or electric field rapidly through stages of preignition breakdown streamers, a region of stable discharge, the CD, with 1 to 10 μA of current, and a condition of arcing. In a pulsed ion source, the time-dependent processes may be somewhat more complex physically and chemically; reactions arise in chronological and positional order within the ion plume. In an IMS equipped with a pulsed CD source, ions once formed at the corona tip may undergo chemical changes in time during swarm drift toward the detector, and this would be reflected in mobility spectra. In one set of studies with a pulsed corona,[32,33] an ion shutter supplemented the pulsed CD to inject ions into the drift region since the corona pulse persisted for nearly 1 ms, too long for ion injection for an IMS drift tube of small dimensions. In the positive polarity, the reaction chemistry in air with ammonia at 2.39 mg/m^3 and water at 80 mg/m^3 was predictable and uncomplicated. Positive ions formed in the pulsed corona included [(H$_2$O)$_n$NH$_4$]$^+$ and [(H$_2$O)$_n$(NH$_3$)NH$_4$]$^+$, as anticipated with any direct current CD or radioactive ion source. Ions formed in negative polarity, however, suggested a distribution of species in a complex and time-dependent manner. Ions arising and decaying in intensity, depending on where the drifting ion swarm was sampled, as a function of shutter delay from the onset of a corona pulse, included anions for [(H$_2$O)$_n$O$_2$]$_2$, [(H$_2$O)$_n$CO$_3$]$_2$, [(H$_2$O)$_n$HCO$_3$]$_2$, [(H$_2$O)$_n$CO$_4$]$_2$, and [(H$_2$O)$_n$NO$_3$]$_2$. Interestingly, the anion for NO$_3$, a bane in the use of CDs for response in negative polarity, was relatively small or not detected, and its absence can be attributed to a counterflow of drift gas sweeping reactive gas neutrals from mixing with ions.

In another pulsed CD design, pulses were submicroseconds long and were short enough that ion formation and injection were achieved with the ion source alone,[34] without grid-based ion shutters as shown in Figure 5.8a. A point-to-plane discharge profile showed a 0.15-μs rise time, a 0.5-μs wide ion pulse, and intensities of 1.6 × 10^{10} ions/pulse. This large number of ions should lead to ion–ion repulsions and peak broadening, and the charge density was trimmed to 10^7 positive ions. This yielded a resolving power of 20 with a 65-mm drift region. While the design is simple and has demonstrated performance, secondary issues such as high-frequency electronic artifacts from the pulses and the engineering to mitigate these may be complications in this design. A particular challenge with this design was the audible snap of the discharge, creating microphonics in the detector–amplifier, where the aperture grid, if flexible, creates a microphone-like behavior. Some of this can be seen in the mobility spectrum (Figure 5.9b) as ripple in the baseline; still, this first demonstration of a pulsed corona source with shutter-free drift tube is promising enough to encourage further engineering optimization.

5.4.2.2 Laser Ionization

Solid samples may be heated and vaporized using laser energy with the additional benefit that compounds may be ionized with the same laser, all at ambient pressure, and characterized in an IMS drift tube. While lasers were used with mobility spectrometers initially for selective photoionization of vapors only,[35,36] ions were observed from laser contact with solids, including metals and salts. This was not simple vaporization by heating and instead was understood as ablation and ionization.[37] In laser desorption ionization (LDI), the laser pulse initiates a mobility spectrum with a

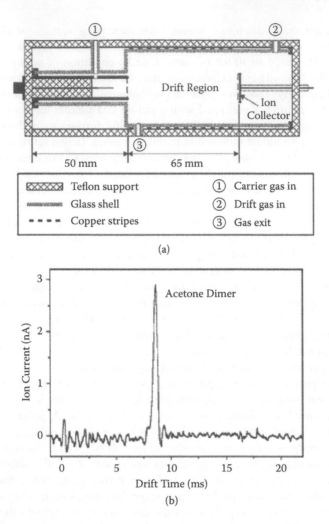

(a)

(b)

FIGURE 5.8 (a) Pulsed corona ion source with high-speed ion injection without an ion shutter and (b) mobility spectrum for acetone. (From An et al., Development of a short pulsed corona discharge ionization source for ion mobility spectrometry, *Rev. Sci. Instrum.* 2005. With permission.)

narrow pulse of photons causing ionization in the gas phase or from a solid substrate. The amount of charge produced is so intense that Coulomb repulsion occurs, and perhaps the kinetics of ion release from a surface is sufficiently slow that peak widths are broad, exceeding several milliseconds (Figure 5.9).[35–37]

Ablation under purified gas atmospheres provided rapid characterization of matter and had the promise to provide information on depths when materials could be eroded.[38] Mobility spectra for a range of metals and nonmetals were distinctive for the materials; however, ablation under conditions of ordinary air atmospheres was complicated by the production of excessive levels of hydrated protons. These were formed through the release of a high-energy electron during the ablation and

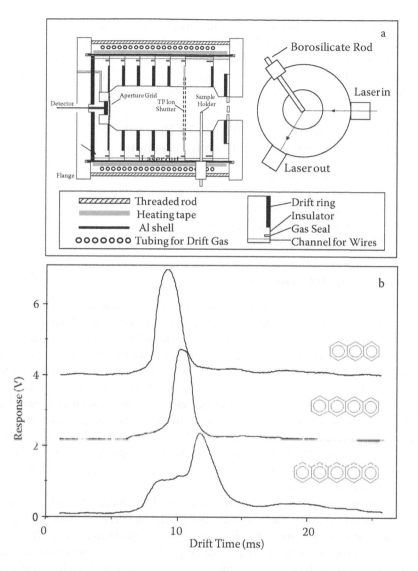

FIGURE 5.9 Mobility spectrometer with laser ablation and ionization of a solid sample (a). Laser ablation and ionization are possible with a solid sample in a purified gas atmosphere. Mobility spectra (b) were generated from thin films of individual polycyclic aromatic hydrocarbons on borosilicate glass. (From Young et al. *Anal. Chim. Acta* 2002, 453, 231–243.)

ionization steps, leading to the cascade of reactions and ending in hydrated protons that obscured, and possibly suppressed formation of, product ions from the solids.

When laser energies are low, measurements of absorbates or thin films of molecules on surfaces could be made without the complications described, and this was demonstrated for polycyclic aromatic hydrocarbons (PAHs).[39,40] An elegant design for laser desorption was used to measure mobility coefficients of 19 aromatic compounds, ranging from benzene to C_{60} fullerene, all without an ion shutter.[39]

The resolving power of this instrument in purified nitrogen was typically 20 to 50 and could reach 75 under optimal conditions. A related work emphasized analytical values showing that PAHs as adsorbates on borosilicate glass could be determined (Figure 5.9) at 40-pg levels or 5.5 to 7 pg/mm^2 using LDI-IMS in air at ambient pressure and 100°C for ion characterization in the drift tube.[40]

5.5 MOBILITY METHODS WITH CONTINUOUS FLOW OF IONS INTO THE DRIFT REGION

Several mobility analyzers commercially available today are designed with a continuous flow or stream of ions from a reaction region into the mobility measurement region, and although resolving power has been considered low, these mobility methods have emerged for laboratory and field measurements. These methods are FAIMS, DMS, DMA, and aIMS methods.

5.5.1 Aspirator IMS Designs

In the concept of aIMS, ions are passed in a flowing gas through a gap comprised of two plates, with an electric field drawing ions toward one of the plates. Ion motion through the gas flow is based on mobility, with ions of highest mobility reaching the plate nearest the point of ion introduction through a slit or a channel or throughout the entire cross section of the gap. Ions of low mobility are carried in the gas flow, reaching the plate further away from the inlet; regardless of displacement, ions must reach a Faraday detector on the plate to be detected. This is seen in Figure 5.10a; ions entering the analyzer region through the whole of the cross section of the gap gave a potentially complex response, particularly for detector elements nearest the front of the analyzer. The bandwidth of response and resolving power could be significantly improved if the aperture was narrowed, and two approaches have been taken with favorable results.

In one approach,[41,42] the ion "beam" is narrowed mechanically with an aperture admitting a thin ribbon of ions into the mobility region (Figure 5.10b), and the field in the gap is varied or scanned by changing voltage applied to the gap. Plots of ion current at a particular voltage $I(V)$ versus V produces an ion breakthrough curve whose first derivative forms a spectrum as commonly displayed. Noteworthy in this effort was the development of a two-flow scheme in which sample flow was entered in a rectangular aperture 100 μm × 5 mm. The flow rate of drift gas was volumetrically greater than the flow of sample gas, and most of the separation region was drift gas; a thin layer of sample flow existed at the wall, and the mean velocity of sample gas was twice that of the drift gas. Computational models were used to explore and optimize parameters; for example, the location of an insulating gap between the detector and the bottom electrode showed significant influence on resolving power, with a maximum value of about 3.8 at a distance of 1 mm from the sample inlet. A description of construction of a small aIMS[42] was accompanied by testing with the best resolving power described yet, 5.5, for a small aIMS operated in air at ambient pressure (Figure 5.10c).

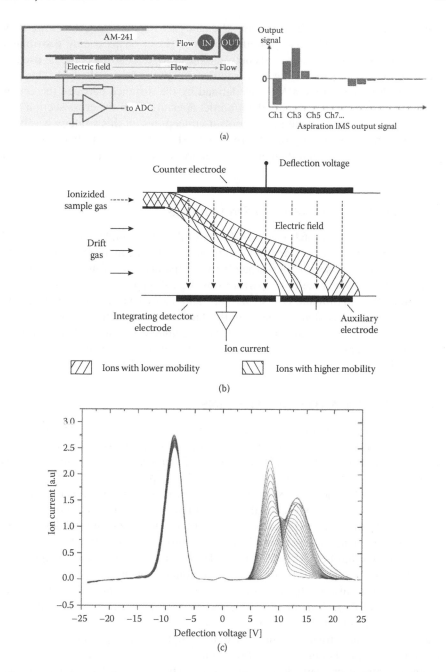

FIGURE 5.10 A design for aIMS (a) where ions enter through nearly the full cross section of the drift tube and a bar plot result from ion discharge on particular detectors. In an effort to improve resolving power, the inlet aperture was narrowed to a fraction of the drift tube flow cross section (b). This resulted in a mobility spectra generated from scanning the voltage (electric field) of the gap (c). (a. Courtesy of Osmo Anttalainen and b, c from Zimmerman et al., *Anal. Chem.* 2008, 80, 6671–6676. With permission.)

In a second approach,[43] ions entering the analyzer region filling the entire flow cross section of an aIMS also passed through wires or strips evenly distributed over the cross section. Electric fields between these wires or strips could be used to neutralize ions in all parts of the cross section except the center channel. Thus, an aperture width entering the aIMS was defined by the distance between the center pair of wires, effectively reducing the ion inlet aperture from the cross section to a narrow ribbon of ions. A large-scale proof-of-concept test of this strategy was built and demonstrated promising results. In a second work, the same team modeled flows in the instrument and demonstrated separation of product ions of dimethyl methyl phosphonate.[44]

5.5.2 DIFFERENTIAL ION MOBILITY SPECTROMETERS

The early designs of microfabricated DMS drift tubes was simple, with ions carried with gas flow from the reaction region or ion source into the analyzer region, filling the 0.5×4 mm cross section of the gap between the analyzer plates.[45] As with the simple aIMS, for which the inlet aperture was the full gap, the same was observed here with DMS, and limitations on peak shape and resolving power were predictably poor. The only significant modification to this has been the injection of ions from the side of the RF region to avoid fringing fields of the analyzer region.[46] This modification did not noticeably improve resolving power yet did apparently allow use of large values for RF, pushing ions to compensation voltages near 100 V.

5.6 SUMMARY AND CONCLUSIONS

The injection of an ion swarm into the drift region of an IMS analyzer constitutes, at the point and time of injection, the very best peak shape possible in the measurement, and after this, peak shape is degraded by diffusion and other effects. This is controlled by the quality of the ion shutter and the shape of the shutter pulse waveform. Consequently, the process of ion injection defines largely the resolving power of drift tubes, including conventional time-of-flight designs and continuous flowing designs. The traditional wire-grid-based ion shutters such as TP and BN designs can provide a convenient means of introducing ions into a drift tube to obtain a mobility spectrum and has been effective in both laboratory and portable IMS analyzers. These designs, even in their best embodiment, contain intrinsic limitations on duty cycle, minimum pulse width, and complexity or cost for assembly of a drift tube. Efforts to avoid these limitations have included multiplex methods applied to the grids, field switching in place of wire grids, and pulsed ion sources. These can bring advantages yet have trade-offs in electronic complexity, computational demands, and sometimes speed of measurement with current technology. None at this moment is commercialized or available in commercial instruments. While limitations and subtle details of performance are much better documented and described than when the second edition of this book was published, wire grids or mesh shutters remain a standard technology for IMS today. Several alternatives exist, and new embodiments of these or entirely new designs may emerge in the next decade, although there is no certainty of improvements.

Advances in continuous ion flow with aIMS methods have occurred as dimensions of the inlet aperture have become identified with limitations in resolving power of aIMS instruments. Two approaches, one pneumatic and one electrical, were developed to enter a thin stream of ions into the cross section of an aIMS. These methods have improved the performance of aIMS instruments and brought additional capabilities into the technical options for IMS.

REFERENCES

1. Tyndall, A.M., *The Mobility of Positive Ions in Gases*, Cambridge Physical Tracts, Editors Oliphant, M.L.E.; Ratcliffe, J.A., Cambridge University Press, Cambridge, UK, 1938.
2. van de Graaff, R.J., Mobility of ions in gases, *Nature* 1929, 124, 10–11.
3. Bradbury, N.E.; Nielsen, R.A., Absolute values of the electron mobility in hydrogen, *Physical Review* 1936, 49(5), 388–393.
4. Author's note: The ceramic two-part spring tensioned ion shutter is unique to PCP, Incorporated, drift tubes at least from the mid-1980s. No open reference to this design is known.
5. Rajapakse, R.M.M.Y.; Eiceman, G.A., Preparation of Bradbury Neilson shutter with temperature conditioning, 2012, in preparation.
6. Tyndall, A.M.; Powell, C.F., The mobility of positive ions in helium. Part I. helium ions, *Proc. R. Soc. Lond. A.* 1931, 134, 125–136.
7. Eiceman, G.A.; Nazarov, E.G.; Stone, J.A.; Rodriguez, J.E., Analysis of a drift tube at ambient pressure: models and precise measurements in ion mobility spectrometry, *Rev. Sci. Instrum.* 2001, 72, 3610–3621.
8. Salleras, M.; Kalmsa, A.; Krenkow, A.; Kessler, M.; Goebel, J.; Meuller, G.; Marcoal, S., Electrostatic shutter design for a miniaturized ion mobility spectrometer, *Sens. Actuators B Chem.* 2006,118, 338–342.
9. Zuleta, I.A.; Barbula, G.K.; Robbins, M.D.; Yoon, O.K.; Zare, R.N., Micromachined Bradbury–Nielsen gates, *Anal. Chem.* 2007, 79(23), 9160–9165.
10. Eiceman, G.A.; Schmidt, H.; Rodriguez, J.E.; White, C.R.; Nazarov, E.G.; Krylov, E.V.; Miller, R.A.; Bowers, M.; Burchfield, D.; Niu, B.; Smith, E.; Leigh, N., Characterization of positive and negative ions simultaneously through measures of K and ΔK by tandem DMS-IMS, 14th International Symposium on Ion Mobility Spectrometry, Château de Maffliers, France, July 16, 2005.
11. Eiceman, G.A.; Vandiver, V.J.; Chen, T.; Rico-Martinez, G., Electrical parameters in drift tubes for ion mobility spectrometry, *Anal. Instrum.* 1989, 18(3–4), 227–242.
12. Tadjimukhamedov, F.K.; Puton, J.; Stone, J.A.; Eiceman, G.A., A study of the performance of an ion shutter for drift tubes in atmospheric pressure ion mobility spectrometry: computer models and experimental findings, *Rev. Sci. Instrum.* 2009, 10, 103103–103110.
13. Spangler, G.E.; Collins, C.I., Peak shape analysis and plate theory for plasma chromatography, *Anal. Chem.* 1975, 47(3), 403–407.
14. Puton, J., Static and dynamic properties of the shutter grid for the ion mobility spectrometer, *Sci. Instrum. (Nauch. Apparat.)* 1989, 4(1), 29–41.
15. Iibeigi, V.; Tabrizchi, M., Peak-peak repulsion in ion mobility spectrometry, *Anal. Chem.* 2012, 84(8), 3669–3675.
16. Sysoev, A.; Adamov, A.; Viidanoja, J.; Ketola, R.A.; Kostiainen, R.; Kotiaho, T., Development of an ion mobility spectrometer for use in an atmospheric pressure ionization ion mobility spectrometer/mass spectrometer instrument for fast screening analysis, *Rapid Commun. Mass Spectrom.* 2004, 18, 3131–3139.

17. Ewing, R.G.; Eiceman, G.A.; Harden, C.S.; Stone, J.A., The kinetics of the decomposition of the proton bound dimers of 1,4-dimethylpyridine and dimethyl methylphosphonate from atmospheric pressure ion mobility spectra, *Int. J. Mass Spectrom.* 2006, 255–256, 76–85.

18. An, X.; Stone, J.A.; Eiceman, G.A., Gas phase fragmentation of protonated esters in air at ambient pressure through ion heating by electric field in differential mobility spectrometry and by thermal bath in ion mobility spectrometry, *Int. J. Mass Spectrom.* 2011, 303, 181–190.

19. Bader, S.; Urfer, W.; Baumbach, J.I., Preprocessing of ion mobility spectra by lognormal detailing and wavelet transform, *Int. J. Ion Mobil. Spectrom.* 2008, 11(1–4) 43–49.

20. Knorr, F.J; Eatherton, R.L.; Siems, W.F.; Hill, H.H., Jr., Fourier transform ion mobility spectrometry, *Anal. Chem.* 1985, 57, 402–406.

21. Tarver, E.E., External second shutter, Fourier transform ion mobility spectrometry: parametric optimization for detection of weapons of mass destruction, *Sensors* 2004, 4(1–3), 1–13.

22. Clowers, B.H.; Siems, W.F.; Hill, H.H.; Massick, S.M., Hadamard transform ion mobility spectrometry, *Anal. Chem.* 2006, 78, 44–51.

23. Szumlas, A.W.; Ray, S.J.; Hieftje, G.M., Hadamard transform ion mobility spectrometry, *Anal. Chem.* 2006, 78, 4474–4481.

24. Kwasnik, M., Caramore, J., Fernandez, F.M., Digitally-multiplexed nanoelectrospray ionization atmospheric pressure drift tube ion mobility spectrometry, *Anal. Chem.* 2009, 81, 1587–1594.

25. Tabrizchi, M.; Jazan, E., Inverse ion mobility spectrometry, *Anal. Chem.* 2010, 82(2), 746–750.

26. Spangler, G.E., Theory for inverse pulsing of the shutter grid in ion mobility spectrometry, *Anal. Chem.* 2010, 82, 8052–8059.

27. Zhou, L.; Collins, D.C.; Lee, E.D.; Lee, M.L., Mechanical ion gate for electrosprayionization ion-mobility spectrometry, *Anal. Bioanal. Chem.* 2007, 388(1), 189–194.

28. Jenkins, A., Ion mobility spectrometers, U.S. Patent 5,200,614; filing date January 16, 1992; issue date April 6, 1993; application number 821,681.

29. Blanchard, W.C.; Nazarov, E.G.; Carr, J.; Eiceman, G.A., Ion injection in a mobility spectrometer using field gradient barriers, i.e., ion wells, *Int. J. Ion Mobility Spectrom.* 2002, 5(3), 15–18.

30. William, C., Blanchard ion detecting apparatus and methods, patent number 6924479; filing date January 24, 2003; issue date August 2, 2005; application number 10/351,107

31. Hoaglund, C.S.; Valentine, S.J.; Clemmer, D.E., An ion trap interface for esi-ion mobility experiments, *Anal. Chem.* 1997, 69, 4156–4161.

32. Hill, C.A.; Thomas, C.L.P., A pulsed corona discharge switchable high resolution ion mobility spectrometer-mass spectrometer, *Analyst* 2003, 128, 55–60.

33. Hill, C.A.; Thomas, C.L.P., Programmable gate delayed ion mobility spectrometry-mass spectrometry: a study with low concentrations of dipropylene-glycol-monomethyl-ether in air, *Analyst* 2005, 130, 1155–1161.

34. An, Y.; Aliaga-Rossel, R.; Choi, P.; Gilles, J.P., Development of a short pulsed corona discharge ionization source for ion mobility spectrometry, *Rev. Sci. Instrum.* 2005, 76, 085105.

35. Lubman, D.M.; Kronlck, M.N., Plasma chromatography with laser-produced ions, *Anal. Chem.* 1082, 54, 1546–1551.

36. Eiceman, G.A.; Vandiver, V.J.; Leasure, C.S.; Anderson, G.K.; Tiee, J.J.; Danen, W.C., Effects of laser beam parameters in laser ion mobility spectrometry, *Anal. Chem.* 1986, 58, 1690–1695.

37. Eiceman, G.A.; Anderson, G.K.; Danen, W.C.; Ferris, M.J.; Tiee, J.J., Laser desorption and ionization of solid polycyclic aromatic hydrocarbons in air with analysis by ion mobility spectrometry, *Anal. Lett.* 1988, 21, 539–552.

38. Eiceman, G.A., Young, D.; Schmidt, H.; Rodriguez, J.E.; Baumbach, J.I.; Vautz, W.; Lake, D.A.; Johnston, M.V., Ion mobility spectrometry of gas-phase ions from laser ablation of solids in air at ambient pressure, *Appl. Spectrosc.* 2007, 61, 1076–1083.
39. Illenseer, C.; Loehmannsroeben, H.G., Investigation of ion molecule collisions with laser-based ion mobility spectrometry, *Phys. Chem. Chem. Phys.* 2001, 3, 2388–2393.
40. Young, D.; Douglas, K.M.; Eiceman, G.A.; Lake, D.A.; Johnston, M.V., Laser desorption-ionization of polycyclic aromatic hydrocarbons from glass surfaces with ion mobility spectrometry analysis, *Anal. Chim. Acta* 2002, 453, 231–243.
41. Zimmermann, S.; Abel, N.; Baether, W.; Barth, S., An ion-focusing aspiration condenser as an ion mobility spectrometer, *Sens. Actuators B Chem.* 2007, 125(2), 428–434.
42. Zimmerman, S.; Barth, S.; Baether, W.; and Ringer, J. Miniaturized low cost ion mobility spectrometer for fast detection of chemical warfare agents, *Anal. Chem* 2008, 80, 6671–6676.
43. Anttalainen, O.; Pitkanen, J., New aspiration IMS cell structure, 17th annual meeting, International Society for Ion Mobility Spectrometry, Ottawa, Ontario, Canada, July 20–25, 2008.
44. Anttalainen, O., New practical structure for second order aspiration IMS, 20th annual conference, International Society for Ion Mobility Spectrometry, Edinburgh, Scotland, 23–28 July 23–28, 2011.
45. Miller, R.A.; Eiceman, G.A.; Nazarov, E.G.; King, A.T., A micro-machined high-field asymmetric waveform-ion mobility spectrometer (FA-IMS), *Sensor Actuators B Chem.* 2000, 67, 300–306.
46. Rorrer III, L.C.; Yost, R.A., Solvent vapor effects on planar high-field asymmetric waveform ion mobility spectrometry, *Int. J. Mass Spectrosc.* 2011, 300, 173–181.

6 Drift Tubes in Ion Mobility Spectrometry

6.1 INTRODUCTION

A drift tube is the central component in making measurements by ion mobility methods and is supported by other components, including power supplies, heaters, ion shutter controllers, gas flows, and detector electronics. Every mobility analyzer of every mobility method will share some or all of the generic designs shown in Figure 6.1, in which the drift tube is defined further as an ion source and reaction region (Chapter 4), a drift (or mobility) region, and a detector (Chapter 7). A rich variety of designs, materials, ion sources, and drift tubes exists today and is supplemented though new methods of mobility measurements. The purpose of this chapter is to present the principles and technology used for obtaining ion mobility spectra in both traditional and emerging mobility methods.

The chapter begins with the drift tube designs that are a legacy of the first commercial ion mobility spectrometer and still found in military and security applications. Analyzers based on these drift tubes are in a reasonably advanced stage of development today and, though limited in resolving power and dynamic range, are nonetheless stable and provide valuable service. In addition, there is an impressive level of innovation on other designs or configurations to measure ion mobilities; often, these are motivated by interests to miniaturize or simplify technology, to improve resolving power, or to combine ion mobility with mass spectrometry (MS).

6.2 TRADITIONAL DRIFT TUBES WITH STACKED-RING DESIGNS

6.2.1 RESEARCH-GRADE AND LARGE-SCALE DRIFT TUBES AT AMBIENT PRESSURE

Drift tubes with drift rings stacked or arranged alternately with insulating rings in a tubular geometry and operated with linear voltage gradients are the best understood of mobility analyzers. While certain features are common to all such drift tubes, there is no conformity in design of rings, control of the electric fields, lengths of reaction and drift regions, choice of ion shutter, method to insulate rings, patterns of internal gas flows, and housing or mechanical support for the drift tube. One basis for classification may be supporting gas atmosphere pressure (ambient or near 1 torr), and a second classification could be the design or dimensions of the drift tube (size, number, material). The following discussion is intended to be instructive but not comprehensive, and only certain drift tubes are included to highlight particular

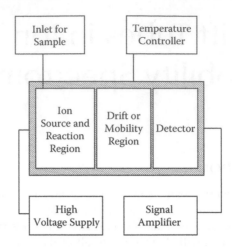

FIGURE 6.1 A block diagram of the main components of a generic ion mobility spectrometer regardless of mobility method. Any ion mobility instrument will be comprised of a drift tube, which includes an ion source, region for mobility characterization, detector and the utilities needed to support the measurement. This usually involves flows of purified gases, power supplies, and temperature control. Options for ion sources were described in Chapter 4, and methods for ion injection were presented in Chapter 5.

features. Discussion includes both commercial- and research-grade designs, with limited treatment, if any, on designs described significantly elsewhere in the book.

Beta VI and succeeding designs. The earliest commercial instrument was the Beta VI (see Chapter 2 and previous editions of this book), which in 1970 shared a common architecture with McDaniel's research-grade drift tube at Georgia Institute of Technology.[1] The drift region was nearly identical to the McDaniel's design and was based on metal rings, each insulated and separated from a neighboring ring by three sapphire balls, stacked and compressed using springs (Figure 6.2a). Unlike the Georgia Tech design, the Beta VI reaction region was configured to accept vapor samples and was notable for the length of the reaction region, which was nearly equal or longer than the drift region. In retrospect, this extended-length reaction region was the best design for ultratrace detection, an early attraction of IMS as a method. This drift tube was placed in a vacuum chamber and was prone, through poor user control of flows, to diffusion of neutrals from the reaction region into the drift region, complicating analytical response. This was ameliorated in later models when the sapphire balls were replaced with Teflon insulating rings.

Baim and Hill. The first drift tube suitable for complete high-speed response in which sample would enter and vent the drift tube rapidly was described in 1982 and intended to blend capillary gas chromatography with an IMS instrument as the detector. This was achieved with a pneumatically "tight" design and unidirectional flow of drift gas.[2] Alternating rings of stainless steel and Macor were fitted together, self-aligning by design as shown in Figure 6.2b. In unidirectional flow, sample neutrals were introduced into the drift gas upflow from the ion source, pushed with drift gas flow into the ion source, and then immediately swept from the drift tube. Neutral vapors from sample were no longer able to diffuse into the drift region or linger

Ion   Drift   Ion
Shutter 2 Rings Shutter 1   Ion Source

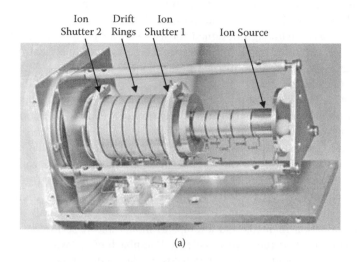

(a)

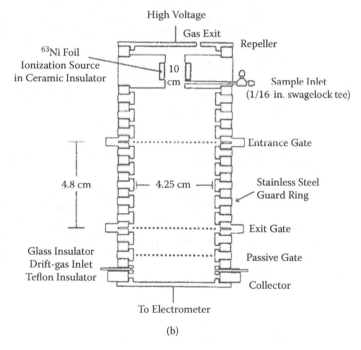

(b)

FIGURE 6.2 The Beta VI instrument (a) originally from Franklin GNO and later PCP, Incorporated, established a general design that has continued to present; it has a beta emitter, reaction region, and drift region with voltage divider. Some pneumatic control was introduced with a unidirectional flow drift tube (b); flow carried sample into and through the ion source region and vented. (Illustration from Baim and Hill, Tunable selective detection for capillary gas chromatography by ion mobility spectrometry, *Anal. Chem.* 1982, 54(1), 38–43. With permission.)

in the reaction region, and this design demonstrated that IMS drift tubes could be fast responding, free of spectral artifacts from uncontrolled ion-neutral associations, and designed for relatively low size and low cost. The drift tube was first controlled by a microprocessor.[3] Although designed for use with a ^{63}Ni foil source, the drift tube could be reconfigured easily for photoionization sources and electrospray ion sources. Rings were closely fitted inside a metal shell, which was used to provide temperature control.

Eiceman et al. The design philosophy of Baim and Hill[2] was retained and simplified further with drift and insulator rings of identical dimensions, and compression/alignment rods were located internal to the stack of rings.[4] This allowed the drift tube to be held together outside a shell and was later modified to include flow barriers and a snap-and-fit design for the rings.[5] The snap-together design with external compression allowed convenient changes of lengths of reaction and drift regions and exchanges of ion sources. This was based on stainless steel drift rings commercially available as washers and insulators of virgin Teflon. Noteworthy on this design was the level of evaluation of fields and attempt to model and describe performance. Similar modeling of drift tubes was made by Baumbach et al., who first proposed that a shell at ground could affect electric fields inside the drift tube.[6]

High resolving power. Impressive drift tube technology with high resolving power was demonstrated by Graseby Dynamics in the late 1980s for modeling and parametric studies of IMS.[7] This drift tube, shown in Figure 6.3a, consisted of a long glass tube with thin copper bands as drift rings attached on the external surface of the tube. The length of the tube was about 50 cm, and internal diameter was about 12 cm. The drift tube was fitted with components directly from the military-grade chemical agent monitor (CAM), including ion source, ion shutter, and aperture/detector assembly. At about 1 cm in diameter, the ion swarm injected into the drift region was only a fraction of the inner diameter of the drift tube, and resolving power of this long drift tube was 150+ in air at ambient pressure. A spectrum is shown in Figure 6.3b.

IONSCAN. A drift tube for bench-scale analyzers appeared in the early 1990s in the IONSCAN family of explosive/narcotic analyzers. This drift tube is shown in Figure 6.4 and embodies a high level of engineering and a thoughtful interface for intended samples. This instrument was a purpose-designed analyzer to detect particulates of high explosives (Section 2.4, Table 2.3); thus, samples were thermally desorbed when pressed with a heated anvil, and vapors were swept into the drift tube. This instrument is impressive as a high-temperature configuration, and the number produced, through several models, was in the thousands. Resolving power is roughly 20 for a drift region of about 10 cm. This instrument was ruggedly designed and built, enough for continuous operation by personnel after minimal training.

Electrospray-based IMS drift tubes. In another pioneering work from the team at Washington State University, difficulties of combining an electrospray ion source with an IMS (i.e., poor resolving power) was attributed correctly to incomplete desolvation of ions. This was solved when the region of the drift tube, normally associated with a reaction region, was branded and extended in length as the desolvation region. A cooled electrospray ionization (ESI) source permitted stable spray, and a counterflow of heated gas dramatically improved resolving power (see Figure 2.4b)

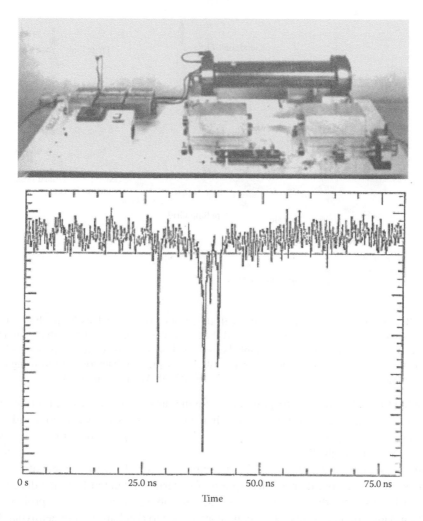

FIGURE 6.3 (a) Photograph of drift tube at ambient pressure with resolving power in excess of 150 built at Graseby Dynamics in the 1980s. (b) A mobility spectrum in negative polarity of a chlorinated phenol obtained with this instrument. (From J. Brokenshire, Courtesy of Graseby Dynamics, Ltd.)

by desolvation of ions and removal of solvent neutral before ions enter a purified gas atmosphere of the drift region.[8-10]

Efforts to join liquid chromatographs (LCs) and IMSs at ambient pressure and with full flow of conventional columns has not been developed since stable response and high-quality spectra can be obtained at flow rates up to about 2 µL/min. Mobility spectrometers with direct flux of ESI spray into the drift tube become inoperable with liquid flows larger than 10 µL/min or liquid compositions with 50% or more water, even with a desolvation region. An interface was described to accept full liquid flow from an LC directly into the drift tube and consisted of three stainless steel rings in a Teflon housing with a stainless steel mesh welded on the third ring,

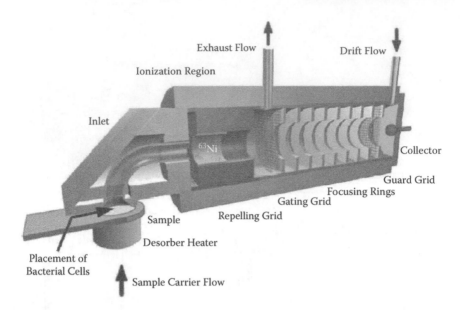

Exhaust Flow

Drift Flow

Ionization Region

Inlet

^{63}Ni

Collector

Guard Grid

Focusing Rings

Gating Grid

Sample   Repelling Grid

Desorber Heater

Placement of
Bacterial Cells

Sample Carrier Flow

FIGURE 6.4 Cross section of inlet and drift tube from a Barringer IONSCAN 400B. The heated anvil and inlet were designed for solids held on a fiber filter. The filter with the sample is placed into and pressed by the movable heated anvil, and desorbed vapors are carried in gas flow into the drift tube. (From Vinopal et al., Finger printing bacterial strains using ion mobility spectrometry, *Anal. Chim. Acta* 2002, 457, 83–95. With permission.)

intended to disrupt the electrospray jet, removing larger aerosols. Liquid formed in the interface was drained through three channels; a small pump was used to aid the flow of trapped liquid from the interface without affecting the pressure or performance in the drift tube.[11]

Side-flow sampling drift tubes. Vapor samples of high chemical complexity, such as those obtained from thermal desorption of materials collected from handbags or human bodies (including synthetic and natural polymers) can produce prolonged contamination of a drift tube. This can be attributed to condensation of semivolatile substances on surfaces of the source and reaction region, which then resemble an exponential dilution chamber with a special circumstance of surface contamination. In an impressive design, vapors desorbed from such samples were entered into and drawn from a drift tube perpendicular to ion flow.[12] The ion source was purged continuously with purified gas and ions swept by flow and field in the direction of the ion shutter, as with other such drift tubes. Similarly, the drift region was provided a purified gas at the detector; flows from both the source and drift regions were removed slightly upfield from the ion shutters. At this point, sample flow was introduced into this same volume of the drift tube and vented with the other flows. Product ions were produced through ordinary reactions, and samples of strong and complex chemical composition could be introduced into this drift tube and clean response restored in seconds. No commercial instrument contains this design feature.

Reverse source flow design. Another innovation in drift tube design, associated also with isolating the reaction region from sample vapors, is a drift tube based on corona discharge (CD). Significantly, reactive gases formed in the corona discharge and were associated with the formation of nitrogen oxide anions. This can be averted if the reactive neutrals are swept from the ionization volume, preserving O_2^- as the main reactant ion.[13] Designs have been described in which ions from the source are extracted from the source region and mixed with sample vapors not permitted to mix with source gas volume.[14]

6.2.2 SMALL AND HANDHELD DESIGNS

A significant decrease in dimensions of drift tubes produced commercially was made with the handheld CAM (Figure 2.3a), which was based on traditional designs with discrete drift rings separated by insulating ceramic standoffs on three internal rods also used to compress the stack of rings. The drift rings had dimensions of 12 mm internal diameter and 25 mm outside diameter; the entire drift tube, including source, was 85 mm long and was placed in a close-fitting protective shell.[15] Sample from a membrane inlet was swept into a 10-mCi ^{63}Ni cylinder and then toward a Tyndall–Powell ion shutter. This gas and drift gas entered at the detector and was removed near the shutter and recirculated through an onboard gas-purifying system. This instrument was applied to a range of consumer or nonmilitary applications with limited success owing to design features optimized for detecting chemical warfare agents.

Several research teams have sought to reduce the size of drift tubes further, and the most adventurous concept was an ion mobility sensor lacking the resolution of a conventional spectrometer.[16] These were to be intentionally small, inexpensive, and limited in resolving power. Another effort at Oak Ridge National Laboratory produced a drift tube that was formed from 25 drift rings with a drift tube length of 35 mm.[17] There was no ion shutter since the ion source was a pulsed ultraviolet laser with a pulse energy of below 0.1 mJ/pulse. This design was functional, yet bandwidths were broad and resolving power was low.

The most successful miniaturization and simplification of traditional drift tube components was the palm-size military instrument now known as the lightweight chemical detector (LCD).[18] In the LCD, the drift rings are attached to a circuit board, which also contains other parts of the electronics as found in consumer electronic products. The structure is similar to other traditional drift tubes, although the drift tube components are formed from small metal squares 15 mm on a side, and drift tube length is 50 mm. There are no insulators apart from air space between the rings. The LCD is operated with a pulse CD, and drift gas flow is a blend of "passive pumping" assisted by a small internal fan. Sample is introduced through a pinhole via an on-demand pulse, originally from a small audio speaker; a temporary stroke on the membrane of the speaker creates a pressure gradient, drawing sample vapors into the ion source volume.

6.2.3 DRIFT TUBES UNDER SUBAMBIENT PRESSURE

Mobility measurements and technology for ion characterization under vacuum or low pressure have become a feature of IMS, largely associated with ion mobility–mass spectrometry (IM–MS) applications for studies in biochemistry, biology, and material sciences. Significant differences in design and operation of drift tubes exist between ambient pressure instruments and those operated at pressures of 1 to 10 torr with helium, for which the dimensions, electric field strengths, and packaging must be adjusted for low pressures. Nonetheless, interesting and elegant drift tube designs can be found in these instruments and strongly merit discussion here. Additional details for these designs may be found in Chapter 9.

ESI ion trap-ion mobility-time-of-flight mass spectrometry (TOF-MS). Ion mobility drift tubes from Clemmer's group at Indiana University have undergone progressive development from the use of an ion trap to gate ions into a mobility spectrometer (Chapter 5) to sophisticated ion trap/ion mobility/TOF-MS methods all combined with ESI sources.[19] The design philosophy was for high resolving power combined with MS to extract information about a sample in plots of ion intensity, drift time, and mass-to-charge ratio (m/z). Large and analytically powerful instruments have been used to identify peptides and to separate charge-state distributions in mixtures such as ubiquitin and myoglobin. In a next increment of design complexity,[20] a quadrupole mass analyzer was placed between the drift tube and TOF-MS. In this ion trap/ion mobility/quadrupole/TOF-MS, the quadrupole mass filter was used to isolate a specific m/z value for collisional activated dissociation and subsequent mass analysis of fragment ions.

Matrix-assisted laser desorption ionization (MALDI) IMS-TOF-MS. Another low-pressure drift tube design was introduced by Russell's group at Texas A&M University and is based on a MALDI source with a drift tube attached to a mass spectrometer for orthogonal ion extraction design. An interface between the drift tube and the MS allows mobility selection and then collision-induced dissociation (CID) with good ion transmission to the MS analyzer.[21] This interface was a stacked-ring ion guide with flexibility to adjust field strength and pressure ratio for controlling ion temperature in CID. The IM-CID-MS method was applied to the fragmentation of peptide ions.

Soft-landing drift tube. A last drift tube to be mentioned in this category is that of Davila and Verbeck, who used a laser ablation source to form ions and a drift tube to select ions for deposition onto substrates for preparative and developmental research of new materials.[22] An inert gas atmosphere (e.g., 8 torr of He) was used to thermalize ions after ablation, and a drift tube was used to isolate selected ions. A unique split-ring design of ion optics at the end of the drift region was controlled to direct ions to a detector or to a substrate for soft landing. Ions at energies below 1 eV were landed onto substrates to explore chemistry of material sciences.

6.2.4 Planar Configurations of Conventional Designs

Cylindrical drift tubes as described commonly require a machine shop to prepare components and technical experience to assemble the many parts into a drift tube. This in addition to the material and production costs can be seen as restrictions in innovations of drift tube designs and development of IMS methods. To speed drift tube fabrication and simultaneously reduce costs of instrument building, a planar design for drift and insulating elements was demonstrated using photoetched circuit boards.[23] Costs to fabricate a drift tube can be estimated as 10% or lower compared with conventional designs; speed, from design through fabrication and to completed instrument, can be days rather than weeks or longer.

In this design, two circuit boards were separated by Teflon gaskets, which defined the drift region as 13 × 25 × 74 mm. The plates contained nine 5 mm wide, evenly spaced drift plates separated by 2.5 mm; these, with a voltage divider, were used to define an electric field in the drift region. The ion shutter had a rectangular shape (10 × 30 mm) to fit inside the cross section of the drift tube, and a stainless steel plate (20 × 10 mm) was used as a detector. The resolving power was 13 and was limited by shutter quality. Previously, Blanchard had used a planar drift tube to move ions in air at ambient pressure with a specialized ion-filtering or mobility-based isolation concept.[24] A planar design is also part of a tandem mobility instrument in development at Excellims (Section 6.6.5).

6.2.5 Drone-Based Drift Tubes

A limitation in all IMS analyzers is the intrinsic properties of any point sensor, namely, the sample must be introduced to the analyzer, which is located at some point that is sometimes at a distance from the sample. In contrast to standoff sensors, which can survey a broad volume of space without moving, a point sensor is defined by a geographically small place. These understandings can be changed if the drift tube can be moved rapidly through the region or volume of interest. This has been achieved with a telemetry-based, drone-portable instrument and employed in monitoring airborne concentrations of vapors of interest. These samples are otherwise inaccessible to a point analyzer. Data handling can be remote, and this was demonstrated using a CAM, a Compaq 386 motherboard with 1-Mb RAM, two H-Cubed Corporation and Tekk Corporation digital radio transmitter and receiver sets, CoSession (Triton Technology) communications software, and a remote, 386-level computer workstation. Altogether, the onboard system weighed less than 7 kg and used less than 30 W of battery power. Approximately 20 spectra per minute were recorded within a typical range of 1 to 2 miles. This was later demonstrated in field trials with analyzers based on the LCD.[25,26]

6.3 HIGH FIELD ASYMMETRIC DRIFT TUBES

6.3.1 General Comments

Field-dependent mobility methods with fields of up to 20,000 V/cm at ambient pressure are impractical with traditional drift tube designs since the required voltage for a 10 cm long drift region would be 200 kV. This practical dilemma was solved in the former Soviet Union in the late 1980s using asymmetric waveforms and parallel plates or cylinders and was reported in early 1993 in a Western journal.[27] Embodiments of these methods have been developed today largely to filter ions before entering a mass spectrometer to decrease chemical noise and select ions of interest for quantitative determinations of specific substances. Attractions of such methods reside in shutter-free designs with continuous flow of ions from the source to the mobility region. The duty cycle for sampling ions from the source is 100%. Signal-to-noise (S/N) ratios can be improved as much as 50-fold when a field asymmetric IMS (FAIMS) is placed between an ESI source and the MS, making these drift tubes powerful complements to MSs for applied pharmaceutical and biomedical measurements. Versions of mobility-dependent drift tubes also exist for stand-alone military applications and in combinations with gas chromatographs.

6.3.2 Cylindrical Field Asymmetric Ion Mobility Spectrometry

The first configuration of field-dependent mobility released from the former Soviet Union and introduced initially into North America was a cylindrical design as shown in Figure 6.5a; in this early configuration, sample molecules were ionized and passed with the gas flow between two concentric tubes (the inner and outer electrodes).[28] The asymmetric electric field, or separation field, was applied between these inner and outer electrodes. In later designs, the ions were separated from sample flow into a flow of purified air.

An early discovery was that the cylindrical design had a focusing effect for ions flowing through the mobility volume,[29] and this arose due to inhomogeneous electric fields in the cross section arising from differences in curvature of the inner and outer tubes. This resulted in ion focusing in the space between the surfaces and enhanced ion peak heights or ion transmission efficiency with increases in separation voltage. Another improvement made with this design by a Canadian team was formation of ions in an external ion source with a counterflow of gas to pass ions into the analyzer and remove solvent or matrix neutrals. Thus, ions from an electrospray ion source could be introduced into the analyzer practically without concern over the introduction of aerosols into the analyzer.[30] The term FAIMS was attached to the cylindrical design, and significant commercial opportunities arose when an ESI-FAIMS configuration was attached to a MS.[31] This initiated a new concept in ion filtering between an ESI source and a MS, and today is commercially available from ThermoFisher Scientific with a hemispherical design.

6.3.3 Microfabricated Differential Mobility Spectrometer

An approach to building small FAIMSs toward low-cost, mass-produced analyzers was developed by Miller et al.;[32] metal electrodes were bonded onto glass or ceramic plates using microfabrication technology methods (Figure 6.5b). Two plates of mirror image were separated by a Teflon gasket and compressed together in a frame to form a drift tube. The size of the gap between the plates was 0.5 mm, and a gas-tight seal or gasket defined the flow channel; dimensions were 5 × 13 mm for the separating plates and 5 × 5 mm for the Faraday detector. Detectors were floated to either +5 V or –5 V so negative and positive ions could be simultaneously detected in one analyzer. Several ion sources have been combined with this drift tube, including a photodischarge lamp,[33] a radioactive ^{63}Ni source,[34] and significantly, an ESI ion source.[35] Notably, AB Sciex has produced a version of these drift tubes for use in ESI tandem MS (Figure 6.6).[36] A Chinese team engineered an improved micro-fabricated Faraday detector based on the original concept introduced by Miller et al.[37]

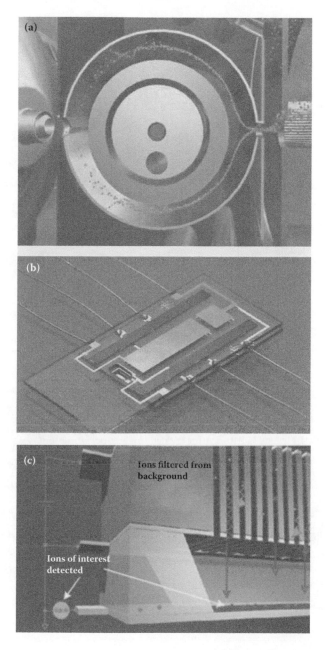

FIGURE 6.5 Drift tubes for FAIMS or DMS including the configuration commercialized by Thermo Fisher Scientific (a) with a cylindrical shape (with permission from Thermo Fisher Scientific); the first small planar design commercialized by Sionex, Incorporated (b) (from Miller et al., A novel micro-machined high field asymmetric waveform ion mobility spectrometer, *Sens. Actuators B* 2000; with permission); and the microfabricated, very small structures of the ultraFAIMS (c) manufactured by Owlstone Nanotechnology (from Owlstone White Paper, 2006).

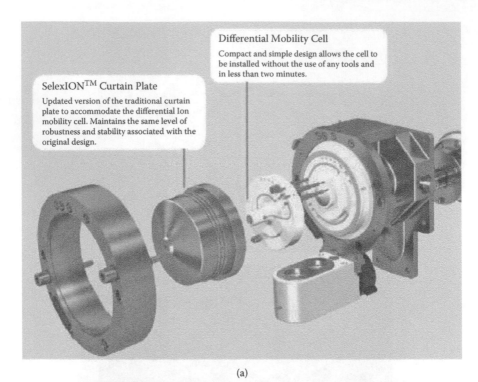

Differential Mobility Cell

Compact and simple design allows the cell to be installed without the use of any tools and in less than two minutes.

SelexION™ Curtain Plate

Updated version of the traditional curtain plate to accommodate the differential Ion mobility cell. Maintains the same level of robustness and stability associated with the original design.

(a)

(b)

FIGURE 6.6 (a) The combination of a DMS with MS commercialized by AB SCIEX with the SelexION technology, which is available as a unit integrated into their lines of mass spectrometers. (Photo courtesy of AB SCIEX.) (b) A photograph of the ultraFAIMS attached to the inlet of a mass spectrometer in recently released technology from Owlstone Nanotech. Both of these designs are intended to separate or filter ions between an electrospray ion source, which often produces a complex mixture of ions from the spray of a liquid sample, and a mass spectrometer.

6.3.4 UltraFAIMS

Another planar differential mobility spectrometric (DMS) instrument, with further miniaturization of the field-dependent mobility concept, was introduced by Owlstone Technology with a "chip" (Figure 6.5c), yielding benefits and some challenges.[38,39] The demand for high voltages in the separating waveform was relaxed with reductions in the dimensions between plates. For example, a separation of 35 μm between plates requires only 270 V to produce 80 kV/cm, and an equivalent field necessitates 4 kV for 500 μm distances as with the small planar DMS noted. At present, the top strength for the electric field is about 75 kV/cm with a frequency of 27 MHz with the ultraFAIMS design. The drift tube is comprised of 50 plates with lengths of 300 μm, yielding about 20 μs ion residence times with flow velocities of 10 m/s. The time at low voltage for an ion in ultraFAIMS is practically insufficient for ion cooling and clustering with polar neutrals as with the 1 MHz DMS or 300 kHz FAIMS instruments described previously. Consequently, positive alpha dependencies, which are based on ion solvation–desolvation processes,[40,41] occur weakly or not at all with the small microFAIMS drift tube. Dispersion plots in early configurations of this technology were largely confined to patterns observed, with negative alpha dependencies affecting analytical resolving power. A strong advantage, however, of the small size and very high electric fields is comparatively facile dissociation or fragmentation of ions. In the newest development to increase analytical information from this tiny analyzer, patterns of ion dissociation and fragmentation as seen in dispersion plots at high fields[39] have been included in identifying chemicals.

6.3.5 Waveforms

In most FAIMS experiments, the waveforms range from 0.2 to 1 MHz and are made with sums of sinusoid waveforms as a matter of convenience and cost. Theoretically, a rectangular waveform would provide improved performance as ions are placed directly and wholly under high or low fields. In nonrectangular shaped waveforms, "ion heating" occurs only above a threshold, and a small portion, of the whole waveform; this means that the desired ion behaviors at field extremes occur inefficiently, with losses in resolving power at a given separating voltage. Improved speeds and lengths of ion heating should lead to improved ion separations, resolving power, and potentially lower demands on field strength. Electronics to produce such rectangular waveforms with 1 MHz and swings of 3 kV are nontrivial and limited by excessive power load, for example, with switching transistors.

At least four teams have sought[42–45] to refine waveforms applied to DMS: Two efforts were made to produce rectangular waveforms and determine improvements over those currently used, and two teams sought to explore theoretically optimum waveforms. The conclusions were as follows:

a. Rectangular waveforms do improve peak separations and allow voltages lower than with the sinusoid added waveforms and thus are an improvement.
b. Seeking an optimum waveform may be ion specific. This would greatly complicate any effort to run optimized separations in a global manner.

Practically, digital waveforms with rectangular profiles are more complex in electrical design and demanding in power tolerance than the design for the sum of sinusoidal waves. This last technology will likely be employed until high-voltage switch technology provides a reliable, economic, and convenient alternative.

6.4 ASPIRATOR DRIFT TUBES

6.4.1 GENERAL COMMENTS

Among the simplest and earliest concepts in mobility analysis is the aspirator IMS (aIMS) method, in which the drift tube functions without an ion shutter, the duty cycle for ion measurement can be true 100%, analyzers have planar geometries, and electronics are comparatively simple.[46,47] While limited in resolving power, handheld aIMS instruments are available commercially from Environics Oy (Figure 6.7) and have been accepted for in-field applications. Causes for low resolving power can be associated with certain design choices, such as open-loop designs in which ambient air is drawn directly into the analyzer to fundamental details such as ion inlet aperture dimensions. Impressive modeling and rebuilding of aIMS drift tubes have occurred in the past decade, showing that improved resolving power is possible while maintaining the attractions of simplicity.

6.4.2 SMALL-SIZE AND LOW-GAS-FLOW DESIGNS

Three drift tube designs had been identified[48,49] for small aIMS instruments; these are termed as follows:

a. First-order integrating design. This has limited resolution and is a fast indicator of the presence of ions with mobilities in certain broad ranges.
b. Multichannel aIMS; several detector channels exist on the ion detection plate. This is used for chemical detection purposes governed in part by the number of channels.
c. Sweep-type aIMS; voltage is changed to control ion arrival at a detector, and one field or more fields could be swept.

A fourth design, a so-called second-order design, was introduced for which ion flow into the analyzer gap is narrow compared to the distance between plates of the drift tube. Two approaches to controlling effective widths of the inlet apertures have been (a) physical structures in the drift tube to introduce a thin band of ions into the main channel of mobility measurement by Zimmermann et al.[50,51] and (b) electric fields to remove ions from all parts but a selected width from the cross section by Anttalainen.[48,49]

Zimmerman modeled and demonstrated an aIMS design with low-cost fabrication by stacking together four inexpensive components. The ion swarm was narrow in width by fluid dynamics and by geometric constraints (i.e., a thin inlet aperture). The aperture for sample flow was about 10% that of the main mobility channel, and modeling showed some affects of unevenness in streams with joining two flows. Still,

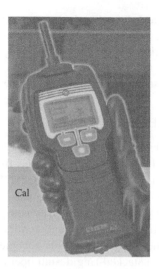

FIGURE 6.7 One aIMS drift tube is commercially packaged as the Chem Pro 100i. (Courtesy of Osmo Anttalainen, Environics Oy.)

the resolving power was 5.5 for the positive reactant ion peak, and separation was evident between peaks of reactant ions and the protonated monomers of dimethyl methylphosphonate (DMMP). Separation was also observed between the protonated monomer and proton-bound dimer of DMMP, although separation among several similar organic compounds was not possible without using pattern recognition methods. In this design, the supporting gas atmosphere was purified in a closed-loop scrubbing flow system.

Anttalainen designed a method to narrow the aperture of ions entering the mobility region of an aIMS drift tube using electric fields between plates located in the cross section of flow. Ions entering the drift region are drawn to and annihilated on plates used to define a thin area of the total cross section, effectively decreasing aperture width. While increases in resolving power were not provided for this new design against the ordinary multichannel aIMS, peak centers for hydrated proton and the product ion of di-isopropyl methylphosphonate (DIMP), likely the protonated monomer, were separated by roughly one peak width. This alone is a large improvement in these small aIMS drift tubes, and growth in small aIMS has become visible both in commercial presence and in appearance of journal articles.[52,53]

6.4.3 DIFFERENTIAL MOBILITY ANALYSIS

Differential mobility analyzers (DMAs) can be seen as a variant of aIMS for which gas flow velocity is sufficiently high to minimize diffusion broadening of ion swarms, and resolving power is high. While DMA methods and technology have been associated with mobility characterization of aerosols or particulates, laboratory designs combined with MSs have been developed for molecule studies[54] and applied in chemical measurements of large polymers[55] and biomolecules formed using electrospray ion sources.[56] A distinction between the classic DMA instrument for particulate characterization and this DMA is the planar design for surfaces[54] similar to other aIMS instruments; critically, efforts are made to maintain laminar flow during the time of ion injection and separation, and large amounts, up to 30 L/min, of purified gas are needed to maintain this condition. The results, however, are favorable for laboratory-based instruments with resolving powers greater than 50.

6.5   TRAVELING WAVE DRIFT TUBES

A new mobility method called traveling wave ion mobility spectrometry (TW-IMS) was released commercially by Waters Corporation as the Synapt™ high-definition mass spectrometry (HDMS) system, including three TW-enabled stacked-ring ion guides in combination with a quadrupole-orthogonal acceleration-TOF mass analyzer.[57] While the resolving power of the mobility component in the Synapt is not large, the TOF-MS is analytically powerful, and commercial availability has made this instrument and subsequent upgrades widely accepted within the biomolecular research community.[58,59] As a stacked-ring design operated in a nitrogen gas atmosphere at 5 torr, discussion of TW-IMS might have been included in Section 6.2.3 with subambient pressure methods; however, the unusual control of electric fields inside the instrument and gas atmosphere merit a separate discussion.

In the drift tube, neighboring rings are connected in opposite phase of a radiofrequency (RF) field that is used to produce a radially confined potential well.[60] A voltage pulse of several volts (e.g., 6 V) is superimposed on one ring (in the Synapt G1) or two rings (in the Synapt G2), causing ion motion. The pulse is moved to a neighboring ring(s) and relaxed, and this pattern is propagated at 300 to 1,300 m/s through the ring structure toward the end of the tube. The process is repetitive, with ions moved by mobility in the direction of the TW. Separation of ion swarms by mobility is achieved by controlling key parameters, including the velocity of the pulse (or wave), the height of the pulse, and gas pressure.

Ion motion in the TW-IMS can be understood through a ratio of ion drift velocity at the steepest wave slope to the wave speed. At given parameters, ions of lower mobility slip over the wave more than those with higher mobility, establishing a basis for mobility separation of ions. At low values of this ratio, ion transit velocity is proportional to the square of the mobility constant K and electric field intensity E. At highest ratio, the transit velocity asymptotically approaches the wave speed, and resolving power depends on mobility, scaling as $K^{1/2}$ in the low ratio and less at higher ratio. Ion injection into the mobility region is made using an ion trap, which can also serve to fragment ions, adding another dimension of selectivity or control in a measurement.

6.6   TANDEM MOBILITY SPECTROMETERS

Tandem methods for chemical measurements are now widely accepted and employed even when tandem is defined in the narrowest sense, such as two-dimensional gas chromatography or tandem mass spectrometry. Tandem or triple quadrupole MSs became commercially available in the early 1980s, and today these instruments are routine concepts as linear quadrupole analyzers or as ion traps. Tandem or sequential methods result in loss of signal current, yet noise is lost more so; thus, S/N is increased.

Since mobility spectrometers are comparatively inexpensive, simple, and with low resolving power compared to MSs, tandem mobility can be achieved for comparatively little additional complexity and cost, particularly in comparison to MSs.

The concept of tandem IMS and its first demonstration date to the mid-1980s under a U.S. Army contract to PCP, Incorporated, and a four-section drift tube of traditional TOF design with a linear or line-of-sight geometry and three regions of electric fields separated by three ion shutters and an ion source volume.[61,62] In this tandem drift tube design (Figure 6.8a), ions formed in the source region were injected using a first ion shutter into a drift region, where ions swarms were separated in drift time. At the end of this first drift region, a second ion shutter was located and synchronized to the first shutter; control of delay could allow ions from only a portion of drift time to be isolated and passed to a second drift region. In this next drift region, vapors could be added to promote selective reactions, or ions could be fragmented by laser irradiation. Ions formed in this second region were then characterized using a third drift region, also preceded with an ion shutter. Although this instrument was a demonstrator unit, the open drift ring structure, much as the original Beta VI design, permitted uncontrolled movement of sample neutrals between mobility regions, complicating the interpretation of results. By the mid-1990s, the need for a dual-shutter tandem instrument was recognized for studies of ion dissociation kinetics; today, this instrument is complimented with vacuum-based IMS–IMS–MS instruments and mobility devices combining different mobility technologies, such as DMS–IMS.

6.6.1 IMS–IMS

6.6.1.1 Kinetic IMS

The lifetimes of gas ions in air at ambient pressure were first measured directly in IMS with a dual-shutter IMS–MS.[63,64] A drift region was used to separate ion swarms and a second ion shutter at the end of the drift region was used to isolate an ion (e.g., a proton bound dimer), which then passed into another drift region before reaching the mass spectrometer. There was time enough to observe decomposition and obtain rate constants. Later, purpose-built IMS–IMS was described and included a Faraday plate detector.[65] A schematic of this drift tube is shown in Figure 6.8b; the instrument is now used in determining the lifetime of ions from explosives in air.[66]

6.6.1.2 Low-Pressure IMS–IMS

An IMS–IMS instrument for use at 3 torr in helium atmosphere has also been described in combination with a MS[67] to improve analytical performance in IM–MS methods. The separating efficiency was indeed improved by a factor of eight, with two-dimensional IMS for mixtures of tryptic peptides. The drift tubes were 100 cm long.

6.6.2 DMS–IMS

A tandem mobility analyzer based on DMS with IMS was designed, built, and operated using a microfabricated DMS analyzer and two miniaturized IMS detectors

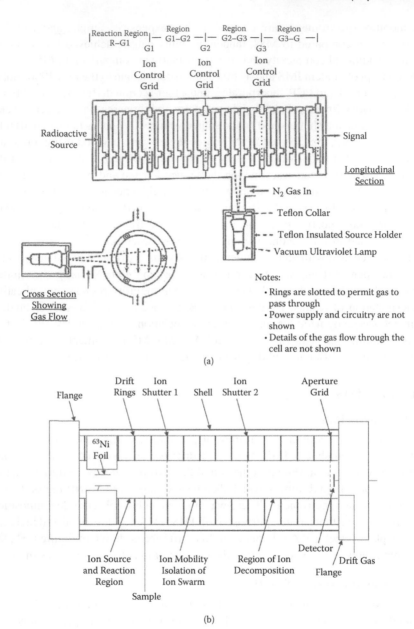

FIGURE 6.8 Schematic of the first tandem ion mobility spectrometer from PCP Incorporated (a) in a configuration with a photodischarge lamp. Three ion shutters were employed to isolate ions of certain mobility for subsequent chemical reactions and then a final section for ion mobility analysis of products. A modern version of this is shown in (b) and is used for the determination of decomposition or fragmentation kinetics of thermalized ions in air at ambient pressure. (From Stimac et al. Tandem Ion mobility spectrometer for chemical agent detection, monitoring and alarm, Contractor report on CRDEC Contract DAAK11 84 C 0017, PCP, West Palm Beach, FL, May 1985.

arranged in an orthogonal configuration (Figure 6.9a), enabling the detection of positive and negative ions simultaneously.[68,69] Ions were passed at a specific compensation voltage (CV) through a DMS drift tube and were drawn from the gas flow into IMS drift tubes of appropriate polarity placed in a twin drift tube geometry. The ion shutters initially were a micromachined design (Figure 5.4) with a 1 cm long conventional drift tube. An anticipated benefit in a DMS–IMS experiment is that ion characterization would occur, with principles of K versus ΔK adding some orthogonal character to the measurement. This was not seen, although resolution over a range of ion masses was level, and this was an improvement over DMS or IMS. Approximately 20 copies of a DMS–IMS instrument were produced commercially by Sionex, Incorporated, from 2005 to 2010.

6.6.3 IMS–DMS

The DMS–IMS design can be inverted, with the IMS drift tube placed before a DMS analyzer, although the scanning speed of a DMS (0.5 to 2 Hz) is a poor direct match with the repetition rate of an IMS drift tube (10 to 30 Hz). In such instances, a dual-shutter IMS design can be successful at isolating an ion swarm for subsequent DMS characterization. A successful combination of a drift tube with an FAIMS-ion trap MS with MS^n capability was recently described[70] and was operated with dual-shutter control in the IMS drift tube (Section 5.3.3). A significant result was the isolation in the drift tube of one peak for tyrosine-tryptophan-glycine, which when introduced into the FAIMS was resolved into two conformer peaks.

6.6.4 DMS–DMS

Although the DMS–IMS measurements described had little orthogonal character owing to similarities in K and ΔK, orthogonality in tandem IMS methods could be added through the introduction of ion–molecule reaction mobility drift tubes. A tandem mobility spectrometer with two sequential (DMS) drift tubes and detectors at ambient pressure was designed to accept a reagent gas in the gap between the two drift tubes.[71] Separate electronic control for each drift tube permitted several modes of operation, including all ion pass, CV scanning, and ion selection over a narrow CV range. Any of these modes could be applied to the drift tubes, allowing a range of analytical measurements analogous to tandem mass spectrometry, although in these studies, ions between mobility regions were clustered, stripped of charge chemically, or entered into a new supporting gas atmosphere rather than fragmented. In one example, the benefit of DMS–DMS was in the the resolution of proton-bound dimers of organophosphates. Ions for both chemicals appeared near 0 V in the CV scale of DMS1 and were baseline resolved in DMS2 after entering into an isopropanol-rich gas atmosphere. This technique is in early stages of development.[71]

6.6.5 Multidimensional IMS

A new concept for IMS, also in early stages of development and meriting discussion, is multidimensional IMS, which is disclosed principally in a patent.[72] In this concept,

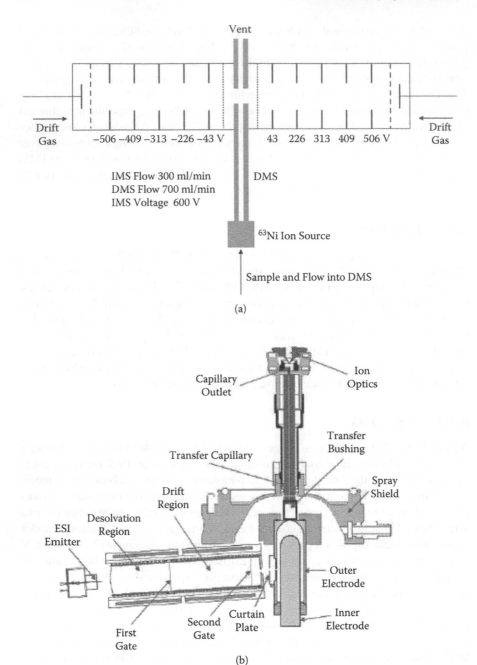

FIGURE 6.9 Two designs of tandem mobility instruments: (a) a DMS–IMS drift tube and (b) an IMS–DMS–MS instrument. (From Pollard et al., *Int. J. Ion Mobil. Spectrom.* 2011, 14(1), 15–22.) These concepts of tandem ion mobility are in research and development stages. (Courtesy G. Eiceman.)

a first drift tube is used for ion swarm separation by mobility, and ions at a particular drift time are orthogonally isolated and introduced into a second drift tube, perpendicular to the first drift tube. This is achieved by rapidly increasing voltages in the first drift tube, causing a field in the direction of the second drift tube. The swarm is then characterized in a second drift tube with temperature or gas compositions differing from the first drift tube. The advantage of this design is that, in principle, this would provide a complete comprehensive two-dimensional mobility measurement. This last concept has not been demonstrated experimentally.

6.7   OTHER DRIFT TUBE DESIGNS

In this section, discussion is given to designs of drift tubes based on unusual or unconventional technology, material sciences, or designs and not mentioned in previous sections. These innovative designs are associated with historic drift tube methods since these are the most extensively developed in the past 40 years.

6.7.1   VOLTAGE DIVIDERS AS FILMS INSIDE TUBES

6.7.1.1   Resistive Ink

A novel approach to drift tube fabrication was described in the early 1980s; these tubes were made of aluminous porcelain tube coated with thick-film resistor material commonly used in microcircuits.[73,74] Separate voltages were applied at the ends of the tube to establish a voltage gradient through the tube from the thin film of resistive ink. The original designers determined mobility coefficients for metal ions in oxygen and later for other chemical systems; performance was found to be the same as stacked-ring designs.[75] This concept was included in small mobility spectrometers (Figure 6.10a) produced commercially by Environmental Technology Group (ETG), Incorporated, in Baltimore, Maryland, during the mid- to late 1980s and was not developed by other teams. Advantages include simplicity, saving of labor and economy of fabrication (although the inks and solvents are expensive), and tolerance of elevated temperatures (the ink is cured near 900°C), although challenges in coating methodology and uniformity can be anticipated.

6.7.1.2   Resistive Glass

Another nonconventional technology with traditional methods of mobility is a monolithic resistive glass tube used as the drift tube (Figure 6.10b). The body of the drift tube was monolithic resistive glass tubes (Burle Electro-Optics, now Photonis, Incorporated) with a resistance of 0.45 GΩ/cm. The glass undergoes a chemical surface treatment, a reduction reaction in hydrogen atmosphere, rendering the surface a resistive layer. Voltages are applied at ends of the tube to establish an electric field throughout the desolvation and drift regions. The tubes were both 3 cm inside diameter and 4 cm outside diameter, with lengths of 12 and 26 cm for desolvation and drift regions, respectively. This technology was distinguished not only by simplicity of construction but also by radial homogeneity of electric fields.[76] While attractive

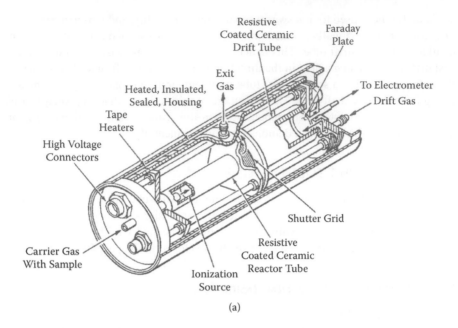

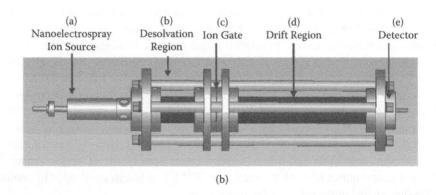

FIGURE 6.10 Drift tube designs with unconventional structures. One drift tube was made of ceramic tubes coated with a semiresistive ink (a) and commercialized by ETG, Incorporated, of Baltimore, Maryland, in the 1980s and early 1990s (From Carrico et al., Simple electrode design for ion mobility spectrometry, *J. Phys. E: Sci. Instrum* 1983. With permission.). Surface-treated glass from Photonics (now Photonis) was employed in a second and recent design (b) (From Kwasnik et al., Performance, resolving power and radial ion distributions of a prototype nanoelectrospray ionization resistive glass atmospheric pressure ion mobility spectrometer, *Anal. Chem.* 2007; with permission).

in simplicity, as was the resistive ink tube described previously, this tube should have improved precision and uniformity of electric fields compared to the resistive ink design. Practical concerns include operation of the tube in air at elevated temperatures, and no systematic description of the tolerance to temperature has been reported or other surface treatment methods described for use with IMS.

6.7.2 OTHER DESIGNS

6.7.2.1 Plastic Drift Tubes

Some designs of drift tubes have been made using epoxy-based printed circuit boards with drift rings[6,77] or in planar designs,[23] and these have been largely demonstrator instruments and lack capability to reach elevated temperatures. An intentional effort to fabricate all parts of a drift tube from inexpensive materials has been described; parts for a complete drift tube were produced using polymer-based fabrication methods.[78] In principle, injection molding could also permit the construction of new internal structures for drift tubes not easily possible with conventional fabrication methods. Plastic components could include electrically conductive, insulating, and static dissipative polymers, and the concept was demonstrated using carbon-loaded nylon and the cyclo-olefin polymer Zeonex. A concern with plastics is release of vapor impurities, and analysis of headspace over Zeonex-encapsulated carbon-loaded nylon for volatile organic compounds suggested levels low enough for use in an IMS drift tube. A snap-together stacked-ring drift tube of 4.25 cm long (Figure 6.11) was tested, and hydrated protons, expected in a clean drift tube, were observed at temperatures up to 50°C. Above this temperature, off-gassing of ammonia caused the formation of proton cluster ions of ammonia. Effects from surface charging were also observed over a 4 h period of continuous operation, although this could be controlled by substituting nylon, a polymer with a significantly lower surface resistivity, for the Zeonex. This concept is at this moment underdeveloped.

6.7.2.2 No-Ring Drift Tube

A drift tube merits mention here for the unconventional drift region, which has implications for drift designs. The drift tube of Irie et al.[79] was formed with a drift region comprised without drift rings. Rather, ions were able to move from the ion shutter to the detector, a distance of 1 to 4 cm, with a voltage of 1 to 5 kV without additional drift rings. Peak width was about 45 μs with a drift time of about 700 μs, yielding a ratio of drift time to a peak width of about 15. Although this ratio is low, the simple drift region demonstrates that mobility can be determined without the addition of drift rings.

6.7.2.3 Grids, Not Rings

At the other extreme of simple designs for rings is a 1970 design of Stevenson et al., who employed only grids in place of rings.[80] Potentials on the grids were established in forward and backward directions through the entire drift length, so the electric field had a triangular pattern with drift length. The grids were spaced at 1 cm, and two sawtooth waveforms of 0–1,000 V and 150 Hz to 150 kHz were placed on alternating rings. A mobility constant can be related to the frequency of the sawtooth waveform, and only an ion of a given mobility is transmitted through the drift region. This drift tube was operated without an ion shutter and constituted a type of linear ion filter. Peak widths at half height were about 2 cm^2/Vs for a peak at 20 cm^2/Vs or a ratio of drift time:peak width of 10. Although this is low by comparison to analytical-grade drift tubes, the ion density was large. Thus, this drift tube could be a type of preparative-scale mobility spectrometer.

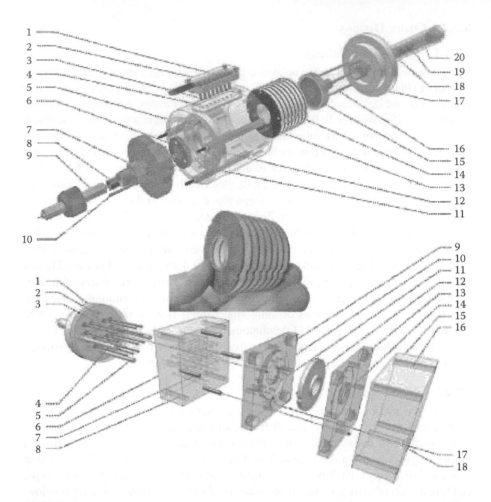

FIGURE 6.11 A mobility spectrometer made from plastic of various insulating or conductive properties. Caption reproduced from Koimtzis et al., *Anal. Chem.* 2011. "Exploded diagrams of the ion mobility spectrometer assembly (Top); the injection molding tool assembly, for the encapsulating insulator (Bottom); and a photograph showing the assembled drift tube (Middle). Key to Top: (1) PCB connector for voltage supply pins. (2) Injection molded voltage supply connector plug. (3) Stainless-steel "snap-fit" connector pins. (4) PTFE Voltage supply socket. (5) Aluminum housing and electrical shield (transparent view). (6) Ion shutters. (7) Graseby HTIMS inlet assembly manifold (includes item 6). (8) ^{63}Ni ionization source. (9) Sheath flow sample inlet. (10) Auxiliary exhaust port. (11) Cartridge heater. (12) Machined PTFE insulating holder to hold and align drift within the assembly. (13) Injection molded electrically conductive electrode. (14) Injection molded insulating spacer. (15) Graseby HTIMS detector and drift flow gas inlet assembly. (16) Drift gas inlet. (17) Machined PTFE lid. (18) Machined and ventilated PTFE insulation. (19) Preamplifier printed circuit board. (20) Preamplifier electrical shielding, copper tube standing at 1.1 kV (transparent view). Bottom: (1) Steel ejector piston. (2) Empty pocked driver pin holder. (3) Ejector pin, these pins "punch" the molded part out of the mold cavity at the end of each cycle. The pins can move through the moving platform block and the moving mold to reach the deepest surface level of the part cavity in (11).

6.7.2.4 New Flow Scheme in Electrospray IMS

A significant record of improvement in drift tube designs with electrospray sources occurred at Washington State University during the 1990s and 2000s. In recent years, teams in Iran have been active in drift tube designs, and a high-resolution electrospray ionization ion mobility spectrometry (ESI-IMS) was described with a reduced diameter in the desolvation region. Another innovation was the addition of desolvation gas near the ion shutter to protect the drift region from aerosols or solvent impurities.[81] The ESI source was located outside the heated volume of the drift tube, and increased temperatures were demonstrated without clogging of the needle or disruption of the ESI process. Increases in desolvation efficiency at higher flow rates and in resolving power (up to 70) were reported with an 11 cm long drift tube and ion shutter pulse of 300 µs. Although this could be seen as a minor modification of technology, the design highlights how processes in a drift tube can be isolated or controlled when design concepts, in this case gas flows, are used in new and flexible schemes. Little more is said here of ESI-IMS owing to the high relevance of ESI through the IM–MS chapter and discussion in this chapter (Section 6.2.1).

6.7.2.5 Twin Drift Tube

As early as the late 1980s, a twin drift tube was demonstrated with the ion source at ground potential and two drift tubes, each polarized with the detector floating at high negative or high positive voltage.[82] The drift tubes were designed in a T arrangement (Figure 6.11), with inlets to the drift tubes in a straight and opposite orientation. Ions from an ion source were carried with gas flow through a cylindrical metal foil (^{63}Ni) and the gas volume between the two drift tubes. Ions were drawn into tubes of opposite polarity; for example, positive ions were pulled in the direction of a detector floating at high negative potential. This instrument demonstrated that positive and negative ions of a substance from a single sample could be conveniently extracted for simultaneous characterization by mobility. This drift tube, as with the high-resolving-power IMS described previously, was developed by Graseby Dynamics under government funding. Although little was disclosed on this tube, the components,

FIGURE 6.11 (continued) (4) Guide pin. (5) Sprue ejector pin (4mm shorter than an ejector pin). (6) Moving platform, a steel block with holes for the various "pins" including dowel pin pockets. This part supports the "moving mold" (11) by the use of four 6 mm bolts (not shown). (7) Dowel pin for the accurate alignment of the mold pair and the moving platform, illustrated by a red arrow. (8) Threaded hole for 6 mm mold-fastener bolt. (9) Counter-bore clearance for the 6 mm mold-fastener bolt. (10) Moving mold. (11) Moving mold part cavity, which carries a negative impression of the left side of the molded part shown in (12). (12) Injection molded part. (13) Fixed mold. (14) Fixed mold part cavity, which carries a negative impression of the right side of the molded part shown in (12). (15) Fixed platform with the polymer injection port and channel. This also supports the fixed mold with four 6 mm bolts. (16) Threaded hole for 6 mm mold-fastener bolt. (17) The blue arrow indicates the flow path of molten polymer through to the part cavity. (18) Polymer injection port." (From Koimtzis et al., Assessment of the feasibility of the use of conductive polymers in the fabrication of ion mobility spectrometers, *Anal. Chem.* 2011, 83(7). With permission.)

and thus dimensions, were derived from spares of the drift tube found in CAM. This design is the basis for the GID-2A Chemical Warfare Agent Detection System.

6.7.2.6 Ion Cyclotron with Mobility

An unusual drift tube was designed with eight regions of four curved 30 mm long drift tubes and four ion funnels (15.2 cm) located between the drift tubes; pressure in the entire drift tube was 2.8 to 3.2 torr in helium at 300 K.[83] Ions were kept in the cyclotron drift tube and were introduced or removed using two drift tubes with Y-shaped geometries. Ions from an ESI source were injected via an ion shutter, and ion swarms moved in the cyclotron by changing electric fields with a frequency that is resonant with the drift time of ions through each region. Those ions with mobilities resonant at a frequency were passed through the ring, while other ions were lost. More than 10 runs around the loop were possible, and the electric field across regions (at fixed frequency) could be scanned to isolate ions of a given mobility. Peak widths were reported as constant even with a significantly increased number of cycles, and resolving powers were calculated as greater than 300. Ions were extracted into a MS using one of the Y-shaped drift tube regions; performance was demonstrated using $[M + 2H]^{2+}$ and $[M + 3H]^{3+}$ ions of a peptide.

6.8 SELECTION OF MATERIALS

In the previous discussions, a large number of drift tube designs and components for drift tubes were described. Little has been described here or in reviews of IMS technology concerning the materials used to construct drift tubes. This is nontrivial, and choices made on materials will govern both the range of operating temperatures possible for a drift tube and the quality of the drift tube response, in ways that are somewhat independent of the design or style of drift tube. In particular, two complications may arise from improper choice of materials or the use of materials under unsuitable operating conditions: Chemicals or impurities will be emitted or desorbed from the materials, causing unwanted contamination or changes in the gas phase ionization chemistry, or compounds in a sample may be adsorbed or lost to surfaces of materials through chemisorption or decomposition. The following discussion has importance for both instrument builders and users of IMS analyzers. Users should be mindful of the issue since commercial instruments have been designed for particular applications and materials included with best compatibility for the application. For example, the CAM contains plastic components that are wholly acceptable for a drift tube operated at ambient temperature; however, the CAM drift tube cannot be operated at temperatures above about 80°C without serious off-gassing of vapors that will interfere with response. In contrast, drift tubes designed for explosives and drug detection will contain materials that do not off-gas at elevated temperature and ideally contain surfaces on the inlet that are not too adsorptive or reactive for hot vapors of polar molecules. Consequently, drugs or thermally labile compounds such as explosives can pass from a sample to the drift tube without undergoing massive condensation or decomposition. Instrument builders should appreciate that reactant ion peaks, even in carefully designed and built instruments, could actually be formed from materials used in the drift tube. The peak expected to be a hydrated proton could actually be protonated monomers or proton-bound dimers arising from vapor impurities released from construction materials, even if thought insignificant, in the drift tube. There is no substitute for care in the selection and use of proper materials for the drift tube.

6.8.1 Conductors

Conducting materials in a drift tube are used in traditional IMS designs for the drift rings, ion shutters, aperture grid, detector plate, gas lines, protective housing, and connecting wires for the drift tube components. Metals might also be found in the ion source assembly (electrospray needle, electrodes in a CD source, and the holder for [63]Ni foil) and in the inlet as a sample transfer line. Stainless steel is the metal preferred for all drift tubes owing to durability; resistance to chemical attack, such as oxidation; and thermal properties of low or null off-gassing. Gold plating has been used with some drift tube designs but is believed unnecessary, only incurring extra cost and complications. Nonetheless, despite favorable chemical properties, stainless steel is a poor conductor of heat and will require extended times to warm and cool. The availability of stainless steel in several grades favors its use as a detector plate, aperture grid, and drift rings, although the difficulty, time, and expense to machine components from bar, rod, or plate stock are considerations. Stamped or formed pieces could reduce the costs of manufacture. For example, drift rings can be made quickly and easily from stainless steel washers available from commercial sources. The washers are faced and trimmed for inner and outer diameters on a lathe, with large savings in time and cost versus complete manufacture from rod stock. In one drift tube design from Baumbach's team in Germany, the drift tube was defined by a Teflon tube, and the drift rings were on the outside of the Teflon tube. In this design, requirements for chemical reactivity could be relaxed, and drift rings could be copper, brass, and even aluminum. Similarly, a high-resolution drift tube described by Brokenshire and manufactured by Graseby Dynamics in the early 1990s consisted of a glass tube with thin copper bands as drift rings attached on the external surface of the tube.[7] Metals share electrical properties, so most considerations will be made on the chemical reactivity, cost, and physical durability of the metal. One special feature is thermal expansion, which can become a consideration when drift tubes are heated; aluminum has the worst coefficient of thermal expansion of metals suitable for IMS drift tubes, and this should factor into designs involving aluminum housings or flanges.

Metal can also be part of an inlet and encounter or make contact with sample. Such metal should be limited to stainless steel or nickel; other common metals such as copper can be excessively reactive or adsorptive toward organic compounds and corrosive gases and should be avoided. Teflon is advisable as much as possible for surfaces of the inlet. Alternatively, metals could be passivated with a thin layer of gold, treated through a proprietary deactivation method, or replaced with glass-lined steel tubing. Inexpensive metals such as copper may be used for prototyping drift tubes, although use should be restricted to low-temperature and nonabsorptive chemicals.

Ion shutters may be made from stainless steel wire, and considerations of thermal expansion may arise within a shutter assembly (see Chapter 4 for detailed discussion of ion shutters). A sophisticated shutter design was based on wires stretched between two ceramic halves. The ceramic pieces were separated by thermally stable springs so that expansion or contraction could occur with variations in temperature without sagging or snapping of the wire from changes in tension. Original descriptions of CD sources and other sources should be consulted for guidance on metals and materials for these specialized components. We have found that the use of standard, tapered, stainless steel syringe needles offers an inexpensive source for CD ionization.

6.8.2 Insulators

Electric fields are established in a drift tube through voltage differences on conducting pieces that are electrically insulated and separated by distances ideally known and fixed. In the earliest drift tube designs of the Beta VI and subsequent generations from the same company, rings were separated by three sapphire balls about 2 mm in diameter symmetrically arranged around the drift ring. In the CAM, the drift rings are held by three stainless steel rods with Teflon sleeves. The rings are insulated by the Teflon, and the spacing between the rings is established by three 4 mm long ceramic tubes with internal diameters large enough to fit over the Teflon-covered rods. In both of these designs, the insulators are Teflon, ceramic, or an inorganic crystalline solid.

The choice of insulator is as important for analytical performance and practical reasons as the conductor for the same reasons: chemical reactivity, physical integrity, cost, and convenience of manufacture. In addition, plastics and other materials are prone to off-gassing with increases in temperature through impurities or small oligomers retained from the production process or through decomposition of the polymer. Another property of the insulator is electrical conductivity, which should be in megaohms per centimeter.

Insulators used in drift tubes include glass, machinable glass, ceramic, machinable ceramic, mica, and plastic materials like polyethylene, polyimide, and Teflon. An insulator of high resistively, high thermal tolerance, and good chemical inertness is Macor™ (55% fluorophlogopite mica and 45% borosilicate glass), which has an added benefit of being machinable. Although Macor can be worked on a lathe and milling machine, precautions are needed to avoid inhalation of dust, and the use of liquid streams may produce a waste liquid that is corrosive to the metal of the lathe and milling machine. Machinable, almost-soft, ceramic is available that can be fired to temperatures over 600°C to become a hardened ceramic. Changes in size after firing can be over 10%, and heating must be uniform and held constant (±2°C), making the application of this ceramic for delicate or fitted parts somewhat difficult.

Polymers such as polypropylene and fluorocarbon resins (the Teflon family of polymers) offer ease of machining and chemical inertness and drift tubes may contain nylon or polypropylene insulating rings as long as temperatures are low. The best of the fluoropolymers regarding temperature stability is PTFE (polytetrafluoroethylene), with an upper temperature of operation more than 260°C. These materials have excellent electrical properties for high impedances of IMS drift tubes and do not exhibit electrical breakdown under high voltage. Sadly, the thermal expansion coefficients for PTFE make deformation a concern, and drift tubes made with PTFE and operated at high temperature must be kept in a durable external shell. Finally,

PTFE is a poor conductor of heat, so the application of heat to a drift tube made from PTFE must be made gradually or localized overheating may occur with melting or decomposition of the plastic. However, the cost and convenience of machining PTFE makes this material attractive for drift tubes even when operated at temperatures as high as 250°C.

6.8.3 MISCELLANEOUS MATERIALS

The same considerations used for insulators and conductors described in the previous sections are applicable to miscellaneous items such as tubing, gaskets, seals, and specialized materials such as adhesives and shells. In low-temperature drift tubes with handheld analyzers, unions for tubing and pump seals and other connections are made using plastics. In research-grade analyzers, connections are usually made with pipe-pipe compression unions or silver-soldered connections. However, material consideration may not be ignored for other parts of a drift tube since even small sources of contamination can render ultrasensitive analyzers, such as IMS drift tubes, unworkable.

Tubing can be used for preheating the drift gas since gas temperature in the drift and reaction regions will affect ion clustering and fragmentation of ions. The appearance of a mobility spectrum will be influenced by gas temperature; however, obtaining a reliable measure of gas temperature is nontrivial. No systematic description of preheating a gas has been given for which the entire gas stream has been monitored by sensors capable of accurately measuring the gas temperature, so that the subject remains unsettled even after 30 years of modern analytical IMS. Any measurements drawn from the drift tube body or flanges will not be reliable for telling ion temperature. One solution for laboratory instruments with a critical need to know temperatures accurately and maintain uniform temperatures is to place the drift tube and tubing and more inside an oven.

In practice, thermocouples have often been attached to the metal housing of a drift tube, providing false measures of temperature of the gas. In a design common to New Mexico State University, several meters of 3.5 mm OD aluminum tubing (a good conductor of heat) are wrapped around the body of the drift tube. Drift gas is passed through the tubing before the gas enters the drift tube near the detector. Preliminary studies made on meter lengths of identical tubing independent of the drift tube showed that gas temperature is not identical to the wall temperature of the tubing even though preheating of gas has occurred, and elevated temperatures are seen with thermocouples placed in the gas effluent. Care must be taken that tubing is sufficiently heated and insulated, or cooling of the gas may occur. Although the gas has been warmed during passage, there is no guarantee that the drift tube is free of thermal gradients. No similar studies were made with stainless steel. Copper tubing should be avoided completely when heated owing to the formation of copper oxides. Under prolonged and elevated temperatures in oxidizing gases such as air, copper oxides may flake and send particulate into the gas streams and places downflow.

Drift tubes under a slight pressure drop must be engineered with seals and gaskets to prevent inflow of surrounding atmosphere into the drift tube. This can be accomplished with thermoplastics or metal-to-metal seals when elevated temperatures are needed or durable elastomers for ambient temperature instruments. Housings and flanges have been universally stainless steel or aluminum.

6.9 SUMMARY AND CONCLUSIONS

A mobility spectrometer may be seen as an assembly of components, and the discussion in this chapter was intended to document the large and increasing diversity in designs of drift tubes in all mobility methods. This should be seen as healthy and promising for a method that is dynamic and growing in value and use. The status of development of IMS today may be compared to similar developments in other methods, such as gas chromatography in the 1960s and 1970s, MS from 1970 to 1990, and LC in the 1990s.

The plain historic linear IMS is today still the method likely to be encountered or used for military and security purposes. The number of instruments already deployed is large, and capabilities still improve; for example, handheld analyzers now exist for explosives detection, and palm-size instruments are available for chemical weapons. As seen here, the types of mobility spectrometers, designs described, and commercial availability have increased greatly since the publication of the second edition of this title. This is made even more diverse owing to a large selection of materials and manufacturing methods. In summary, the technological developments in ion mobility methods have grown during the past decade in a complementary manner to the applications and journal articles (Chapter 2).

REFERENCES

1. Cohen, M.J.; Karasek, F.W., Plasma chromatography—a new dimension for gas chromatography and mass spectrometry, *J. Chromatogr. Sci.* 1970, 8, 330–337.
2. Baim, M.A.; Hill, H.H., Jr., Tunable selective detection for capillary gas chromatography by ion mobility spectrometry, *Anal. Chem.* 1982, 54, 38–43.
3. Baim, M.A.; Schuetze, F.J.; Frame, J.M.; Hill, H.H., Jr., A microprocessor controlled ion mobility spectrometer for selective and nonselective detection following gas chromatography, *Am. Lab.* 1982, 14, 59–70.
4. Eiceman, G.A.; Leasure, C.S.; Vandiver, V.J.; Rico, G., Flow characteristics in a segmented closed-tube design for ion mobility spectrometry, *Anal. Chim. Acta* 1985, 175, 135–145.
5. Eiceman, G.A.; Nazarov, E.G.; Stone, J.A.; Rodriguez, J.E., Analysis of a drift tube at ambient pressure: models and precise measurements in ion mobility spectrometry, *Rev. Sci. Instrum.* 2001, 72, 3610–3621.
6. Soppart, O.; Baumbach, J.I., Comparison of electric fields within drift tubes for ion mobility spectrometry, *Meas. Sci. Technol.* 2000, 11, 1473–1479.
7. Brokenshire, J.L., High resolution ion mobility spectrometry, 1991 joint meeting FACSS/Pacific Conference and 27th Western Regional ACS meeting, Anaheim, CA, October 6–11, 1991.
8. Wu, C.; Siems, W.F.; Asbury, G.R.; Hill, H.H., Jr., Electrospray ionization high-resolution ion mobility spectrometry-mass spectrometry, *Anal. Chem.* 1998, 70, 4929–4938.
9. Lee, D.S.; Wu, C.; Hill, H.H., Detection of carbohydrates by electrospray-ionization ion mobility spectrometry following microbore high-performance liquid-chromatography, *J. Chromatog. A* 1998, 822, 1–9.

10. Wu, C.; Siems, W.F.; Klasmeier, J.; Hill, H.H., Separation of isomeric peptides using electrospray ionization/high-resolution ion mobility spectrometry, *Anal. Chem.* 2000, 72, 391–395.
11. Tadjimukhamedov, F.K.; Stone, J.A.; Papanastasiou, D.; Rodriguez, J.E.; Mueller, W.; Sukumar, H.; Eiceman, G.A., Liquid chromatography/electrospray ionization/ion mobility spectrometry of chlorophenols with full flow from large bore LC columns, *Int. J. Ion Mobil. Spectrom.* 2008, 11(1–4), 51–60.
12. Schellenbaum, R.L.; Hannum, D.W., Laboratory evaluation of the PCP large reactor volume ion mobility spectrometer (LRVIMS), Technical Report No. SAND89-0461*UC-515, Sandia National Laboratories, Albuquerque, NM, March 1990.
13. Bell, A.J.; Ross, S.K., Reverse flow continuous corona discharge ionization, *Int. J. Ion Mobil. Spectrom.* 2002, 5(3), 95–99.
14. Tabrizchi, M.; Khayamian, T.; Taj, N., Design and optimization of a corona discharge ionization source for ion mobility spectrometry, *Rev. Sci. Instrum.* 2000, 7, 2321–2328.
15. Turner, R.B.; Brokenshire, J.L., Hand-held ion mobility spectrometers, *Trends Anal. Chem.* 1994, 13(7), 281–286.
16. Baumbach, J.I.; Berger, D.; Leonhardt, J.W.; Klockow, D., Ion mobility sensor in environmental analytical chemistry: concept and first results, *Int. J. Environ. Anal. Chem.* 1993, 52(1–4), 189–93.
17. Xu, J.; Whitten, W.B.; Ramsey, J.M., A miniature ion mobility spectrometer, *Int. J. Ion Mobil. Spectrom.* 2002, 2, 207–214; also see Xu, J.; Whitten, W.B.; Ramsey, J.M., Space charge effects on resolution in a miniature ion mobility spectrometer, *Anal. Chem.* 2000, 72, 5787–5791.
18. Snyder, A.P.; Harden, C.S.; Shoff, D.B.; Ewing, R.G.; Katzoff, L.; Bradshaw, R.; Turner, R.B.; Adams, J.N.; Taylor, S.J.; FitzGerald, J., Miniature ion mobility spectrometer monitor, in *Proceedings of the ERDEC Scientific Conference on Chemical and Biological Defense Research*, Aberdeen Proving Ground, MD, November 15–18, 1994, 145–151.
19. Hoaglund, C.S.; Valentine, S.J.; Clemmer, D.E., An ion trap interface for ESI-ion mobility experiments, *Anal. Chem.* 1997, 69, 4156–4161.
20. Hoaglund-Hyzer, C.S.; Clemmer, D.E., Ion trap/ion mobility/quadrupole/time-of-flight mass spectrometry for peptide mixture analysis, *Anal. Chem.* 2001, 73, 177–184.
21. Gillig, K.J.; Ruotolo, B.; Stone, E.G.; Russell, D.H.; Fuhrer, K.; Gonin, M.; Schultz, A. J., Coupling high-pressure MALDI with ion mobility/orthogonal time-of-flight mass spectrometry, *Anal. Chem.* 2000, 72, 3965–3971.
22. Stephen, D.J.; Birdwell, D.O.; Verbeck, G.F., Drift tube soft-landing for the production and characterization of materials: applied to Cu clusters, *Rev. Sci. Instrum.* 2010, 81(3), 034104–034110.
23. Eiceman, G.A.; Schmidt, H.; Rodriguez, J.E.; White, C.R., Krylov, E.V., Stone, J.A., Planar drift tube for ion mobility spectrometry, *Instrum. Sci. Technol.* 2007, 35(4), 365–383.
24. Blanchard, W.C., Using non-linear fields in high pressure spectrometry, *Int. J. Mass Spectrom. Ion Proc.* 1989, 95, 199–210.
25. Arnold, N.S.; Meuzelaar, H.L.C.; Dworzanski, J.P.; Cole, P.C.; Snyder, A.P., Feasibility of drone-portable ion mobility spectrometry, U.S. Army Chemical Research Development and Engineering Center Scientific Conference on Chemical Defense Research, Arberdeen Proving Grounds, MD, November 1991.
26. Cao, L.; Harrington, P.B.; Harden, C.S.; McHugh, V.M.; Thomas, M.A., Nonlinear wavelet compression of ion mobility spectra from ion mobility spectrometers mounted in an unmanned aerial vehicle, *Anal. Chem.* 2004, 76, 1069–1077.

27. Buryakov, I.A.; Krylov, E.V.; Nazarov, E.G.; Rasulev, U.Kh., A new method of separation of multi-atomic ions by mobility at atmospheric pressure using a high-frequency amplitude-asymmetric strong electric field, *Int. J. Mass Spectrom. Ion Proc.* 1993, 128, 143–148.

28. Carnahan, B.; Day, S.; Kouznetsov, V.; Tarrasov, A., Development and applications of a traverse field compensation ion mobility spectrometer, in *Fourth International Workshop on Ion Mobility Spectrometry*, Cambridge, UK, July 1995.

29. Guevremont, R.; Purves, R.W., Atmospheric pressure ion focusing in a high-field asymmetric waveform ion mobility spectrometer, *Rev. Sci. Instrum.* 1999, 70, 1370–1383.

30. Purves, R.W.; Guevremont, R., Electrospray ionization high-field asymmetric waveform ion mobility spectrometry-mass spectrometry, *Anal. Chem.* 1999, 71, 2346–2357.

31. Purves, R.W.; Guevremont, R.; Day, S.; Pipich, C.W.; Matyjaszczyk, M.S., Mass spectrometric characterization of a high-field asymmetric waveform ion mobility spectrometer, *Rev. Sci. Instrum.* 1998, 69(12), 4099–4105.

32. Miller, R.A.; Eiceman, G.A.; Nazarov, E.G.; King, A.T., A novel micro-machined high field asymmetric waveform ion mobility spectrometer, *Sens. Actuators B* 2000, 67, 300–306.

33. Eiceman, G.A.; Nazarov, E.G.; Tadjikov, B.; Miller, R.A., Monitoring volatile organic compounds in ambient air inside and outside buildings with the use of a radio-frequency-based ion-mobility analyzer with a micromachined drift tube, *Field Anal. Chem. Tech.* 2000, 4, 297–308.

34. Eiceman, G.A.; Tadjikov, B.; Krylov, E.; Nazarov, E.G.; Miller, R.A.; Westbrook, J.; Funk, P., Miniature radio-frequency mobility analyzer as a gas chromatographic detector for oxygen-containing volatile organic compounds, pheromones and other insect attractants, *J. Chromatogr. A* 2001, 917, 205–217.

35. Levin, D.S.; Miller, R.A.; Nazarov, E.G.; Vouros, P., Rapid separation and quantitative analysis of peptides using a new nanoelectrospray—differential mobility spectrometer-mass spectrometer system, *Anal. Chem.* 2006, 78, 5443–5452.

36. Schneider, B.B.; Covey, T.R.; Coy, S.L.; Krylov, E.V.; Nazarov, E.G., Chemical effects in the separation process of a differential mobility/mass spectrometer system, *Anal. Chem.* 2010, 82,1867–1880.

37. Tang, F.; Wang, X.; Xu, C., FAIMS biochemical sensor based on MEMS technology, in *New Perspectives in Biosensors Technology and Applications*, Serra, A.P. (Ed.) ISBN: 978-953-307-448-1, InTech, Shanghi, China, 2011, pp. 1–32.

38. White Paper, Owlstone Nanotech, OWL-WP-1 v3.0 21-3-06. Cambridge UK, 13 pp, copyright 2006, available at http://info.owlstonenanotech.com/rs/owlstone/images/FAIMS%20Whitepaper.pdf.

39. Shvartsburg, A.A.; Smith, R.D.; Wilks A.; Koehl, A.; Ruiz-Alonso, D.; Boyle, B., Ultrafast differential ion mobility spectrometry at extreme electric fields in multichannel microchips, *Anal. Chem.* 2009, 81, 6489–6495.

40. Krylov, E.; Nazarov, E.G.; Miller, R.A.; Tadjikov, B.; Eiceman, G.A., Field dependence of mobilities for gas-phase-protonated monomers and proton-bound dimers of ketones by planar field asymmetric waveform ion mobility spectrometer (PFAIMS), *J. Phys. Chem. A* 2002, 106, 5437–5444.

41. Krylova, N.; Krylov, E.; Eiceman, G.A.; Stone, J.A., Effect of moisture on the field dependence of mobility for gas-phase ions of organophosphorus compounds at atmospheric pressure with field asymmetric ion mobility spectrometry, *J. Phys. Chem. A* 2003, 107, 3648–3654.

42. Papanastasiou, D.; Wollnik, H.; Rico, G.; Tadjimukhamedov, F.; Mueller, W.; Eiceman, G.A., Differential mobility separation of ions using a rectangular asymmetric waveform, *J. Phys. Chem. A* 2008, 112, 3638–3645.

43. Krylov, E.V.; Coy, S.L.; Vandermey, J.; Schneider, B.B.; Covey, T.R.; Nazarov, E.G., Selection and generation of waveforms for differential mobility spectrometry, *Rev. Sci. Instrum.* 2010, 81, 24101–24112.

44. Prieto, M.; Tsai, C.W.; Boumsellek, S.; Ferran, R.; Kaminsky,I.; Harris, S.; Yost, R.A., Comparison of rectangular and bisinusoidal waveforms in a miniature planar FAIMS, *Anal Chem.* 2011, 83, 9237–9243.

45. Shvartsburg, A.A.; Smith, R.D., Optimum waveforms for differential ion mobility spectrometry (FAIMS), *J. Am. Soc. Mass Spectrom.* 2008, 19, 1286–1295.

46. Sacristan, E.; Solis, A.A., A swept-field aspiration condenser as an ion-mobility spectrometer, *IEEE Trans. Instrum. Meas.* 1998, 47, 769–775.

47. Ebert, H., Aspirations apparat zur bestimmung des Ionengehaltes der atmosphare, *Phys. Z.* 1901, 2, 662–666.

48. Anttalainen, O., New practical structure for second order aspiration IMS, 20th annual conference, International Society for Ion Mobility Spectrometry, Edinburgh, Scotland, July 24–28, 2011.

49. Anttalainen, O.; Pitkaenen J., New aspiration IMS cell structure, 17th Annual Conference on Ion Mobility Spectrometry, Ottawa, Canada, July 20–25, 2008.

50. Zimmermann, S.; Barth, S.; Baether, W., A miniaturized low-cost ion mobility spectrometer for fast detection of trace gases in air, *IEEE Sensors 2008*, 740–743.

51. Barth, S.; Baether, W.; Zimmermann, S., System design and optimization of a miniaturized ion mobility spectrometer using finite element analysis, *IEEE Sensors* 2009, 9, 377–382.

52. Tuovinen, K.; Paakkanen, H.; Hänninen, O., Determination of soman and VX degradation products by an aspiration ion mobility spectrometry, *Anal. Chim. Acta* 2001, 440, 151–159.

53. Raatikainen, O.; Pursiainen, J.; Hyvonen, P.; Von Wright, A.; Reinikainen, S.-P.; Muje, P., Fish quality assessment with ion mobility based gas detector, *Mededelingen—Faculteit Landbouwkundige en Toegepaste Biologische Wetenschappen* (Universiteit Gent), 2001, 66, 475–480.

54. Rus, J.; Moro, D.; Sillearo, J.A.; Royuela, J.; Casado, A.; Estevez-Molinero, F.; de la Mora, J.F., IMS-MS studies based on coupling a differential mobility analyzer (DMA) to commercial API–MS systems, *Int. J. Mass Spectrom.* 2010, 298, 30–40.

55. Ude, S.; de la Mora, J.F.; Thomson, B.A., Charge-induced unfolding of multiply charged polyethylene glycol ions, *J. Am. Chem. Soc.* 2004, 126, 12184–12190.

56. Hogan, C.J., Jr.; de la Mora, J.F., Ion mobility measurements of nondenatured 12–150 kDa proteins and protein multimers by tandem differential mobility analysis-mass spectrometry (DMA-MS), *J. Am. Soc. Mass Spectrom.* 2011., 22, 158–172.

57. Pringle, S.D.; Giles, K.; Wildgoose, J.L.; Williams, J.P.; Slade, S.E.; Thalassinos, K.; Bateman, R.H.; Bowers, M.T.; Scrivens, J.H., An investigation of the mobility separation of some peptide and protein ions using a new hybrid quadrupole/traveling wave IMS/oa-TOF instrument, *Int. J. Mass Spectrom.* 2007, 261, 1–12.

58. Williams, J.P.; Bugarcic, T.; Habtemariam, A.; Giles, K.; Campuzano, I.; Rodger, P.M.; Sadler, P.J., Isomer separation and gas-phase configurations of organoruthenium anti-cancer complexes: ion mobility mass spectrometry and modeling, *J. Am. Soc. Mass Spectrom.* 2009, 20(6), 1119–1122.

59. Inutan, E.D.; Wang, B.; Trimpin, S., Commercial intermediate pressure MALDI ion mobility spectrometry mass spectrometer capable of producing highly charged laserspray ionization ions, *Anal. Chem.* 2011, 83, 678–684.

60. Shvartsburg, A.A.; Smith, R.D., Fundamentals of traveling wave ion mobility spectrometry, *Anal. Chem.* 2008, 80(24), 9689–9699.

61. Stimac, R.M.; Wernlund, R.F.; Cohen, M.J.; Lubman, D.M.; Harden, C.S., Initial studies on the operation and performance of the tandem ion mobility spectrometer, Pittcon. 1985, New Orleans, LA, March 1985.

62. Stimac, R.M.; Cohen, M.J.; Wernlund, R.F., Tandem ion mobility spectrometer for chemical agent detection, monitoring and alarm, Contractor Report on CRDEC Contract DAAK11 84 C 0017, PCP, West Palm Beach, FL, May 1985, AD B093495.

63. Ewing, R.G., Kinetic decomposition of proton bound dimer ions with substituted amines in ion mobility spectrometry, dissertation, New Mexico State University, Las Cruces, December 1996.

64. Ewing, R.E.; Eiceman, G.A.; Harden, C.S.; Stone, J.A., The kinetics of the decompositions of the proton bound dimers of 1,4-dimethylpyridine and dimethyl methylphosphonate from atmospheric pressure ion mobility spectra, *Int. J. Mass Spectrom.* 2006, 76–85, 255–256.

65. An, X.; Stone, J.A.; Eiceman, G.A., Gas phase fragmentation of protonated esters in air at ambient pressure through ion heating by electric field in differential mobility spectrometry and by thermal bath in Ion Mobility Spectrometry, *Int. J. Mass Spectrom.* 2011, 303(2–3), 181–190.

66. Rajapakse, R.M.M.Y.; Stone, J.A.; Eiceman, G.A.; Design and performance of ion mobility spectrometer for kinetic studies of ion decomposition in air at ambient pressure, 2012, in preparation.

67. Valentinea, S.J.; Kurulugamaa, R.T.; Bohrera, B.C.; Merenbloomaa, S.I.; Sowell, R.A.; Mechref, Y.; Clemmer, D.E., Developing IMS-IMS-MS for rapid characterization of abundant proteins in human plasma, *Int. J. Mass Spectrom.* 2009, 283, 149–160.

68. White, C.R., Characterization of tandem DMS-IMS2 and determination of orthogonality between the mobility coefficient (K) and the differential mobility coefficient (ΔK), MS thesis, New Mexico State University, Las Cruces, NM, May 2006.

69. Eiceman, G.A.; Schmidt, H.; Rodriguez, J.E.; White, C.R.; Nazarov, E.G.; Krylov, E.V.; Miller, R.A.; Bowers, M.; Burchfield, D.; Niu, B.; Smith, E.; Leigh, N., Characterization of positive and negative ions simultaneously through measures of K and ΔK by tandem DMS-IMS, ISIMS 2005, Château de Maffliers, France, July 2005.

70. Pollard, M.J.; Hilton, C.K.; Li, H.; Kaplan, K.; Yost, R.A.; Hill, H.H., Ion mobility spectrometer-field asymmetric ion mobility spectrometer-mass spectrometry, *Int. J. Ion Mobil. Spectrom.* 2011, 14(1), 15–22.

71. Menlyadiev, M.; Stone, J.A.; Eiceman, G.A., Tandem ion mobility measurements with chemical modification of ions selected by compensation voltage in differential mobility spectrometry/differential mobility spectrometry instrument, *Int. J. Ion Mobil. Spectrom.* 2012, 15, 123–130.

72. Wu, C., Multidimensional ion mobility spectrometry apparatus and methods, U.S. Patent 7,576,321 B2, Aug. 8, 2009.

73. Linuma, K.; Takebe, M.; Satoh, Y.; Seto, K. Design of a continuous guard ring and its application to swarm experiments, *Rev. Sci. Instrum.* 1982, 53, 845–850.

74. Linuma, K.; Takebe, M.; Satoh, Y.; Seto, K., Measurements of mobilities and longitudinal diffusion coefficients of Na^+ Ions in Ne, Ar, and CH_4 at room temperature by a continuous guard-ring system, *J. Chem. Phys.* 1983, 79, 3893–3899.

75. Carrico, J.P.; Sickenberger, D.W.; Spangler, G.E.; Vora, K.N., Simple electrode design for ion mobility spectrometry, *J. Phys. E Sci. Instrum.* 1983, 16, 1058–1062.

76. Kwasnik, M.; Gonin, M.; Fuhrer, K.; Barbeau, K.; Fernandez, F.M., Performance, resolving power and radial ion distributions of a prototype nanoelectrospray ionization resistive glass atmospheric pressure ion mobility spectrometer, *Anal. Chem.* 2007, 79, 7782–7791.

77. Bathgate, B.; Cheong, E.C.S.; Backhouse, C.J., A novel electrospray-based ion mobility spectrometer, *Am. J. Phys.* 2004, 72(8), 1111.

78. Koimtzis, T.; Goddard, N.J.; Wilson, I.; Thomas, C.L., Assessment of the feasibility of the use of conductive polymers in the fabrication of ion mobility spectrometers, *Anal. Chem.* 2011, 83(7), 2613–2621.
79. Irie, T.; Mitsui, Y.; Hasumi, K., A drift tube for monitoring ppb trace water, *Jpn. J. Appl. Phys.* 1992, 31, 2610–2615.
80. Stevenson, P.C.; Thomas, R.A.; Lane, S., Ion-mobility spectrometer for radiochemical applications, *Nucl. Instrum. Methods* 1970, 89, 177–187.
81. Khayamian, T.; Jafari, M.T.; Design for electrospray ionization-ion mobility spectrometry, *Anal. Chem.* 2007, 79(8), 3199–3205.
82. Atkinson, R.; Clark, A.; Taylor, S.J., Ion mobility spectrometer comprising two drift chambers, U.S. Patent 7994475, February 28, 2008.
83. Kurulugama, R.; Nachtigall, F.M.; Lee, S.; Valentine, S.J.; Clemmer, D.E., Overtone mobility spectrometry: part 1. Experimental observations, *J. Am. Soc. Mass Spectrom.* 2009, 20, 729–737.

7 Ion Detectors

7.1 INTRODUCTION

The process of converting a swarm of gas phase ions into an electrical signal that provides both arrival time information and amplitude information is known as ion detection. The ion detector is typically located after the ion mobility separation cell and has a number of ideal requisites to accomplish the transduction of mobility-separated ions in a manner that minimizes loss of sensitivity and IMS resolving power.

Sensitivity in an IMS is a function of the transport efficiency of vapors from the sample into the ionization source, the efficiency of the ionization processes, the transfer of the analyte ions through the spectrometer, and finally the conversion of ions into an electronic signal as a function of arrival time. Ideally, ion detectors for high sensitivity have unity ion detection efficiency with low noise. The response should be stable from minute to minute, hour to hour, day to day, and month to month, enabling routine, dependable, and quantitative operation of the ion mobility spectrometry (IMS) instrument. In addition to high sensitivity and stability, an ion detector should have a wide mass range response, detecting ions of any size or mass equally without variation or discrimination. It should have a wide dynamic range with a high saturation level so that samples containing both low and high abundances of ions are accurately measured. With time-of-flight IMS, peak widths are often less than 1 ms, so the rise time of the detector should be on the order of 10 μs or faster.

Resolving power (the ratio of the peak position to the width of the peak at half maximum) in IMS is a function of the broadening of the ion pulse as it is created and migrates through the IMS. Fast response times with short recovery times are essential for ion detectors to accurately track the leading and trailing edges of a swarm of mobility-separated ions. For IMS instruments to be applied for practical and routine operation, detectors should also have long lifetimes with low maintenance requirements. If they do need to be replaced, they should be inexpensive and easy to install.

The primary operational parameter that determines the type of ion detector required depends on the pressure at which the ions are detected. In general, detectors used for ambient pressure IMS are different from those used under vacuum conditions. Thus, this chapter is divided into two sections describing the ion detectors used at ambient pressure and those used at low pressures.

7.2 AMBIENT DETECTION OF MOBILITY-SEPARATED IONS

7.2.1 Faraday Cup and Faraday Plate Detectors

At ambient pressure, at which the majority of IMS instruments operate, ion detection is accomplished by converting ions impinging on the detector into current by

155

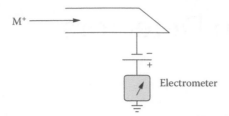

FIGURE 7.1   Schematic of a Faraday cup. (From Wikipedia.)

neutralization on a collector electrode. Named for Michael Faraday, this common ion detector is called a Faraday cup for vacuum detection and a Faraday plate for ambient pressure detection. A schematic of a Faraday cup is shown in Figure 7.1. As ions impinge on the internal surface of the cup, current flows in the metal as the ions are neutralized. When positive ions are neutralized on the cup, electrons flow toward the cup, and when negative ions are neutralized, electrons flow toward the electrometer.

The Faraday cup was originally designed to capture ions in a vacuum. Under vacuum conditions, ions can impinge on a metal surface with sufficient energy to knock out other ions or electrons. If these secondary ions or electrons escape neutralization on the collector, the efficiency of ion detection will be decreased. The cup design was used to capture secondary ions and electrons, providing a quantitative counting method for ions in low-pressure or vacuum conditions.

Ion collectors at ambient pressure, however, do not require a cup design because the impinging ion on the collector does not have sufficient energy to produce secondary ions; thus, the potential loss of secondary ions is not a problem for quantification. Under ambient pressure conditions, ions can be collected with a simple flat-plate design called a Faraday plate. Faraday plates can be easily interfaced to an IMS as a flat, terminal plate after the ion separator.

For 100% collection efficiency, the simple relationship between current measured and the number of ions collected is

$$\frac{N}{t} = \frac{I}{e} \tag{7.1}$$

where the number of ions N impinging on the surface of the Faraday plate per second t is equal to the current I in amperes divided by the elementary electron charge e of 1.60×10^{-19} C. Ion currents detected in IMS are on the order of a few nanoamperes (10^{-9} amperes). One nA corresponds to about 6×10^{9} ions/s. Johnson noise (electronic noise due to the random thermal movement of electrons in the Faraday plate and its concomitant electrical connections) at ambient temperature (25°C) is about 1×10^{-12} amperes, so the detection limit of an ambient pressure ion collector is about 2×10^{6} ions/s. For a single IMS peak, which can be 0.2 ms wide, the detection limit is around 100 ions.

For drift tube IMS systems, efficient collection of the ions is not the only consideration for ion detection. As a swarm of ions approaches the Faraday plate, it induces a charge on the plate that starts current to flow in the detector. That is, as a swarm of positively charged ions approaches a Faraday plate, electrons in the metal plate

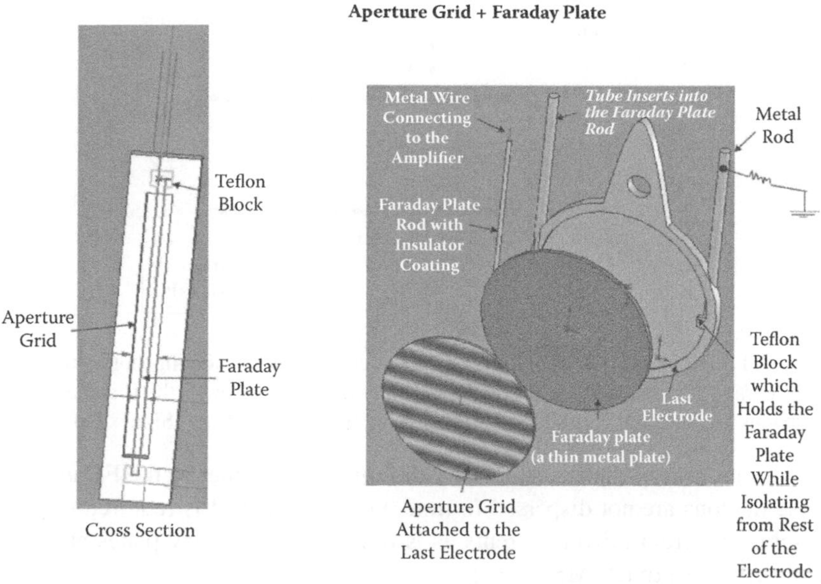

Aperture Grid + Faraday Plate

FIGURE 7.2 Faraday plate assembly with aperture grid. (Drawn by Manuja Lamabadusuriya.)

will be attracted to the surface of the plate. This flow of electrons from the body of the metal plate to the surface of the plate nearest the approaching ion swarm will be measured and recorded as an ion current by the electrometer. This current is induced by the approaching ions *before* the ion swarm reaches the surface of the Faraday plate. In other words, the approaching charge will be "mirrored" on the Faraday plate, causing the detector to respond before the packet arrives at the Faraday plate. As a result, the apparent width of the arriving ion swarm is increased, decreasing the measured resolving power of the spectrometer.

To resolve the mirror image problem, an "aperture grid" is place before the Faraday plate to shield the effect of the incoming charge from the Faraday plate. Figure 7.2 shows a drawing of a Faraday plate assembly with the aperture grid. In this design, the Faraday plate is a thin metal foil attached with a shielded lead directly to the amplifier and sandwiched between an electrically isolated metal screen that serves as the aperture grid and a metal plate that serves as the last IMS guard ring. The last electrode and the aperture grid are at some appropriate electrical potential above and below the Faraday plate, respectively. Once ions pass through the aperture grid, they are focused onto the Faraday plate by the aperture grid and the terminal electrode.

While time-dispersive ion mobility devices of the type used for drift tube IMS require aperture grids prior to the Faraday plate to preserve the resolving power of the instrument, ion filters and scanning mobility spectrometry such as differential mobility spectrometry (DMS), field asymmetric IMS (FAIMS), differential mobility analysis (DMA), and aspiration IMS (aIMS) do not require an aperture grid and can efficiently detect ions with a simple Faraday plate. In these devices, ions do not travel as a discrete swarm, and the exact arrival time of the ions is not critical. Figure 7.3 shows a schematic of a typical differential ion mobility spectrometer (DIMS) in

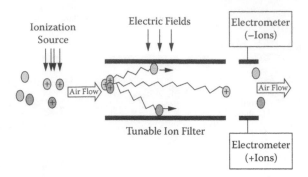

FIGURE 7.3 Differential ion mobility spectrometer (DIMS) showing both positive and negative Faraday plates used for detecting the ions that pass through the tunable ion filter. Because ions are detected continuously, no aperture is required. (From Sionex Corporation.)

which ions are transported continuously through the ion filter to the Faraday plate. Because the ions are not dispersed in time, no aperture grid is required to prevent band spreading from mirror currents as is required for Faraday plates attached to time-of-drift (drift time) IMSs.

7.2.2 Novel Ambient Pressure Ion Detectors

7.2.2.1 Capacitive–Transimpedance Amplifier

One recent approach to increased sensitivity in IMS instruments operating at atmospheric pressures has been the use of a microfaraday array coupled with capacitive–transimpedance amplifier (CTIA).[1] Figure 7.4 provides a picture of a

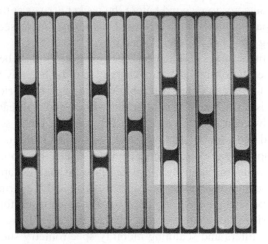

FIGURE 7.4 Visual image of individual detector elements. Pixels are 5, 2.45, and 1.60 mm long and 145 μm wide, with 10-μm spacing. There is also a 10-μm guard electrode present between each detector element. (From Babis et al., Performance evaluation of a miniature ion mobility spectrometer drift cell for application in hand-held explosives detection ion mobility spectrometers, *Anal. Bioanal. Chem.* 2009, 395, 411–419. With permission.)

microfaraday array used in this type of detector. Each of the plates in the microfaraday array is an independently addressable channel of a multiplexer, which selectively transfers the output of the CTIA to an analog-to-digital converter for recording. The primary advantage of the faraday array is that channels receiving a more intense beam of ions can be singly read and reset often, while channels receiving low-intensity beams can be integrated for longer periods. A single CTIA detector was capable of detecting 90 singly charged ions, and when multiple readouts were employed with ensemble averaging, the detection limit was reduced to 8 singly charged ions. When a microfaraday array was coupled to a traditional IMS instrument, detection limits for HMX were reduced to 6×10^{-15} g.[2]

7.2.2.2 Radiative Ion–Ion Neutralization

An alternative to charge detection devices for measuring ambient pressure IMS spectra uses the concept of radiative ion–ion recombination (RIIR).[3] Figure 7.5 shows a schematic diagram of an IMS with RIIN detection. The IMS cell consisted of 30 electrodes, 8 in the reaction region and 22 in the drift region, separated by a Bradbury Nielsen ion gate. The secondary electrospray ionization (SESI) needle was operated in the negative mode at a –3-kV bias above the IMS tube to produce analyte ions. The analytes were detected by a photomultiplier tube (PMT) through the light produced on neutralization with positive ions generated from a corona discharge ionization (CDI) needle biased at +3.25 kV and positioned at the end of the IMS cell. A second ion gate was positioned near the end of the IMS cell and was maintained in the "open" state. Vaporous samples were introduced into the reaction region through a heated sample tube, wherein heated gas flowed over an analyte of interest that had been deposited onto glass wool.

Figure 7.6 depicts the first ion mobility spectrum recorded in which charge detection devices were not used. In this spectrum, negative reactant and product ions were combined with cations to produce light measured with a PMT. PMTs are sealed with a glass window, where the photons enter the tube, so that the inside of the tube can be evacuated. Thus, PMTs can be used under atmospheric pressure conditions. The potential for detecting light from radiative ion–ion neutralization is that of ion counting at atmospheric pressure. In addition, it may be possible to achieve

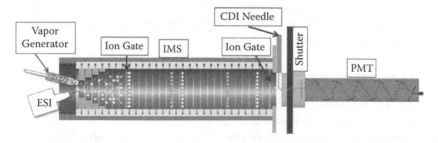

FIGURE 7.5 Schematic of IMS-RIIN instrument. Light emitted from ion-ion neutralization is detected as the response. (From Davis et al., Radiative ion-ion recombination: a new gasphase atmospheric pressure ion detector mechanism, *Anal. Chem.* 2012, May 11 [Epub ahead of print]. With permission.)

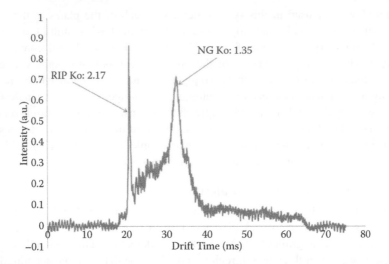

FIGURE 7.6 The first ion mobility spectrum generated from the light emitted from ion-ion recombination. The two peaks are a sharp reactant ion peak with a K_o of 2.17 cm²/V*s and a product ion peak of nitroglycerin with a K_o of 1.35 cm²/V*s. (From Davis et al., Radiative ion-ion recombination: a new gas-phase atmospheric pressure ion detector mechanism, *Anal. Chem.* 2012, May 11 [Epub ahead of print]. With permission.)

species-dependent wavelengths of light produced, providing qualitative information on the analyte ions.

7.2.2.3 Cloud Chamber Detector

In a cloud chamber detector, ions at atmospheric pressure are electrically focused into a cloud chamber filled with cold water or octane vapors. The presence of the ion serves as the nucleus for the formation of small droplets that can scatter light from a laser beam passing through the cloud chamber. When mobility-separated ions entered the cloud chamber, perturbation in the laser light due to the formation of ion-nucleated particles was detected by a PMT. When the chamber was supersaturated with water, the scattered light intensity increased in the presence of ions, but when the chamber was supersaturated with octane, the intensity decreased in the presence of ions. Mobility spectra of difluorodibromomethane have been reported using cloud chamber detection.[4]

7.3 LOW-PRESSURE DETECTION OF MOBILITY-SEPARATED IONS

Low-pressure ion detectors for IMS are typical of those used in mass spectrometry. In fact, whenever low-pressure ion detection methods are used with IMS, the IMS is interfaced to a mass spectrometer. Mass spectrometers after IMSs serve as mass filters or mass dispersive devices prior to low-pressure ion detection. Mass discrimination is not necessary, however, as in the case of a quadrupole mass spectrometer operating in the radio frequency (RF)–only regime, where complete ion mobility spectra can be obtained. A more detailed discussion of the advantages and disadvantages of ion mobility coupled with mass spectrometry can be found in Chapter 9. The discussion in this chapter, however, focuses on describing ion detectors commonly used under low-pressure conditions traditionally used for mass spectrometry.

The Faraday cup described in Figure 7.1 was the earliest mass spectral detector in which ion detection was accomplished by direct charge (current) measurement. The Faraday cup is a fixed detector in which mass spectrometers must be scanned to focus ions into the cup. Because mass spectrometric ion beams can be as low as a few fA (1 fA = 6,242 ions/s), 10^9 to 10^{13} electronic amplification is required. The high input impedance with large feedback resistance required for Faraday cup amplification produces a slow, stable signal but with high electronic noise. Limited by noise and speed, Faraday cup detectors are relatively insensitive and too slow for application to scanning or time-dispersive mass spectrometry.

Secondary electron multiplier (SEM) detectors replaced Faraday cup detectors for scanning mass spectrometers. The electron multiplier is based on the concept of a photomultiplier except that there is no glass membrane, so ions (electrons) can enter the amplification region of the detector. Because electron multipliers are not sealed and are open to the atmosphere, they must be operated under vacuum conditions and therefore cannot be used directly in atmospheric pressure IMS.

The first stage of a SEM detector is called a conversion dynode. Here, either positive or negative ions are focused onto a surface coated with a material having a low work function so that when the ion impinges on the surface, electrons are emitted from the surface into the vacuum of the surrounding area. Electrons produced from the conversion dynode are called secondary electrons and are accelerated in an electric field to the first dynode of the electron multiplier. These secondary electrons gain energy from the electric field, and when they impinge on the first dynode sputter multiple additional electrons from the dynode's surface. This amplification process of electron acceleration, dynode impact, and multiple electrons sputtering from the dynode's surface per electron impacted is repeated through a series of discrete dynodes to generate an overall amplification greater than 10^6 electrons/ion. A schematic diagram of a discrete-dynode SEM is shown in Figure 7.7, pictorially demonstrating the electron multiplication process.

A more efficient SEM can be constructed with an electron-emissive dynode made from a resistive film on an insulating surface. With this design, a bias voltage is placed on the entrance to the tube, setting up a voltage gradient on the surface of the dynode. Electrons are multiplied through many random collisions on the surface as they cascade through the electric field generated by the dynode surface. SEMs with this type of dynode are called continuous-dynode electron multipliers (CDEMs).

A multichannel plate (MCP) is a type of CDEM in which a series of microchannels on a disk-shaped device are coated with an electron-emissive material to generate 10^2 to 10^4 amplification as the electrons cascade through the microchannels. MCPs can be stacked to increase amplification or focused onto a fluorescent surface for ion-beam imaging. Because of the short electron pulse widths (~1 ns) obtained with MCPs, they are the ion detector of choice for time-of-flight mass spectrometers.

Whether an electron multiplier has a discrete-dynode or continuous-dynode design, it can be operated in both analog and pulse-counting modes. For analog detectors,

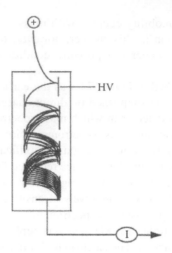

FIGURE 7.7   Discrete-dynode electron multiplier. (From Babis et al., Performance evaluation of a miniature ion mobility spectrometer drift cell for application in hand-held explosives detection ion mobility spectrometers, *Anal. Bioanal. Chem.* 2009, 395, 411–419. With permission.)

this secondary ion current is converted to voltage through a current-to-voltage converting amplifier and then digitized for data analysis. Alternatively, each approximately 10^6 electrons/ion pulse is counted as a discrete event. Noise events typically produce less-intense electron pulses ($<10^5$ electrons/ion) and thus can be discriminated against. While the pulse-counting mode is more sensitive than the analog mode, the analog mode has a wider dynamic range. Thus, both methods are in common use, pulse counting when the signal is weak and analog when the signal is strong. Both discrete-dynode- and continuous-dynode-type electron multipliers are used for the detection of mobility-separated ions when IMSs are interfaced to mass spectrometers.

7.4   SUMMARY

Overall, ion detectors for IMS are quite simple. For ambient pressure devices, Faraday plate detectors are used. While quantitatively efficient, their limit of detection is determined by Johnson (thermal) noise from the connecting electronics of the device. For time-dispersive devices, the band broadening must be minimized by the use of an aperture grid coupled to the Faraday plate. For continuous IMS devices such as those for DMS, FAIMS, aIMS, and DMA, aperture grids are not required. When mass spectrometers are used after IMS, low-pressure ion detection devices are used. These detectors amplify ions though high-energy collisions with surfaces of low work functions to produce a cascade of ions, amplifying response. Two of these ion detectors most commonly used today in mass spectrometry are the continuous-dynode electron multipliers and the MCP. Both of these detectors can measure ion current as an analog signal for a large dynamic response range, or the detectors can be operated in an ion-counting mode to detect very low signals.

REFERENCES

1. Babis, J.S.; Sperline, R.P.; Knight, A.K.; Jones, D.A.; Gresham, C.A.; Denton, M.B., Performance evaluation of a miniature ion mobility spectrometer drift cell for application in hand-held explosives detection ion mobility spectrometers, *Anal. Bioanal. Chem.* 2009, 395, 411–419.
2. Denson, S.; Denton, B.; Sperline, R.; Rodacy, P.; Gresham, C., Ion mobility spectrometry utilizing micro-faraday fringe array detector technology, *IJIMS* 2002, 5(3), 100–103.
3. Davis, E.J.; Seims, W.F.; Hill, H.H., Jr., Radiative ion-ion recombination: a new gas-phase atmospheric pressure ion detector mechanism, *Anal. Chem.* 2012, May 11 [Epub ahead of print].
4. Kendler, S., Ion detection in IMS devices: a new concept, in Israel Institute for Biological Research Conference, Ness-Ziona, Israel, 2010.

8 The Ion Mobility Spectrum

8.1 INTRODUCTION

The ion mobility spectrum has many forms that share one common feature: The ion current intensity is measured as a function of an ion's mobility in a gas. As with other types of spectrometry, the ion mobility spectrum is obtained by correlating a change in a spectrometer's parameter with a physical property of the ions. In light spectrometry, the number of photons is recorded as a function of photon energy; in mass spectrometry, the number of ions is recorded as a function of mass, and in ion mobility spectrometry (IMS), the number of ions is recorded as a function of an ion's collision cross section, which is related to its mobility. The type of IMS depends on the instrumental parameter that is scanned to produce the intensity versus mobility spectrum. To understand the many types of mobility spectra, we must first consider the relation among mobility, electric field, and pressure.

8.2 MOBILITY, ELECTRIC FIELD, AND PRESSURE

The mobility of an ion swarm is defined as the ratio of the swarm's average velocity to the electric field through which it migrates:

$$K = \frac{v_d}{E} \qquad (8.1)$$

where K is the ion mobility coefficient (or ion mobility constant), v_d is the ion swarm's velocity, and E is the electric field through which the ion swarm migrates. The mobility coefficient is traditionally reported in units of square centimeters per volt per second ($cm^2V^{-1}s^{-1}$).

In addition to the electric field, temperature and pressure affect the velocity of an ion swarm. At constant pressure, an ion swarm's velocity will increase with increasing temperature, and at constant temperature, an ion swarm's velocity will decrease with increasing pressure. To correct for the variation in an ion swarm's velocity as a function of temperature T and pressure P, mobilities are commonly reported as reduced mobilities K_0 at standard number density ($N_0 = 2.687 \times 10^{19}$ cm^{-3}) using standard temperatures ($T_0 = 273.15$ K) and pressures ($P_0 = 760$ torr) according to the following relation:

$$K_0 = K \frac{N}{N_0} = K \frac{P}{P_0} \frac{T_0}{T} \qquad (8.2)$$

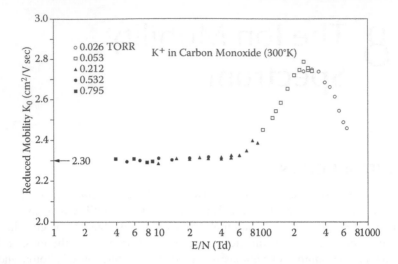

FIGURE 8.1 Dependence of K_o on E/N. (From Thomson et al., Mobility, diffusion, and clustering of potassium(+) in gases, *J. Chem. Phys.* 1973, 58, 2402–2411. With permission.)

The reduced mobility of an ion swarm, however, varies as a function of the electric field and buffer gas pressure through which it migrates in a complex manner. Figure 8.1 shows the dependence of the mobility of an ion swarm as a function of E/N where K_o is the reduced mobility of an ion swarm, E is the electric field, and N is the number density of the buffer gas.[1] When E is given in volts and N in particles/cubic centimeter, the ratio is reported in units of townsends (Td). One townsend is equal to 1×10^{-17} V cm^2 and calculated by the following relation:

$$Td = \frac{E}{N} * 10^{17} V \text{ cm}^2 \tag{8.3}$$

For example, at 20°C and atmospheric pressure, $N = 2.5 \times 10^{-19}$ cm^{-3}. Thus, for a townsend of 1 the electric field would be 250 V/cm.

In Figure 8.1, the reduced mobility is plotted for potassium ions in a buffer gas of carbon monoxide from Td = 1 to Td = 1,000.[1] Note that the reduced mobility is constant up to a Td of about 50. The reduced mobility increases from Td = 50 to Td = ~300. For values higher than Td = 300, the reduced mobility decreases as a function of increasing Td. While the absolute values of K_o vary as a function of the ion and the buffer gas, the shape of the mobility-versus-Townsend curve as shown in Figure 8.1 is similar for all ions and buffer gases. Thus, there are three distinct regions: (1) a low-field region where reduced mobility is constant; (2) an intermediate-field region where the reduced mobility increases with increasing Td; and (3) a high-field region where mobility decreases with increasing Td. In the low-field region, the energy that an ion receives from the electric field is much less than that it receives for the thermal energy of the buffer gas. In the intermediate field region, the energy that an ion receives from the thermal energy of the buffer gas is similar to that received from the electric field; in the high-field region, the energy an ion receives from the electric field is much greater than that received from the thermal energy of the buffer gas. Note that while mobility is constant, increasing or decreasing as a function of increasing Td, the velocity of the ion swarm always increases with increasing electric field. In no case would the velocity of an ion decrease if the electric field was increased at constant pressure. The region in which the mobility is constant as a function of Td is that region in which drift tube ion mobility spectrometry (DTIMS), aspiration ion mobility spectrometers (aIMS), and differential mobility analysis (DMA) are normally operated, while the region where mobility changes as a function of Td is where the ion mobility techniques called differential mobility spectrometry (DMS) and field asymmetric ion mobility spectrometry (FAIMS) occur. As a side note, the condition where Td << 1 is that region where liquid phase ion mobility and electrophoresis occur, while the condition where Td >> 1,000, is that region where mass spectrometry occurs.

8.3   ION MOBILITY SPECTRA

The separation of ions by gas phase mobility has been accomplished using three different approaches: temporal, spatial, and focusing. Temporal separation records arrival times for a pulse of ions traveling axially through an electric field; spatial methods track unique transit paths of ions as they travel orthogonally to the electric field lines; and focusing methods vary the electric field to bring different ions into focus on a Faraday plate or orifice into a mass spectrometer. Temporal separations are typical of those used by DTIMSs and traveling wave ion mobility spectrometers (TW-IMSs); spatial separations are achieved by aIMS, while focusing separation is achieved using DMA, DMS, and FAIMS. The figures discussed next provide examples of ion mobility spectra obtained from each type of ion mobility separation.

Figure 8.2 shows a mobility spectrum in which the ion current is plotted as a function of arrival time obtained from an IMS DTIMS used for explosive detection.[2] In this negative-polarity spectrum, the intensity of the current at the Faraday plate is plotted as a function of the time after the ion shutter is opened to introduce an ion swarm into the drift region of the IMS. When an ion swarm arrives at the electrode, an increase in current produces a peak representing the arrival time of the swarm. The arrival time axis is normally recorded in milliseconds, and the measured mobility is determined by the relation

$$K = \frac{L}{Et_d} \frac{cm^2}{Vs} \qquad (8.4)$$

where L is the length of the drift section of the tube in centimeters, E is the strength of the electric field in volts per centimeter, and t_d is the arrival time of the ion swarm in seconds.

Three ion peaks are seen in this mobility spectrum. The first, at a drift time of about 6 ms, is called the reactant ion peak (RIP) and is present in the spectrum even when a sample is not. The next two ion peaks are ion peaks of 4,6-dinitro-o-cresol, a common contaminant from smoke, with a reduced mobility of 1.59 V cm^{-1} s^{-1} and 2,4,6-trinitrotoluene (TNT) with a reduced mobility of 1.54 V cm^{-1} s^{-1}. For

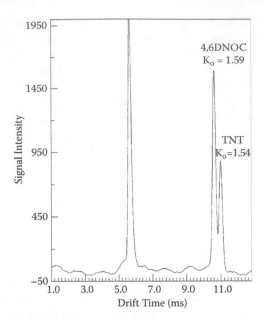

FIGURE 8.2 Drift tube ion mobility spectrum of 2,4,6-trinitrotoluene (TNT) and 4,6-dinitro-o-cresol (4,6DNOC). (From Wu et al., Construction and characterization of a high-flow, high-resolution ion mobility spectrometer for detection of explosives after personnel portal sampling, *Talanta* 2002, 57, 123–134. With permission.)

detection of TNT, a window is set to monitor the expected arrival time. When current is observed in that time window, an alarm for TNT is sounded.

Ion mobility spectra for the TW-IMS look similar, but because the electric field is applied as successive waves, sweeping the ion toward the detector, mobilities and collision cross sections cannot be calculated directly from TW-IMS but must be calibrated to DTIMS instruments.

Figure 8.3 shows a readout from an aIMS.[3] Because the spectrometer contains a series of discrete electrodes, the spectrum is represented as a histogram of response at each electrode. Resolving power is low with aIMS; thus, the pattern of the histogram is used to detect the presence of target compounds rather than discrete peaks.

On the other hand, the DMA instrument, which operates similarly to the aIMS, contains only one collector electrode. Ions are focused onto the Faraday plate as a function of scanning the orthogonal voltage through which the ions traverse. Spectra are plotted as the ion current on the Faraday plate as a function of $1/K$, where K is the mobility of the ion. In the DMA literature, Z is used to denote ion mobility instead of K. As with the aIMS, resolution power is low; however, monomers can be separated from dimers. For example, Figure 8.4 shows a typical ion mobility spectrum of a 9.2-kDa polystyrene resin obtained from a DMA.[4]

Spectra from DMS and FAIMS (both of these methods are collectively referred to as differential ion mobility spectrometry, DIMS) are similar to one another, although the magnitude of the parameters may differ. Figure 8.5 shows a typical FAIMS

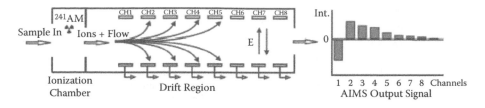

FIGURE 8.3 Ion mobility spectrum from an aspiration-type ion mobility spectrometer. Spectrum is an array of Faraday plates producing a histogram-type display.[3]

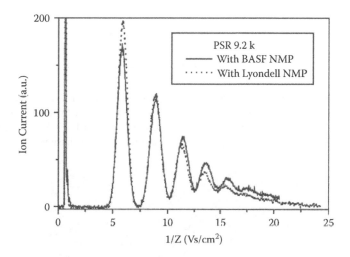

FIGURE 8.4 DMA spectrum of 9.2-kDa polystyrene. The sharp peak to the left is from small background ions due to the electrospray process. The next peak to the right beginning at a 1/Z value of about 5 is the monomer, followed by the dimer, trimer, and so on.[4] The resolving power of this spectrum is approximately 2. (From Ku et al., Mass distribution measurement of water-insoluble polymers by charge-reduced electrospray mobility analysis, *Anal. Chem.* 2004, 76, 814–822. With permission.)

spectrum in which the intensity of the ion current impinging on the detector is plotted as a function of the compensation voltage C_v as the C_v is scanned from −30 to 0 V. The detector used for this analysis was an LCQ XL linear ion trap in which the FAIMS response was for a mass-selected ion at 477 Da. This FAIMS extracted ion trace shows the separation of isobars. A monomer of polyethylene glycol (PEG) was clearly separated by FAIMS from the target isobaric compound loperamide. These two isobars would not have been separated by mass spectrometry alone.[5] FAIMS and DMS spectra can also be detected with a simple Faraday plate detector. Similar to DTIMS, a widow can be set for the detection of a target compound to produce a small and versatile IMS instrument for monitoring and detection.

When the compensation voltage is varied as a function of the dispersion voltage, a two-dimensional (2D) DIMS spectrum can be achieved that can more completely characterize an ion mixture.[6] Figure 8.6 is a CV-DV scan for DIMS at an elevated

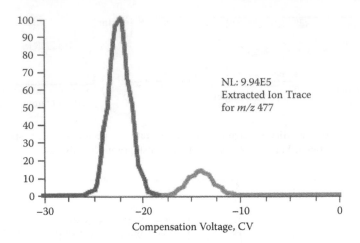

FIGURE 8.5 FAIMS spectrum of loperamide in the presence of polyethylene glycol. The resolving power of this spectrum is about 4. (From Horner and Phillips, The use of FAIMS to separate loperamide from PEG prior to ms analysis using an LTQ XL, Thermo Fisher Scientific Application Notes 2008, Application Note 395. With permission.)

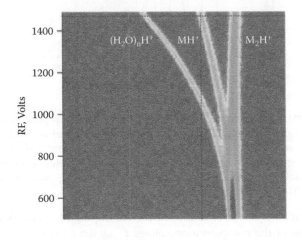

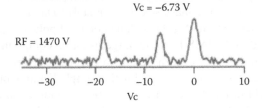

FIGURE 8.6 Dispersion/compensation spectrum for DMMP at 1.55 atm. The spectrum shown below is the compensation voltage spectrum at a dispersion voltage of RF plus 1,470 peak voltage. (From Nazarov et al., Pressure effects in differential mobility spectrometry, *Anal. Chem.* 2006, 78, 7697–7706. With permission.)

pressure of 1.55 atm. This graph provides a full spectrum of information about the behavior of ions in DIMS instruments. The peak value for the radio-frequency (RF) voltage (called the dispersion voltage) is given on the y-axis of the two-dimensional plot and is varied from 700 to 1,500 V. There are three ions present in this mixture: the reactant ion, which is the hydronium ion $[(H_2O)_nH^+]$; the monomer ion, MH^+ of dimethyl methyl phosphonate (DMMP); and the dimer ion, M_2H^+ of DMMP. When the dispersion voltage is low, say 800 V, it is clear from the 2D ion mobility spectrum that the three ions are not separated with the low dispersion voltage and appear at the same compensation voltage. As the dispersion voltage increases, the dispersion of the three ions increases, such that the lighter ion, the reactant ion, is moved to more negative compensation voltage, while the monomer of DMMP is also moved to more negative C_v voltages, while the larger dimer ion appears unaffected by the dispersion voltage. While dispersion increases with increasing dispersion voltage, sensitivity decreases. So, as with most analytical methods, you cannot maximize both resolution and sensitivity but must optimize the voltages to achieve optimal resolution and sensitivity at a level required for the analysis. When the dispersion voltage was held constant at 1,470 V, baseline resolution of all three ions was achieved.

8.4   IMS AS A SEPARATION DEVICE

The value of IMS as an analytical method rests not only with its high sensitivity, but also with the ability to obtain qualitative information from mobility. In one way, IMS can be thought of as a separation technique similar to chromatography or electrophoresis. In fact, in the early days of IMS, the technique was called plasma chromatography because of the selective interaction the ions have with the neutral, near-stationary phase of the buffer gas, providing both separation of and qualitative information about the sample. Thus, this section of the chapter uses common chromatographic terms to describe ion mobility spectra. Describing IMS in chromatographic terms provides a method by which IMS can be compared with other analytical instruments.[7]

As with chromatography, the position of the peak in IMS provides qualitative information. The location of the ion swarm as it exits a drift region is dependent on the type of instrumentation used: For drift tube instruments, it is the arrival time of the ion swarm at the Faraday plate or mass spectrometer orifice; for DMS, it is the compensation voltage required to create a stable path through the instrument, and for aspiration-type instruments, it is the location of the Faraday plates as a function of the strength of the electric field. All of these qualitative measurements can be related to the mobility of the ion swarm, although in some cases this relation is complex and not well understood. Nevertheless, the relationships of K, K_0, and Ω to ion mobility spectra have been described elsewhere in this book and serve as the qualitative basis of IMS. Until the fundamental relation of ion–molecule interactions can be understood sufficiently to model ion behavior in IMS instruments, IMS standards will serve to calibrate the various IMS platforms.

As a separation device, IMS depends on both the position and the width of the ion peak. The position of the peak is determined by the ion's mobility, while the width of the peak is a complex function of the introduction method, diffusion, homogeneity of the electric field, and the engineering of the mobility cell. The "goodness" of an ion mobility spectrum can be discussed in terms of three analytical figures of merit: resolving power, separation selectivity, and resolution.

8.4.1 Resolving Power

The ability of an analytical instrument to identify a component of a complex mixture rests to a large degree with its resolving power, that is, the efficiency of the method in separating a number of components into unique compartments of the analytical space. For DTIMS, resolving power R_p has been defined as the ratio of the arrival time t_d to the full width at half maximum w_h of the arriving ion swarm.

$$R_p = \frac{t_d}{w_h} \qquad (8.5)$$

For the ion mobility spectrum shown in Figure 8.7, the ion produced from methamphetamine had an arrival time of 24.94 ms and a peak width at half height of 0.154 ms.[8] Thus, the resolving power for this DTIMS was 24.94/0.154 = 161.

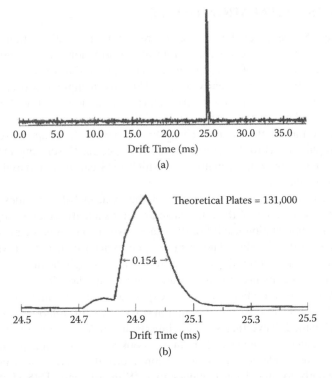

FIGURE 8.7 Measuring resolving power in IMS. (a) The mass-selected ion mobility spectrum of methamphetamine obtained with an atmospheric drift tube IMS interfaced to a quadrupole mass spectrometer.[8] (b) Expanded scale of spectrum A. (From Wu et al., Electrospray ionization high-resolution ion mobility spectrometry-mass spectrometry, *Anal. Chem.* 1998, 70(23), 4929–4938. With permission.)

Resolving powers are related to chromatographic efficiencies in terms of theoretical plates by the simple relation

$$n = 5.55R_p^2 \tag{8.6}$$

where n is the number of chromatographic theoretical plates, and R_p is the IMS resolving power. For the case shown in Figure 8.7, an IMS resolving power of 161 is equivalent to 144,000 theoretical plates. It should be noted here that IMS is not a chromatographic technique, and the mechanisms of separation are quite different. Nevertheless, it is convenient to have a direct method to compare the efficiencies of these separation techniques.

The previous discussion refers only to the low-field drift tube type of IMS. For other types of IMS, the concept is the same, but the specific measurements may differ. For the TW-IMS, the definition of R_p is the same as for the drift tube IMS, t_d/w_h. However, for FAIMS and DMS, it is $R_p = CV/\Delta CV_h$, where CV is the compensation voltage at the peak's maximum height, and ΔCV_h is the full width of the compensation voltage at half the height of the peak at its maximum. For DMAs, the resolving power is defined as $V/\Delta V_h$. In general, resolving powers of commercial IMSs range from a low of about 2 for aIMS to a high of about 80 for the linear IMS. Typically, commercial IMS drift tubes have a resolving power from 20 to 40, that is, the number of theoretical plates is 2,220–8,880.

8.4.2 Separation Selectivity

Separation selectivity is defined as the relative positions of two ion swarms in the mobility spectrum as defined by the following relation:

$$\alpha = \frac{X_2}{X_1} \tag{8.7}$$

where X_1 is the position of swarm 1 on the mobility spectrum, and X_2 is the position of swarm 2 so that α is always equal to or greater than 1.00. When the resolving power of an IMS is too low to generate a desired resolution of two ions, it may be possible to change α to resolve the ions. For example, the mobility spectrum in Figure 8.5 has a very low resolving power of around 2, but the α value is sufficiently high so that both peaks, the ions for loperamide and polyethylene glycol, were baseline resolved. For this case, $\alpha = C_{v2}/C_{v1} \sim -22.5 \text{ V}/14 \text{ V} = 1.60$. With IMS systems of high resolving powers, ions with much smaller separation selectivity values can be resolved.

Separation selectivity in IMS can be affected in a number of ways. The most direct approach is to modify or change the buffer gas (see also Chapter 11). For example, carbon dioxide as a buffer gas, due to its higher polarizability, can separate ions that cannot be separated with nitrogen alone. Four common buffer gases used in IMS are helium, argon, nitrogen (air), and carbon dioxide, with respective polarizabilities of 0.205, 1.641, 1.740, and 2.911 debye. One separation that has demonstrated the effects of buffer gases on resolution is that of chloroaniline from iodoaniline.[9] In helium, for which the size of the ion is dominant, chloroaniline, the smaller ion with a high charge density, has the higher mobility; in carbon dioxide, for which polarization is important, iodoaniline, the large ion with a lower charge density, has the higher mobility. Thus, in helium chloroaniline can be separated from iodoaniline, with chloroaniline eluted first, while in carbon dioxide, chloroaniline can also be separated from iodoaniline, but the iodoaniline is eluted first.

Other examples of buffer gas selectivity include the resolution of arginine from phenylalanine in carbon dioxide but not in nitrogen;[10] the separation of lorazepam from diazepam in both helium and carbon dioxide but not in argon and nitrogen;[11] the resolution of carbohydrate isomers such as methyl-β-D-mannopyranoside from methyl-α-D-mannopyranoside in carbon dioxide but not nitrogen;[12] and the resolution of heroin from THC (tetrahydrocannabinol) using nitrous oxide as the buffer gas in a commercial IMS instrument.[13]

Modification of the buffer gas either by mixing buffer gases or by the addition of trace amounts of a modifier can also influence IMS separation selectivity. The most common mixed buffer gas used in IMS is air, although other mixtures may provide altered α values. If IMS separation cannot be accomplished by mixing or changing the buffer gas, the addition of modifiers to the buffer gas may provide enough selective ion–molecule interactions to effect resolution. For example, Figure 8.8 shows the separation of valinol from serine when 2-butanol was added to the nitrogen buffer gas.[14] The addition of chiral modifiers has led to the gas phase ion mobility separations of enantiomers such as D- and L-methionine, S- and R-atenolol, L- and D-tryptophan, and L- and D-methyl glucopyranoside.[15]

Separation selectivity can also be altered by the type of ion that is formed for the compound of interest. Cation or anion adducts can have a significant effect on the separation. In the positive ion mode, most ions formed from electrospray or any chemical ionization methods are protonated cations. However, if Na^+ is added to the electrospray solution, sodiated adducts may be formed as the primary response ion. For example, the protonated ions of two isomers, methyl-β-glucopyranoside and methyl-α-glucopyranoside, cannot be baseline separated in DTIMS, but the sodiated adducts of the same isomers can be baseline separated.[16] Figure 8.9 demonstrates separation selectivity induced by cation adduction. These IMS spectra are of the methyl-β-D-galactopyranoside and the methyl-α-D-galactopyranoside isomers adducted with cobalt acetate, silver, and lead acetate. The cobalt acetate adducts had a separation factor of 1.02, the silver adducts had a separation factor of 1.05, and the lead acetate adducts had a separation factor of 1.07.

8.4.3 Resolution

The resolution of an ion mobility separation depends on both the resolving power R_p and the separation selectivity α of the IMS system. Resolution R determines how well two peaks, with drift times of t_1 and t_2 and peak widths of w_1 and w_2, can be separated from one another and is defined as

$$R = \frac{2(t_2 - t_1)}{w_2 + w_1} = \frac{R_p}{1.74}\left(\frac{\alpha - 1}{\alpha}\right) \tag{8.8}$$

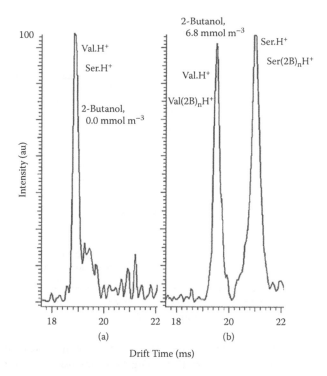

FIGURE 8.8 Separation of a mixture of valinol and serine by introducing 2-butanol into the buffer gas. (a) IMS spectra of the mixture in nitrogen only showing one overlapping peak for both compounds at 18.9 ms. (b) Resolution of the mixture by intruding 1.7 mmol m⁻³ of 2-butanol modifier into the buffer gas. (From Fernandez-Maestre et al., Using a buffer gas modifier to change separation selectivity in ion mobility spectrometry, *Int. J. Mass Spectrom.* 2010. With permission.)

Using this relation, a resolution of 1.00 provides a separation with a 2% overlap between the two ions. If we know the resolving power of the spectrometer, then we can determine the relative mobilities that can be separated. Table 8.1 provides a comparison of selectivity, resolving power, and resolution. For example, if the resolving power of a particular instrument is only 10, then the relative mobility values have to be greater than 1.2 for the IMS instrument to achieve a resolution of 1 between the two ions. In the example of Figure 8.5, the relative separation of the two components was 1.60; thus, a resolving power of 2 was sufficient to produce the baseline resolution of the two components. On the other hand, for complex sample mixtures such as environmental or biological samples, many ions may have similar mobilities. For relative mobility differences of 1% ($\alpha = 1.01$), resolving powers of greater than 174 are required to separate these ions with a 2% overlap.

8.5 QUANTITATIVE ASPECTS TO RESPONSE

IMSs perform successfully in several applications where only threshold response is necessary or important, and examples include the detection of explosives and

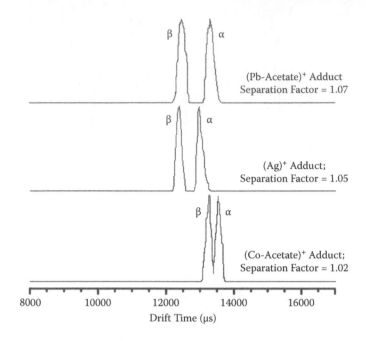

FIGURE 8.9 Selectivity induced by cation adduction. Ion mobility separation of methyl-β-D-galactopyranoside from its isomer methyl-α-D-galactopyranoside using different cation adducts. (From Dwivedi et al., Rapid resolution of carbohydrate isomers by electrospray ionization ambient pressure ion mobility spectrometry-time-of-flight mass spectrometry (ESI-APIMS-TOFMS), *J. Am. Soc. Mass Spectrom.* 2007, 18, 1163–1175. With permission.)

TABLE 8.1
Comparison of IMS Resolution, Selectivity, and Resolving Power

Resolution	Selectivity	Required Resolving Power
1.00	1.20	>10
1.00	1.10	>19
1.00	1.05	>36

screening suspect objects for drugs of abuse. In these applications, the detection of even trace amounts of the target analyte will trigger an alarm, and no further quantitative information is necessary. In other instances, knowledge of both the presence of a substance and the quantitative amounts are required. Examples of such a quantitative use of an IMS instrument include the determination of chemical warfare agents and the screening of air for volatile organic compounds; in both of these examples, human mortality or vitality is associated with dose–response relationships, and the level of dose is essential information. In the past decade, improvements in threshold applications of IMS have arisen from better sampling methods, more sensitive and rugged instrumentation, and better data-processing techniques.

These are seen with advances in the detection of explosives with handheld analyzers as well as with fixed or stationary analyzers. Parallel advances have occurred with instrumentation and procedures requiring quantitative determinations. These have occurred with the availability of improved drift tubes that exhibit fast response, low memory effects, and reproducible delivery of sample to the ion source. Together, these have resulted in improved precision, sensitivity, and linear range compared to previous generations of analyzers. In the following sections, the principles underlying quantitative behavior for IMS analyzers are discussed.

8.5.1 EFFECT OF SAMPLE CONCENTRATION ON RESPONSE

The first step in IMS response, ionization of neutral analyte molecules, fundamentally establishes a starting point from all other events that occur subsequently in a drift tube and establishes the boundaries for quantitative response. Central to all discussion of quantitative aspects of IMS is the limited reservoir of charge available for ion formation. The amount of charge available for formation of product ions is practically limited by the strength of the ionization source, by the overall sequence of reactions leading to the formation of reactant ions, and finally by the reaction rate for the formation of product ions. Under typical operating conditions with a ^{63}Ni source, the upper limit for ion density in a ^{63}Ni source is 10^9 to 10^{10} ions/cm^3/s, and this establishes the maximum charge available for the formation of product ions. In practice, this value may be reduced due to ion losses with poor designs of electric fields and flow patterns. Furthermore, chemical losses in an actual measurement may arise through the presence of impurities, competing reaction pathways, and decomposition or fragmentation of the target product ion.

At low concentrations of the analyte, reactant ions are consumed proportionally to the density of the vapor neutrals per Equation 8.9:

$$\text{Rate} = k[M]\left[H^+(H_2O)_n\right] \tag{8.9}$$

The rate constant k in Equation 8.9 is roughly the collision rate constant, which controls the formation of product ions. As shown in Figure 8.10, the density of product ions, for a sample at a constant concentration, is dependent on both residence time of the sample in the source and its concentration. The depletion of reaction ions is naturally rooted in kinetics of Equation 8.9 and parallels inversely the formation of product ions. The relationship is virtually stoichiometric. For a fixed time of residence of sample molecules in the reaction region, the number of product ions and their peak intensity in the mobility spectrum will be quantitatively proportional to the concentration of sample molecules. Also, increases in residence time for [M] in the source region will lead to an increase in the number of product ions as long as the source is not saturated with sample vapor. Once the reservoir of reactant ion charge is totally consumed, no additional increase in product ion intensity will be seen even with additional increase in the concentration of sample molecules. In practice, this limits the upper end of the linear range associated with ion sources at ambient pressure.

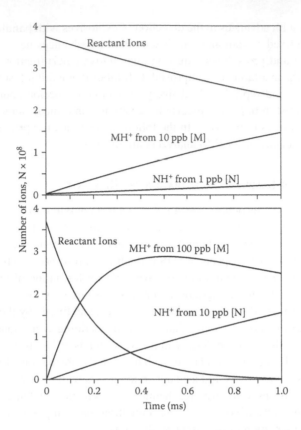

FIGURE 8.10   Plots of ion number with reaction time or time of mixing for ions and neutrals at ambient temperature and pressure. The curves show response to two analytes of comparable proton affinity. The analytes differed by 10-fold in concentration, illustrating the kinetic basis of forming product ions.

One of the best-recognized quantitative patterns in mobility spectra is the relationship between protonated monomers and proton-bound dimers, common to several types of compounds, including esters, ketones, alcohols, amines, and organophosphorus compounds. As the vapor concentration of the analyte M is increased in an ion source (Figure 8.11), a protonated monomer peak first appears with a corresponding decrease in the reactant ion peak intensity. A second peak appears with a further increase in analyte concentration and is the proton-bound dimer; as this occurs, intensity declines for peaks of both the reactant ion and the protonated monomer. When the analyte vapors are removed from the ion source and the analyte concentration decreases, these patterns are reversed: Intensity declines for the proton-bound dimer and increases for the protonated monomer peak. Eventually, the protonated monomer peak intensity declines, and the reactant ion peak is restored to the original intensity. The dynamics of these changes are shown in Figure 8.11. While this pattern is commonly associated with positive polarity IMS, similar patterns may be observed with negative ions such as $M·Cl^-$ and $M_2·Cl^-$, as observed with volatile halogenated anesthetics[17] and other chemicals that form long-lived negative ion clusters.

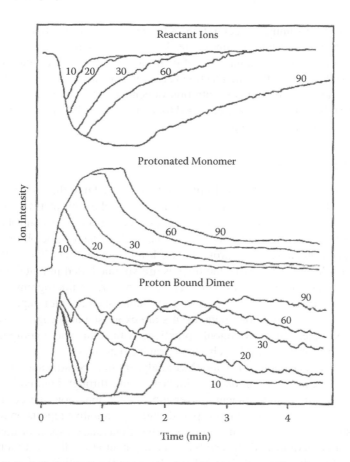

FIGURE 8.11 Plots of ion intensities from vapor sampling of a moving plume of a chemical vapor. The numbers are time in seconds used to generate the plume. These plots show the dynamics in an IMS analyzer.

When the reactant ion population is completely exhausted, the proportional relationship between product ion intensity and neutral vapor density no longer holds, and this negates conventional calibrations. Thus, further quantitative value in the response of the analyzer is lost with existing methods and interpretations. Another troubling phenomenon as concentrations of sample are increased in the source or reaction region is the diffusion of analyte molecules from the source region into the drift region. In the drift region, neutrals of sample may associate with product ions of the same chemical, leading to the formation of cluster ions of the kind $M_nH^+(H_2O)_x$, where n can be 3 or larger. Cluster ions in the ion swarm will rapidly undergo loss and then addition of neutral adducts while the swarm traverses the drift region,[18] and the observed reduced mobilities for such ions are weighted averages of the reduced mobilities of all the individual or cluster ions that participate in the equilibrium. A practical implication of this is that drift times of the product ions can vary across a large range, which is established by concentration of sample neutrals in the drift region. Since distribution of sample neutrals in the drift region will be irregular and dependent on sample concentration, attempts to assign identity to ions using K_o values become difficult or impossible. In summary, depletion of reactant ions should be avoided, and a residual level of peak intensity for the reactant ions should be maintained to ensure that the reaction region is not saturated, leading to quantitatively unclear behavior and concentration-dependent drift times for product ions. Most modern IMS drift tubes are designed to avoid this condition, but user carelessness can circumvent good designs.

8.5.2 Detection Limits

In practice, two definitions for the limit of detection (LOD) or the minimum detectable level (MDL) may be made when considering application of IMS technology. When carrying out continuous monitoring of ambient air or a controlled atmosphere, MDL is the lowest concentration of the analyte that can be observed in the mobility spectrum above a background noise level (signal/noise [S/N] > 3). Typical units for such MDLs for volatile compounds are volume based in units of parts per million, parts per billion, milligrams per cubic meter, or micrograms per cubic meter. When samples are provided as discrete items such as filter paper swipes or as solid-phase microextraction fibers, MDLs are expressed as picograms/sample or nanograms/sample. This is commonly practiced with semivolatile compounds and when IMS drift tubes are used as chromatographic detectors.

The excellent detection limits for IMS analyzers may be attributed to the high efficiency of chemical ionization at ambient pressure through high collision rates, long residence times, and efficient conversion of each ion–molecule collision into an activated intermediate. During the past decade, quantitative reports on MDLs for IMS measurements have become common. One elementary aspect of quantitative response with a traditional drift tube (i.e., linear potential gradient with ion shutter) is that only a small portion of the total ion charge in the reaction region is utilized. The ion shutter is open to passage of ions between the reaction and drift regions for only 0.1–0.3 ms in a 20-ms duty cycle; thus, only about 1% of all possible ions are sampled, and even fewer are detected. Increases in the strength of the ionization source, the width of the ion shutter, or the duty cycle of the shutter have limited effectiveness, such that ion–ion repulsions in the drift tube become noticeable at levels above about 10^5 ion/mm^3, and peak heights decline as the peak widths undergo broadening from coulombic repulsions.[19]

One of the earliest reports on MDLs in IMS was for dimethyl sulfoxide (DMSO); a peak for 10^{-10} moles (about 8 ng) of DMSO was seen using an IMS-MS with a gas chromatographic (GC) inlet.[20] Shortly thereafter, a GC-IMS-MS was used to detect 10^{-8} g of musk ambrette,[21] and an even better MDL of 10^{-10} g was reported for an isolated or stand-alone IMS. However, the method of sample delivery with such IMS configurations involved depositing a solution on a wire and inserting the wire into the heated IMS inlet, where the sample vapors were desorbed. This made exact weight determination of the sample somewhat problematic. In addition, the studies did not exploit fully the definition of detection limit[22] regarding the S/N. The successful combination of capillary columns and fast-responding drift tubes has allowed reliable determinations of low MDLs. Baim and Hill reported[23] an MDL of 6 pg for the methylester of 2,4-D (dichlorophenylacetic acid) and an MDL of 28 pg for di-n-hexyl ether[24] using 3 for the S/N value. Organophosphorus compounds with strong proton affinities exhibited detection limits of about 10 pg, based on the protonated monomer ion.[25] The earliest rigorous evaluation of IMS as a stand-alone instrument for MDLs was for the detection of *sec*-butylchloro-diphenyl oxides in biological tissues, where an MDL of about 20 pg was reported.[26]

Continuous sampling of vapors provided other instances for which IMS was deemed attractive due to inherent sensitivity to an analyte, as typified by nickel carbonyl in air, where reported detection limits were 0.2–0.3 ppb.[27] In another report of actual MDLs for continuous monitoring, Karpas reported detection limits of 10 ppb for bromine in the ambient air of a chemical plant.[28] Other detection limits for continuous monitoring included volatile halogenated anesthetics with MDLs from 0.1 to 1,000 ppb[18] and for nicotine with an MDL of 50 ppb.[29] Perhaps the area of greatest interest in detection limits for IMSs has been that of explosive detection (see also Chapter 12). Karasek first reported the detection of 10 ng of TNT with a S/N of greater than 1,000 from which an MDL in the low-picogram range was predicted by extrapolation.[30] Lawrence and Neudorfl showed that ethylene glycol dinitrate (EGDN) undergoes ionization with O_2^- reactant ions to EGDN*NO_3^-, which is an inefficient use of analyte vapor, but that ionization could be enhanced through the use of Cl^- as the reactant ion to form EGDN*Cl^-.[31] Detection limits for EGDN with and without chloride reactant ions were 500 and 30 pg, respectively. This was presumably due to the simplification of ionization routes with a single reactant ion, chloride, rather than the complicated mixtures often seen with air and perhaps to the formation of a long-lived chloride adduct. The ultimate detection limits for TNT and RDX in air were quoted as 3 ppt (v/v) in air or 0.3 ppt with signal averaging on a large-volume reactor.[32] These low MDLs were attained with a scrupulously clean drift tube in purified nitrogen supporting atmosphere and should not be expected for routine screening instrumentation. Throughout this discussion, the MDLs have been given on an absolute basis for specific analyzers and do not include any sample preparation or preenrichment. For example, the group at Sandia reported 0.03-ppt MDLs for explosives in air when a sampling tube was used to enrich vapors.[32]

Field instruments, unlike most laboratory instruments, operate at near-ambient temperatures and are equipped with membrane inlets to maintain clean atmospheres for internally recirculated gases. Sample vapors must pass through a membrane that exhibits selectivity governed by the molecular structure of the membrane, the presence of functional groups on the analyte, the temperature of the membrane, and the vapor pressure of the analyte. The membrane efficiency will rarely be 100%; consequently, MDLs with membrane-equipped instruments will be poorer than those where sample is deposited directly into the reaction region. An MDL of about 10 ppb was reported for hydrazine and monomethylhydrazine (rocket propellants) in air using handheld IMSs and continuous sampling of a vapor stream.[33] When such handheld, membrane-equipped analyzers received vapors in short bursts (through a flow injection sample inlet), MDLs were expressed as 2 ng for aniline[34] and 5 ng for dialkyl phthalates.[35] The membrane degrades detection limits by a factor of 10 to 100, suggesting that the yields of molecules that permeate across the membrane barrier are between 1% and 10%.

8.5.3 Reproducibility, Stability, and Linear Range

The reproducibility of signal intensities and drift times had, before 1990, not been considered widely in reports or journal articles on IMS, possibly since detection limits were the main concern. Consequently, there is only a relatively brief record available in the literature on the repeatability of IMS measurements, which is a key to any quantitative analytical method. The few examples that are available are concerned with short-term repeatability, and the relative standard deviation (RSD) for peak areas in these is between 5% and 25%.[34] In one study with a handheld IMS analyzer, reproducibility was 6 to 27% RSD for 5 to 2,500 ng of dialkylphthalates, as shown in Table 8.2.[35] Measurements of hydrazine vapors at 10 to 200 ppb using the same instrument showed precision of about 3 to 16% RSD for these high-reactive and -adsorptive chemicals.[33]

A fourth example is for the determination of phenol through thermal desorption from filter paper. The repeatability was 42 to 0.7% RSD in the range of 10 to 10,000 ng, respectively; experimental results are shown in Table 8.3 and include errors from sample handling.[36] All of these studies occurred under carefully controlled laboratory conditions.

TABLE 8.2
Quantitative Response of a Mobility Spectrometer for Thermal Desorption of Phthalate Esters from Filter Paper

Peak Area Amount (µg)	Mean ($N \times 10^3$)	Standard Deviation (% RSD)	Peak Area Amount (µg)	Mean ($N \times 10^3$)	Standard Deviation (% RSD)
0.0050	2.01	1.49	0.005	2.36	2.27
	1.40	(27)		2.02	(8.0)
	1.05			2.42	
0.05	1.95	1.79	0.05	2.90	3.15
	1.71	(7.0)		3.28	(6.0)
	1.70			3.27	
0.25	4.61	3.90	0.25	5.92	5.49
	3.91			5.39	(6.0)
				5.16	
0.50	10.5	8.42	0.50	8.19	7.55
	6.86	(18)		6.25	(12)
	6.88			8.22	
1.0	8.42	13.4	1.0	12.5	11.0
	16.2	(26)		9.75	(10)
	15.5			10.8	
2.5	23.3	23.6	2.5	19.9	20.2
	24.1	(2.0)		22.5	(9.0)
	23.2			18.1	

Standard deviation includes the sample preparation and thermal desorption steps.[38]

TABLE 8.3

Precision of Five Replicate Determinations of Phenol by
Vapor Desorption-Ion Mobility Spectrometry

Phenol Amount (ng)	ΣPeak Area (area counts)	% Relative Standard Deviation
0	1,235	22.9
10	3,934	31.7
20	3,395	42.1
30	5,313	36.6
40	5,526	36.6
50	6,486	13.8
60	9,481	11.1
70	10,426	14.8
80	12,800	9.0
90	14,252	12.9
100	17,495	7.6
200	30,389	10.6
300	38,872	10.0
400	39,904	11.5
500	43,049	6.9
600	47,225	7.0
700	51,181	4.3
800	53,888	7.4
900	54,753	8.0
1,000	66,087	3.4
2,000	74,663	3.8
3,000	81,964	1.4
4,000	84,101	1.4
5,000	89,207	6.1
6,000	96,092	1.8
7,000	98,844	1.4
8,000	103,834	2.2
9,000	103,831	0.7
10,000	100,320	11.7

A measure of routine repeatability in IMS can be seen in GC-IMS measurements using the Volatile Organic Analyzer (VOA; see Chapter 5). The repeatability with the VOA for drift time (as $1/K_o$) was better than 0.1% RSD for most chemicals and at worst 0.4% RSD. Peak areas were repeatable, with an average of 10% RSD for three measurements for a mixture of 18 volatile organic compounds and a calibrant as shown in part in Table 8.4.[37] This compared favorably with a laboratory based GC-IMS, for which a selected compound in a mixture of organophosphorous compounds was determined at 88 to 1,390 pg, and repeatability was roughly 6 and 25% RSD for four to six measurements in each of two reagent gases, water and DMSO.[25]

TABLE 8.4
Quantitative Response of the Volatile Organic Analyzer for the International Space Station

| | Concentration (mg/m³) | | | | | | | |
	Target	1	2	3	Average	Std	% RSD	% Error
Methanol	170	200	194	204	199	5	2.5	17.3
2-Propanol	717	833	597	672	701	121	17.2	2.3
1-Butanol	702	442	421	393	419	25	5.9	40.4
Ethanal	202	223	216	205	215	9	4.2	6.3
m,p-Xylenes	1,793	1,625	1,943	1,839	1,802	162	9.0	0.5
o-Xylene	633	587	633	727	649	71	11.0	2.5
Toluene	414	425	561	440	475	75	15.7	14.8
2-Butanone	480	298	269	270	279	16	5.9	41.9
Ethyl acetate	279	199	315	236	250	59	23.7	10.4
Dichloromethane	177	197	191	191	193	3	1.8	9.0

The long-term stability of calibration in IMS is rarely documented, although the single open-source example, the VOA, suggested that calibration curves may be stable for months within repeatability. Otherwise, supporting numerical evidence is difficult to obtain. One creative solution to calibration of an IMS analyzer in a changing matrix was that of Dam for monitoring stack emissions.[38] In this IMS monitoring system, which had been installed and operated at a DuPont chemical production facility for over a decade without serious shortcomings in stability, an inlet flow system to the IMS was equipped to receive a flow of standard levels of the target chemical while monitoring for the target chemical. This constituted a type of flowing standard addition calibration and was able to compensate for quantitative complications from changes in the matrix. Military applications of IMS require that handheld units are designed to remain in sealed containers with shelf lives of a decade and when unwrapped are expected to power up to operating levels within minutes. One of the rare documented instances of stability over a period of weeks is the comparison of the calibration of a GC-IMS system to hydrazines before and after a flight of a handheld unit on flight STS-37 of the U.S. space shuttle. The repeatability of calibration of hydrazines over a 6-week interval was 5–10%, which was better than the measured RSD in laboratory exercises.[33]

Sensitivity is defined as the increase in response per unit concentration or as the slope of the plot of detector response versus concentration. Of interest here is the quantitative response for different compounds over a range of concentrations. Commonly, linear ranges of 10–100 have been reported for IMS, and working ranges can be near or larger than 1,000. A typical response curve for a mobility spectrometer is shown for phenol in Figure 8.12.[36] Just below the threshold of response (4 ng), only a very slight increase in the product ion peak area is observed. Between 40 and 100 ng, peak area increases linearly from 6 to 16 units (Figure 8.12a), that is, doubling in intensity when the concentration is doubled. As the amount of phenol is increased

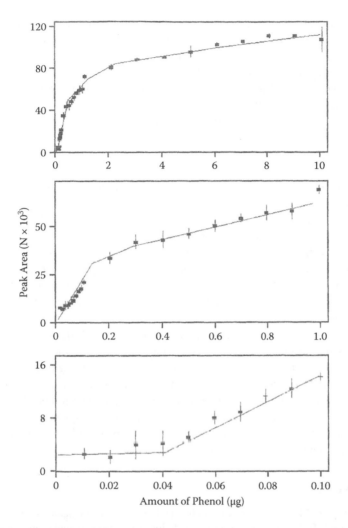

FIGURE 8.12 Quantitative response, including precision for phenol, with IMS determination of phenol on filter paper. Phenol was applied to the filter paper as a solution, dried, and thermally desorbed into the IMS drift tube.

to 200 ng, the slope of the response curve changes, and peak area goes from 30 to 65 units for 200 to 1,000 ng (Figure 8.12b), that is, doubling in intensity when the concentration is quadrupled. From 2,000 to 10,000 ng, the peak area increases from 70 to 105 units (i.e., a 500% increase in concentration resulted in a 50% increase in peak area). The amount of phenol cited here was placed on the filter paper; thus, the amounts of phenol in the gas phase should be substantially lower than these amounts owing to inefficiency in the thermal desorption step and efficiency of passage of phenol through the membrane inlet of the handheld analyzer. Nonetheless, the trends in these curves are more significant here than absolute mass calibrations. The plot in Figure 8.12c is what is commonly described as the calibration curve for

IMS and shows a linear range from 40 to 1,000 ng, a second linear region from 1,000 to 2,000 ng, and a third range above 2,000 ng in which the ion source is nearly saturated (and reactant ions depleted). Thus, the dynamic or working range is from 40 to 10,000 ng. Repeatability in the plots illustrates that discrimination near the threshold is possible, with 10–40% RSD indicating that a fine distinction can be made between 60 and 80 ng. In the middle range of response, the RSD was 5–10%, and this was reduced to about 2% at the upper range of response.

8.6 SUMMARY

In summary, ion mobility separations occur by a variety of methods. In all cases, there is an instrumental scan parameter that controls the separation of the ion. For example, in the drift tube spectrometers, it is the arrival time of the ion; for the aspiration spectrometers, it is the position of the faraday plates; for the mobility analyzers, it is the strength of the orthogonal voltage; and for the DMSs, it is the compensation voltage. The relative value of these scan parameters for two ions is called the separation factor α, and the resolving power of a spectrometer is determined by the ratio of the scan parameter to the width of the scan parameters for a packet of ions.

The resolution of two ion peaks in IMS depend on both the resolving power of the instrument and the separation selectivity of the buffer gas and ion interactions. Separation selectivity can be controlled by adjusting the drift gas polarizability, by the addition of modifiers to the drift gas, and by controlling the type of ions and ion adducts produced in the ionization process. Thus, the large number of instrumental parameters that exist for IMS makes the method complex as an analytical technique, but when parameters are optimized, IMS becomes a powerful and flexible analytical measurement device for a large number of analytes.

REFERENCES

1. Thomson, G.M.; Schummers, J.H.; James, D.R.; Graham, E.; Gatland, I.R.; Flannery, M.R.; McDaniel, E.W., Mobility, diffusion, and clustering of potassium(+) in gases, *J. Chem. Phys.* 1973, 58, 2402–2411.
2. Wu, C.; Steiner, W.E.; Tornatore, P.S.; Matz, L.M.; Siems, W.F.; Atkinson, D.A.; Hill, H.H., Jr., Construction and characterization of a high-flow, high-resolution ion mobility spectrometer for detection of explosives after personnel portal sampling, *Talanta* 2002, 57, 123–134.
3. Makinen, M.A.; Anttalainen, O.A.; Sillanpaa, M.E.T., Ion mobility spectrometry and its applications in detection of chemical warfare agents, *Anal. Chem.* 2010, 82, 9594–9600.
4. Ku, B.K.; Fernandez de la Mora, J.; Saucy, D.A.; Alexander, J.N., Mass distribution measurement of water-insoluble polymers by charge-reduced electrospray mobility analysis, *Anal. Chem.* 2004, 76, 814–822.
5. Horner, J.; Phillips, J., The use of FAIMS to separate loperamide from PEG prior to ms analysis using an LTQ XL, Thermo Fisher Scientific Application Notes 2008, Application Note 395.
6. Nazarov, E.G.; Coy, S.L.; Krylov, E.V.; Miller, R.A.; Eiceman, G.A., Pressure effects in differential mobility spectrometry, *Anal. Chem.* 2006, 78, 7697–7706.

7. Asbury, R.G.; Hill, H.H., Jr., Evaluation of ultrahigh resolution ion mobility spectrometry as an analytical separation device in chromatographic terms, *J. Microcol. Sep.* 2000, 12(3), 172–178.

8. Wu, C.; Siems, W.F.; Asbury, G.R.; Hill, H.H., Electrospray ionization high-resolution ion mobility spectrometry-mass spectrometry, *Anal. Chem.* 1998, 70(23), 4929–4938.

9. Asbury, G.R.; Hill, H.H., Using different drift cases to change separation factors (alpha) in ion mobility spectrometry, *Anal. Chem.* 2000, 72(3), 580–584.

10. Asbury, R.G.; Hill, H.H., Jr., Separation of amino acids by ion mobility spectrometry, *J. Chromatogr. A* 2000, 902(2), 433–437.

11. Matz, L.M.; Hill, H.H., Jr.; Beegle, L.W.; Kanik, I., Investigation of drift gas selectivity in high resolution ion mobility spectrometry with mass spectrometry detection, *J. Am. Soc. Mass Spectrom.* 2002, 13(4), 300–307.

12. Dwivedi, P.; Bendiak, B.; Clowers, B.H.; Hill, H.H., Rapid resolution of carbohydrate isomers by electrospray ionization ambient pressure ion mobility spectrometry-time-of-flight mass spectrometry (ESI-APIMS-TOFMS), *J. A. Soc. Mass Spectrom.* 2007, 18(7), 1163–1175.

13. Kanu, A.B.; Hill, H.H., Jr., Identity confirmation of drugs and explosives in ion mobility spectrometry using a secondary drift gas, *Talanta* 2007, 73(4), 692–699.

14. Fernandez-Maestre, R.; Wu, C.; Hill, H.H., Jr., Using a buffer gas modifier to change separation selectivity in ion mobility spectrometry, *Int. J. Mass Spectrom.* 2010.

15. Dwivedi, P.; Wu, C.; Matz, L.M.; Clowers, B.H.; Siems, W.F.; Hill, H.H., Jr., Gas-phase chiral separations by ion mobility spectrometry, *Anal. Chem.* 2006, 78(24), 8200–8206.

16. Dwivedi, P.; Bendiak, B.; Clowers, B.H.; Hill, H.H.H., Jr., Rapid resolution of carbohydrate isomers by electrospray ionization ambient pressure ion mobility spectrometry-time-of-flight mass spectrometry (ESI-APIMS-TOFMS), *J. Am. Soc. Mass Spectrom.* 2007, 18, 1163–1175.

17. Eiceman, G.A.; Shoff, D.B.; Harden, C.S.; Snyder, A.P.; Maritinez, P.M.; Fleischer, M.E.; Watkins, M.L., Ion mobility spectrometry of halothane, enflurane, and isoflurane anesthetics in air and respired gases, *Anal. Chem.* 1989, 47, 403–407.

18. Preston, J.M.; Rajadhyax, L., Effect of ion-molecule reactions on ion mobilities, *Anal. Chem.* 1988, 60(1), 31–34.

19. Spangler, G.; Collins, C.I., Peak shape analysis and plate theory for plasma chromatography, *Anal. Chem.* 1975, 47(403–407).

20. Cohen, M.J.; Karasek, F.W., Plasma chromatography—a new dimension for gas chromatography and mass spectrometry, *J. Chromatogr. Sci.* 1970, 8, 330–337.

21. Karasek, F.W.; Keller, R.A., Gas chromatograph/plasma chromatograph interface and its performance in the detection of musk ambrette, *J. Chromatogr. Sci.* 1972, 10, 626–628.

22. Long, G.L.; Winefordner, J.D., Limits of detection: a closer look at IUPAC definitions, *Anal. Chem.* 1983, 55, 712A–714A.

23. Baim, M.A.; Hill, H.H., Jr., Determination of 2,4-dichlorohenoxyacetic acid in soils by capillary gas chromatography with ion mobility detection, *J. Chromatogr.* 1983, 279, 631–642.

24. St. Louis, R.H.; Siems, W.F.; Hill, H.H., Jr., Evaluation of direct axial sample introduction for ion mobility detection after capillary gas chromatography, *J. Chromatogr.* 1989, 479, 221–231.

25. Eiceman, G.A.; Harden, C.S.; Wang, Y.-F.; Garcia-Gonzalez, L.; Schoff, D.B., Enhanced selectivity in ion mobility spectrometry analysis of complex mixtures by alternate reagent gas chemistry, *Anal. Chim. Acta* 1995, 306, 21–33.

26. Tou, J.C.; Boggs, G.U., Determination of sub parts-per-million levels of sec-butyl chlorodiphenyl oxides in biological tissues by plasma chromatography, *Anal. Chem.* 1976, 48, 1351–1357.

27. Watson, W.M.; Kohler, C.F., Continuous environmental monitoring of nickel carbonyl by Fourier transform infrared spectrometry and plasma chromatography, *Environ. Sci. Technol.* 1979, 13, 1241–1243.
28. Karpas, Z.; Pollevoy, Y.; Melloul, S., Determination of bromine in air by ion mobility spectrometry, *Anal. Chim. Acta* 1991, 249, 503–507.
29. Eiceman, G.A.; Sowa, S.; Lin, S.; Bell, S.E., Ion mobility spectrometry for continuous on-site monitoring of nicotine vapors in air during the manufacture of transdermal systems, *J. Hazard. Mater.* 1995, 43, 13–30.
30. Karasek, F.W.; Denney, D.W., Detection of 2,4,6-trinitrotoluene vapours in air by plasma chromatography, *J. Chromatogr.* 1974, 93, 141–147.
31. Lawrence, A.H.; Neudorfl, P., Detection of ethylene glycol dinitrate vapors by ion mobility spectrometry using chloride reagent ions. *Anal. Chem.* 1988, 60, 104–109.
32. Schellenbaum, R.L.; Hannum, D.W., Laboratory evaluation of the PCP large volume ion mobility spectrometer (LRVIMS), *Sandia Rep.* March 1990, SAND89-0461*UC515.
33. Eiceman, G.A.; Salazar, M.R.; Rodriguez, M.R.; Limero, T.F.; Beck, S.W.; Cross, J.H.; Young, R.; James, J.T., Ion mobility spectrometry of hydrazine, monomethylhydrazine, and ammonia in air with 5-nonanone reagent gas, *Anal. Chem.* 1993, 65, 1696–1702.
34. Eiceman, G.A.; Garcia-Gonzalez, L.; Wang, Y.-F.; Pittman, B.; Burroughs, G.E., Ion mobility spectrometry as flow-injection detector and continuous flow monitor for aniline in hexane and water, *Talanta* 1992, 39, 459–467.
35. Poziomek, E.J.; Eiceman, G.A., Solid-phase enrichment, thermal desorption, and ion mobility spectrometry for field screening of organic pollutants in water, *Environ. Sci. Technol.* 1992, 26, 1313–1318.
36. Smith, G.B.; Eiceman, G.A.; Walsh, M.K.; Critz, S.A.; Andazola, E.; Ortega, E.; Cadena, F., Detection of *Salmonella typhimurium* by hand-held ion mobility spectrometer: a quantitative assessment of response characteristics, *Field Anal. Chem. Technol.* 1997, 4, 213–226.
37. Limero, T.F.; Reese, E.; Trowbridge, J.; Hohman, R.; James, J.T., The Volatile Organic Analyzer (VOA) aboard the international space station, Society of Automotive Engineers 2002, Paper Offer Number 021CES-317.
38. Dam, R., Analysis of toxic vapors by plasma chromatography, in *Plasma Chromatography*, Editor Carr, T.W., Plenum Press, New York, 1984, pp. 177–214.

9 Ion Mobility–Mass Spectrometry

9.1 COMBINING MOBILITY WITH MASS

The addition of mass information to mobility information opens new horizons for both ion mobility spectrometry (IMS) and mass spectrometry (MS). The combination of mass and mobility within one spectrum provides information on ion structure that is not possible with either method alone. By adding ion mobility information to mass information, the size of an ion as well as its mass may be measured. For example, MS is often blind to low levels of ions in complex mixtures due to chemical noise of the matrix and to isomeric/isobaric components of the mixture. Thus, the addition of size information to mass information expands the parametric space that can be used for ion detection.

Of course, the primary advantage of adding IMS to MS is that ion mobility provides a rapid preseparation step prior to MS. Separation based on ion size as well as mass provides a two-dimensional (2D) analysis of each ion, generating an increase in peak capacity over that possible by MS alone. As in MS, a scan rule for IMS using the ratio of cross section to charge Ω/z instead of mass/charge ratio m/z can be written as follows:

$$\frac{\Omega}{z} = 0.265 \left(\frac{\pi}{kT} \right)^{1/2} \left(\frac{1}{m} + \frac{1}{M} \right)^{1/2} \frac{t_d E}{L} \frac{760}{P} \frac{T}{273} \frac{e}{N_o}$$

where Ω is the collision cross section of the ion in A^2, z is the number of charges on the ion, k is Boltzmann's constant, T is the temperature of the buffer gas, m is the mass of the ion, M is the mass of the buffer gas, t_d is the drift time of the ion in milliseconds, E is the electrical field in the ion mobility drift tube, L is the length of the drift tube in centimeters, P is the pressure of the buffer gas (in torr) in the drift region, e is the electron charge, and N_o is the number density under standard temperature and pressure conditions.

Unlike true orthogonal separation methods, IMS and MS are related through ion size. That is, in general, ions of larger mass have larger sizes. Assuming that biomolecular ions are rigid spheres, an ion's collision cross section Ω increases as a function of mass m according to the following relation: $\Omega \propto m^{2/3}$.

But, biomolecules are not rigid spheres; they have different shapes that produce different collision cross sections Ω. To a first approximation, when Ω is plotted as a function of m, unique mobility-to-mass correlation curves (often called trend lines) are generated for different classes of compounds. Trend lines were first observed by

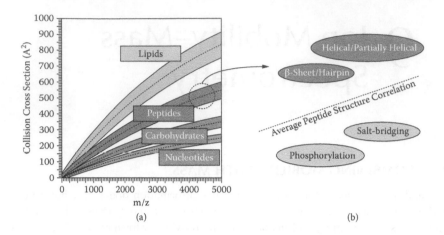

(a)   (b)

FIGURE 9.1 Mobility–mass correlation curves (trend lines). (From McLean et al., Ion mobility–mass spectrometry: a new paradigm for proteomics, *Int. J. Mass Spectrom*. 2005, 240, 301–315. With permission.)

Karasek[1] in the 1970s and evaluated in more detail by Karpas[2] in the late 1980s, but their specific advantages for application to biological matrixes have been recently articulated by Woods.[3]

Figure 9.1 shows a 2D plot of ion cross section Ω as a function of mass-to-charge ratio *m/z*.[4] Within this plot, mobility–mass correlation for four classes of biomolecules (lipids, peptides, carbohydrates, and nucleotides) are observed. Ions of a particular class, however, do not fall exactly on a trend line for that class but rather form a trend band in which compounds within a class can be found. For example, as illustrated in the expanded region of Figure 9.1, collision cross sections for specific peptide ions vary around an average peptide trend line. Structural information about the peptides can be obtained from the position of the peptide with respect to the average trend line as noted by Russell et al.[5] and Clemmer et al.[6] For example, peptides containing β-sheet and helical structures are found above the average peptide trend line, while phosphorylated and bridged peptides are found below the average peptide trend line.

An important qualitative indicator for ions based on the position that they occupy in mobility–mass correlation space is the specific cross section (SCS). Analogous to specific volume in three dimensions, this value is the 2D ratio of the measured collision cross section in $Å^2$ to the mass in daltons of the ion for which the collision cross section was measured (Ω/m). This quantity is reported in $Å^2$/daltons and is essentially the slope of the trend lines for a given class of compounds. Ions that fall into a specific class will have similar Ω/m values. As noted above with respect to peptides, small differences in Ω/m values for compounds in a class from the average Ω/m for a class provide information on the structure of ion.

As described in previous chapters, there are many different types of ion mobility methods. These include drift tube ion mobility spectrometry (DTIMS), traveling wave ion mobility spectrometry (TW-IMS), differential mobility spectrometry (DMS), differential mobility analysis (DMA), and aspiration ion mobility spectrometry (aIMS). All of these IMS methods have been interfaced to MSs.

In addition to the number of types of spectrometers available for measuring ion mobility, numerous instruments are available for measuring mass. Thus, IMS-MS is combinatorial in nature. IMS instruments have been interfaced to quadrupole (Q) MSs, ion trap MSs, time-of-flight mass spectrometers (TOF-MSs), Fourier transform ion cyclotron resonance MSs, and most recently a Waters Synapt G-2 MS containing a quadrupole ion filter with a TW-IMS coupled with a TOF-MS. With these various IMS-MS configurations, high-resolution IMS-MS, IMS-MS^2, IMS-MS^3, IMS-MS-IMS-MS, IMS-MS^2-IMS-MS, and IMS-MS-IMS-MS^2 spectrometry has been demonstrated. In the following sections, each type of ion mobility-MS is described.

9.2 LOW-PRESSURE DRIFT TUBE ION MOBILITY SPECTROMETRY–MASS SPECTROMETRY

Traditional or DTIMS measures the velocity of an ion swarm as it drifts through a constant electric field. When the electric field is low (<10 Td), the ion's velocity is constant and can be used to measure the collision cross section. All other types of ion mobility methods must be calibrated with known standards to determine cross sections, and in some cases the relationship between mobility and cross section is still not well understood. Over the years, DTIMS has been conducted under a variety of experimental conditions. For example, the pressure of the buffer gas has been varied from a few torr to several atmospheres; the temperature of the buffer gas has varied from below zero to 400°C; and the electric field through which the ions migrate has varied from high to low fields. The two major operating regimes for DTIMS are low and ambient pressures of the buffer gas. Thus, the following discussion of DTIMS coupled with MSs is divided into these two pressure regimes.

In the early days of ion mobility, before 1970, drift tubes were operated at low pressures (<10 torr) to interface with MSs. Many of the early experiments in ion mobility MS were conducted by interfacing low-pressure DTIMS instruments to a variety of MSs, including magnetic sector MSs,[7] TOF-MSs,[8–10] and QMSs.[11,12] The purpose of these experiments was primarily to investigate ion–molecular interactions in specific buffer gases. The concept of using ion mobility to evaluate ion structure was introduced by Jarrold et al.[13] and Bowers et al.[14] Modifying an early design by Bohringer and Arnold,[15] they measured mobility by injecting mass-selected ions into a low-pressure drift tube. From these early studies of ion-neutral reactions of ion clusters and metal ions, the use of mass-selected ion mobility studies has become important for the investigation of proteins, peptides, nucleic acids, and carbohydrates.

9.2.1 ELECTROSPRAY LOW-PRESSURE ION MOBILITY MS

A giant step forward in ion mobility–MS occurred in the late 1990s when Clemmer et al. applied IMS coupled to TOF spectrometry for the analysis of peptides from protein digests.[16] The creative novelty of this approach was to use an ion trap as the IMS gate.[17] In this manner, they could continuously accumulate ions from an electrospray ion source and periodically inject them into a low-pressure IMS. Figure 9.2 shows an early design of this instrument.

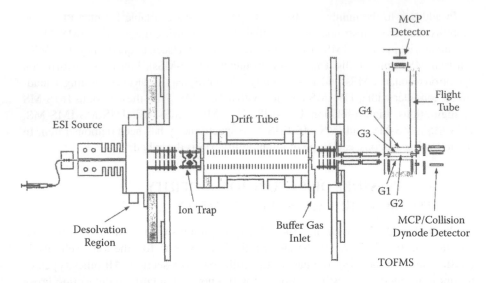

FIGURE 9.2 Schematic diagram of electrospray ionization-ion trap-low-pressure ion mobility spectrometer-time-of-flight mass spectrometer (ESI-IT-IMS-TOF). (From Henderson et al., ESI/ion trap/ion mobility/time-of-flight mass spectrometry for rapid and sensitive analysis of biomolecular mixtures, *Anal. Chem.* 1999, 71, 291. With permission.)

In this design, an aqueous sample is continuously electrosprayed through a 3.2-mm diameter orifice into an ion trap aperture such that the ion beam is accumulated for approximately 100 ms. The ion trap in this example was held between 10^{-4} and 10^{-3} torr with radio-frequency (RF) fields between 2,000 and 2,600 V at a frequency of 1.1 MHz. The trap consisted of two end-cap electrodes and a center ring electrode. These concentrated packets of ions were confined to a small volume in the center of the trap that is aligned with the drift tube entrance aperture. Ions are then injected through a 1.6-mm diameter exit aperture into the ion mobility drift tube. Ions are injected by turning off the RF field and supplying a 0.5-µs pulse of a direct current (DC) voltage of –100 to –200 V. The trap was biased about 30 V relative to the ion mobility drift tube. The ion mobility drift tube in this example was 40.4 cm in length with 80-µm diameter entrance and exit apertures. Thus, the pressure in the drift tube could be maintained at 2 to 3 torr of He, significantly above the operating pressures of the trap and the TOF-MS. The ions are then separated by ion mobility as they drift through the buffer gas in the drift tube under the influence of a weak electric field (8 to 10 V/cm). Note that this is a drift field of approximately 12.2 Td, which is on the high end of the low-field condition in which collision cross sections can be accurately calculated.

After mobility separation, the ions exit the ion mobility drift tube and are shaped into a ribbon trajectory with a combination of Einzel and DC-quadrupole lenses so that the ion beam traverses the 25-cm long focusing region and passes through a 1.6 × 12.7 mm slit into the source region of a TOF-MS.

The TOF-MS is positioned orthogonal to the IMS tube so that the ion trajectories in the TOF are at right angles to their trajectories in the IMS. Orthogonal introduction

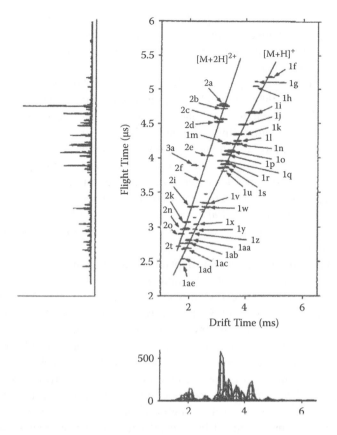

FIGURE 9.3 Contour plots of nested drift time (bottom) and flight time (left) data for a mixture of peptide ions that were formed by direct electrospray of a tryptic digest of cytochrome c. Projection of the data along the bottom and left axes shows ion mobility and time-of-flight distributions, respectively. (Taken from Henderson et al., ESI/ion trap/ion mobility/time-of-flight mass spectrometry for rapid and sensitive analysis of biomolecular mixtures, *Anal. Chem.* 1999, 71, 291. With permission.)

of ions to TOF-MSs helps keep the ion energies homogeneous during ion injections into the field free region of the TOF. The advantage of IMS coupled with TOF-MS is in matching the time regimes of both systems. IMS spectra are developed on the millisecond time frame, while mass spectra are developed on the microsecond time scale. Thus, thousands of mass spectra can be "nested" inside an ion mobility spectrum. Several of these MS-nested IM spectra can be averaged to improve sensitivity and produce 2D ion mobility MS spectra of the type shown in Figure 9.3.

Figure 9.3 provides an example of a typical spectrum from an ion mobility TOF-MS. This particular spectrum is of a mixture of peptides generated from a tryptic digest of cytochrome c. Note that for this 2D spectrum the mass flight time is plotted along the *y*-axis, and the mobility drift time is plotted along the *x*-axis. The spectrum along the *y*-axis is the mass spectrum of the sample, while the spectrum along the *x*-axis is the ion mobility spectrum. Within the 2D spectrum, each line represents an individual ion swarm produced from the electrospray process. The data points in the spectrum appear in the shape of a line because the resolving power for mobility spectrometry is much less than that for MS. Nevertheless, ions are clearly separated in mobility–mass.

The major feature of this spectrum is that the data form families of ions. The basis of these ion families is that of multiple charging. For the same mobility time, there is a family of ions with a light mass and one with a heavier mass. The ions that form the light masses are those with a single charge, while those that form the upper family are doubly charged. Thus, an ion with a heavy mass can have a similar mobility to that of a lighter mass ion because the charge state is twice as strong. As discussed previously, ion mobility separates based on the Ω/z ratio in a manner analogous to MS, which separates ions based on m/z ratio. The 2D spectrum produces data for each ion that we call its SCS (Ω/m). Thus, the ions with a larger cross section per mass form the basis for the lower family of ions, while ions of smaller Ω/m form the upper family of ions.

The resolving power of an IMS is dependent on the number of collisions an ion has with buffer gas atoms or molecules and on the potential through which the ion travels. Thus, for reduced pressure instruments, increasing the drift tube length increases resolving power. Both Smith and Bowers have used long IMS tubes to improve resolving power. When longer tubes are used, larger diameters are also necessary due to the high radial diffusion rates of ions at reduced pressures. To refocus the ions into the MS, Smith and coworkers developed an electrodynamic ion funnel.[18] These ion funnels are now used uniformly in IMS instruments operated at reduced pressure. They have been used not only to interface IMS tubes to MSs but also to refocus ions as they travel through the long drift tubes. One disadvantage of incorporation of an ion funnel into an IMS tube is that it creates multiple ion paths (i.e., those ions near the wall of the tube travel further to reach the focal point than those ions in the center of the drift tube). This multiple-path process in IMS decreases resolving power as well as the accuracy by which collision cross sections can be determined. The major advantage of ion funnels is improved sensitivity.

In summary, the advantages of low-pressure IMS are that ion trap injection can replace ion gates to improve sensitivity, and ion-focusing devices such as the ion funnel can be employed at low pressure to offset ion diffusion. Due to the complexity and size of the instrument, however, no commercially available low-pressure IMSs with static homogeneous electric fields are interfaced to MSs. Nevertheless, their performance in research laboratories has clearly demonstrated many advantages of coupling IMS to MS.

9.2.2 Matrix-Assisted Laser Desorption Ionization Low-Pressure DT Ion Mobility MS

A similar but modified approach for low-pressure ion mobility MS was taken by Russell's group.[19] Figure 9.4 shows the schematic of an early matrix-assisted laser desorption ionization (MALDI) IMS-TOF-MS that used a short IMS drift tube at a pressure of 1 to 10 torr of He, resulting in a mobility resolving power of about

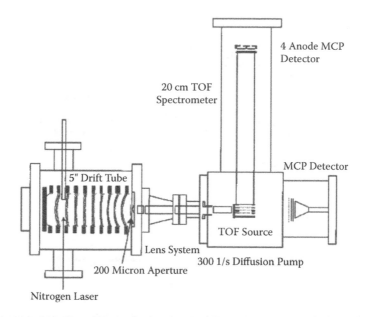

FIGURE 9.4 Schematic of MALDI ion mobility TOF-MS. (From Gillig et al., Coupling high-pressure MALDI with ion mobility/orthogonal time-of flight mass spectrometry, *Anal. Chem.* 2000, 72(17), 3965–3971. With permission.)

25 coupled to a small TOF-MS with a resolving power of 200. In this approach, the MALDI ion mobility MS was evaluated for proteomics using a tryptic digest of proteins. As discussed previously, the electrospray of a protein digest produces a distribution of charge states. MALDI, on the other hand, produces only singly charged ions. The production of a single charge state reduces the complexity of the spectrum, and the separation of isomers with IMS provides additional information not possible with MS alone. With MALDI, all of the ionization occurs in a single laser pulse, so sample is not wasted, and a near 100% duty cycle can be achieved.

As shown in Figure 9.4, ions are formed directly in the low-pressure ion mobility cell operating between 1 and 10 torr. The ion mobility tube and the TOF chamber were separated by a 200-μm diameter pinhole leak. The laser was a focused nitrogen laser (337 nm) with a pulse width of about 4 ns and a frequency of 20 Hz. Thus, ion mobility spectra were taken every 50 ms. The plume of ions formed by laser ionization is quickly cooled, limiting initial spatial spreading of the ion pulse. No ion gate or ion trap is required for MALDI ion mobility MS. At the interface between the IMS and the MS, the field strength is increased to focus mobility-separated ions through the aperture and into the TOF.

Today, MALDI ion mobility MS instruments have longer drift cells and use both ultraviolet (UV) and infrared (IR) laser ionization.[3] In addition, the longer drift cells have nonlinear electric fields, which help to focus the ions into the center of the tube and increase the transmission efficiency of the ion, reducing both resolving power and cross-section accuracy. MALDI ion mobility MS has been used for the analysis of very complex biological samples. Figure 9.5 is a typical MALDI ion mobility MS

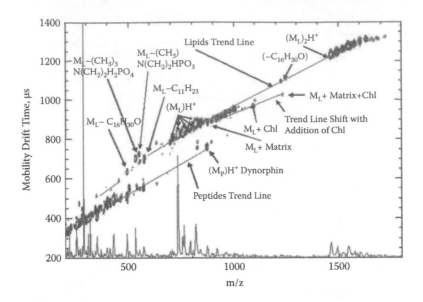

FIGURE 9.5 MALDI ion mobility (TOF)MS spectrum showing separate trend lines for peptides and lipids. (Reprinted from Woods et al., Lipid/peptide/nucleotide separation with MALDI-ion mobility-TOF MS, *Anal. Chem.* 2004, 76(8), 2187–2195. With permission.)

spectrum illustrating one of the major advantages of IMS when coupled with MSs for the analysis of complex biological samples. Figure 9.3 demonstrated that IMS could separate charged states, but because MALDI produces only singly charged ions, the two families of ions shown in Figure 9.5 are of different compound classes rather than different charge states.

Figure 9.5 is also different from Figure 9.3 in that the axes are plotted differently. In Figure 9.5, the mass data are plotted as the independent data along the *x*-axis, and the mobility data are plotted as the dependent data along the *y*-axis. Thus, when reading 2D ion mobility MS data, caution must be used to determine which scheme has been adopted for plotting data. One is not better than another, but it is common to find ion mobility MS data plotted in both ways.

The data shown in Figure 9.5 are from a mixture of sphingomyelin (ML), dynorphin (MP), and chlorisondamine (Chl). The primary feature of this 2D spectrum is the presence of two families of ions. Because this spectrum was obtained with a MALDI ionization source, all the ions are singly charged. So, unlike the ion families discussed for Figure 9.3, these trend lines do not arise from multiple charges but from structure similarities among compound classes.

Here are shown two trend lines: one for lipids and one for peptides. As discussed in Figure 9.1, there are a number of trend "bands" in which various classes of compounds appear. Thus, the location of an ion within the 2D spectrum provides information on its identity. With higher IMS resolving powers than possible with low-pressure IMS, specific trend lines for classes of compounds can be identified within the trend bands.

The 2D image shown in Figure 9.5 is an example of ion mobility MS spectra that have mobility information on the y-axis and mass information on the x-axis. As shown in Figure 9.5, lipids have larger SCSs (Ω/m) than peptides. Structurally, this can be visualized as lipid structures being more floppy, more unfolded, than those of peptides. Peptides are wrapped and folded back on themselves, making a tighter, more compact structure. In general, SCSs of compound classes follow the order lipids > carbohydrates > peptides > nucleotides.

One important application of 2D ion mobility MS spectra is the shift in trend lines as a result of tagging a component. For example, in Figure 9.5 chlorisondamine complexes with certain lipids by wrapping around the amine to form a tighter, denser, Chl-lipid complex. Those lipids that complex with chlorisondamine can easily be identified by the trend line shift that occurs with the addition of Chl. This approach has been used to identify certain reactive components in a mixture by tagging them with compounds that differ in SCS and thus shift the trend line for the tagged species to a new position.

9.2.3 TRAVELING WAVE ION MOBILITY MS

Figure 9.6 shows a schematic of an ion mobility MS called the Synapt, the only low-pressure IMS–MS that is commercially available. Known as the TW-IMS, this mobility cell is embedded in a Q-TOF-type MS with multiple capabilities. The figure depicts the ions traveling from left to right. On the far left is the ESI source, which introduces aqueous samples from a high-performance liquid chromatographic (HPLC) instrument or by direct infusion. As the electrosprayed ions enter the MS, they are bent in a "Z" manner to eliminate the solvent and focus the ions into a traveling wave ion transfer lens. From here, the ions enter a QMS, where a mass can

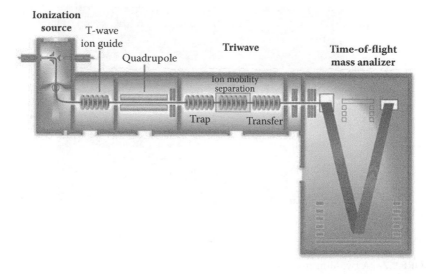

FIGURE 9.6 Schematic of a traveling wave ion mobility–mass spectrometer. (Available at http://pubs.acs.org/cen/coverstory/86/8637cover.html/. With permission.)

be selected and sent to the triwave assembly. The triwave assembly is made up of three traveling wave cells that can be operated in a variety of modes. The first cell of the triwave assembly can serve as an ion transfer cell or as a collision-induced dissociation (CID) cell. From the CID cell, fragment ions are transferred to the second traveling wave cell, which can be operated as a transfer cell or as an IMS for mobility separation of the fragment ions.

After mobility separation, ions are transferred to a third traveling wave cell similar to the first cell of the triwave assembly and operating as an ion transfer lens or as a CID cell. Finally, the ions are transferred into a high-resolution TOF-MS. Thus, many types of analyses can be performed with this single instrument. It can operate simply as an IMS-TOF or a Q-TOF, but it can also operate as a Q-IMS-TOF or, in its most powerful mode, as a Q-IMS-CID-TOF.

One of the driving forces for the development of this commercially available low-pressure IMS was the observation that doubly charged peptide ions could be separated from singly charged peptide ions by IMS, as demonstrated by Figure 9.3. The formation of only 2^+ and 1^+ ions from the ESI of a tryptic digest was expected since trypsin cleaves proteins only after basic amino acids. Thus, the peptide that is left has an N-terminal end that can accept a proton and a basic amino acid on the C-terminal end that can also accept a proton, but these are the only two sites on the peptide that can be protonated. Thus, $(M + H)^+$ and $(M + 2H)^{2+}$ are the ions that would be expected. If the $(M + H)^+$ can be separated from the $(M + 2H)^{2+}$ ions, then sequencing a protein becomes much simpler. Figure 9.7 shows a simple TW-IMS-TOF 2D spectrum of a protein digest demonstrating the ability of the Synapt to achieve the goal of charge state separation.

The resolving power of the TW-IMS in the first-generation Synapt was only about 10. In the second generation, the TW-IMS resolving power has been significantly

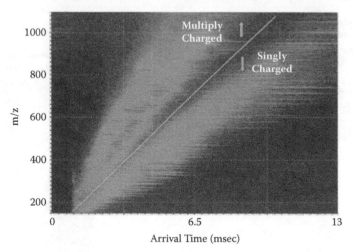

FIGURE 9.7 A two-dimensional plot of ion arrival time versus *m/z* for a protein digest mixture obtained using TW-IMS.[39] (Reprinted from Pringle et al., An investigation of the mobility separation of some peptide and protein ions using a new hybrid quadrupole/travelling wave IMS/oa-ToF instrument, *Int. J. Mass Spectrom.* 2007, 261, 1–12. With permission.)

improved to 40, a value similar to other IMS instruments. The Synapt G2 (G2 for generation 2) is capable of isomer and conformer separation, making the Synapt G2 one of the most powerful MSs for the separation and analysis of complex mixtures such as those encountered in metabolomics and proteomics.

9.3 ATMOSPHERIC PRESSURE DRIFT TUBE ION MOBILITY–MASS SPECTROMETRY

Mobilities at atmospheric pressure were first reported by A. J. Dempster[20] using the ion mobility device of Frank and Pohl.[21] These experiments demonstrated that ion mobility was inversely proportional to pressure up to at least 1,000 atm. Ion mobility at high pressures, however, was essentially ignored until Cohen demonstrated its potential as an ion separator, naming the analytical technique plasma chromatography (PC).[22] Cohen then interfaced PC to MS to produce a commercial research tool for mass identification of mobility-separated ions.[23] This atmospheric pressure (or more precisely ambient pressure) ion mobility drift tube has been the backbone of the analytical applications of IMS. Since its development as an analytical tool in the 1970s, ambient pressure IMS has been interfaced to a variety of MSs for mobility–mass information. The following sections describe several ion mobility MS systems using a variety of MS interfaces to standard ambient pressure ion mobility drift tubes.

9.3.1 DTIMS–Quadrupole Mass Spectrometry

The first MS to which an ambient pressure tube was interfaced was a QMS by Franklin GNO Corporation and applied by Karasek (Figure 9.8).[24]

Developed before the invention of ESI, this instrument was primarily used for the mobility–mass detection of volatile compounds from direct injection of vapors and from gas chromatography. Later, an electrospray version of IMS was interfaced to a QMS for the evaluation of aqueous samples.[25] Because quadrupole mass spectra must be obtained by scanning through the quadrupole voltages at a relatively slow rate, 2D data from IMS–QMS are not readily available. Instead, the normal mode of operation for an IMS–QMS is to use the single-ion monitoring mode of the MS and provide mass-selected ion mobility spectra. Nevertheless, when monitoring isomer and enantiomer separations by ion mobility or targeting a single compound of a complex mixture, IMS–QMS provides a sensitive and mass-selective method of detection.

9.3.2 DTIMS Time-of-Flight Mass Spectrometry

A more natural fit for IMS as described previously is the nesting capabilities of IMS-TOF-MS. Figure 9.9 is a schematic diagram of an ambient pressure ESI ambient pressure IMS interfaced to a TOF-MS (ESI-DTIMS-TOF-MS).

In Figure 9.9, ions travel from left to right. Sample compounds are introduced into the desolvation region of the IMS, where the electrified spray is propelled down the electric field into a counterflow of a heated buffer gas. The purpose of the heated buffer gas is to evaporate the electrospray solvent and sweep neutral vapors away from

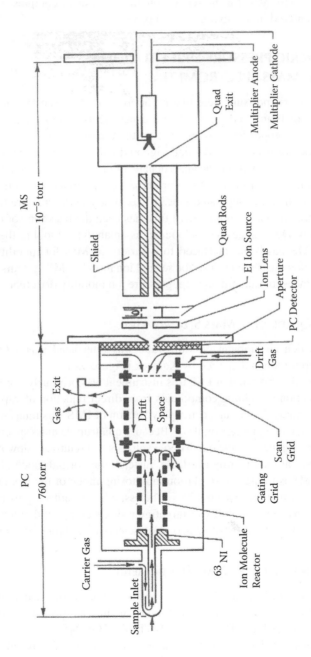

FIGURE 9.8 First ambient pressure ion mobility–mass spectrometer. (From Karasek, Kim, and Hill, Mass identified mobility spectra of p-nitrophenol and reactant ions in plasma chromatography, *Anal. Chem.* 1976, 48(8), 1133–1137. With permission.)

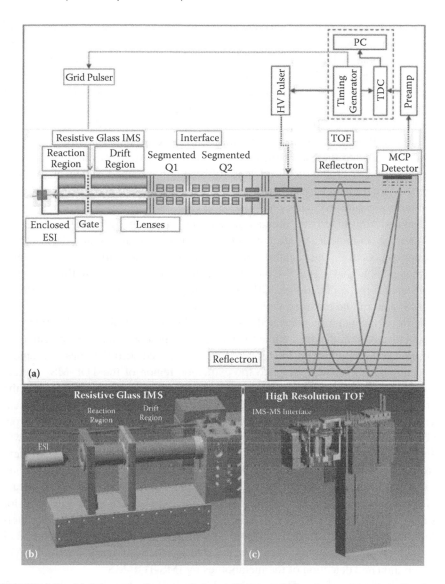

FIGURE 9.9 (a) Schematic diagram of glass tube ion mobility spectrometer interfaced to a time-of-flight (TOF) mass spectrometer via a 300-μm inner diameter pinhole leak followed by two segmented quadrupole ion guides and a set of focusing ion lenses. Three-dimensional schematic of the resistive glass tube ion mobility spectrometry (b) coupled with high-resolution TOF mass spectrometer (c). (From Kaplan et al., Resistive glass IM–TOFMS, *Anal. Chem.* 2010, 82(22): 9336–9346. With permission.)

the ion mobility drift region of the spectrometer. The instrument shown in Figure 9.9 is the only commercially available ambient pressure IMS interfaced to a TOF-MS. Once the electrospray solvent has been evaporated from electrosprayed droplets, molecular ions free of solvent clusters continue to migrate in the electric field toward a closed Bradbury Nielson ion gate. Periodically, the ion gate is opened for a few tenths of a millisecond to permit a pulse of ions into the drift region of the IMS tube. With this instrument, the IMS tube was constructed from a continuous resistive glass tube where a potential was applied to the one end and the voltage dropped linearly throughout the length of the tube, producing a linear electric field of about 250 V/cm. At ambient pressure of 760 torr and a temperature of 426 K, the number density is 1.8×10^{19} cm^{-3}, producing an E/N of 1.4 Td (1 townsend = $E/N \times 10^{17}$), well within the low-field condition of IMS, in which Ω values can be accurately determined. After injection into the drift region, the ion swarms migrate through the buffer gas in a 20-cm drift tube. At the end of the drift tube, ions are moved, electrodynamically, through a 300-μm pinhole leak into the differentially pumped region of the MS, where they are trapped with a segmented quadrupole transfer lens.

Ambient pressure IMS tubes are almost always interfaced to MSs through pinhole leaks to minimize band broadening and loss of resolving power at the pressure interface. Pinhole leaks are not without problems; clustering and declustering reactions can occur on the vacuum side of the interface, depending on pressure and voltage applied to that region.[26] Control of the pressure and voltage in this region of the MS enable the application of CID of the analyte ions. Thus, either IMS-MS or IMS-MS[2] data can be obtained with this instrument.

The RF of the quadrupole located just behind the pinhole leak is used to trap ions and hold them within the quad while the buffer gas is pumped away. The potential on each segment in the quadrupole transfer lens is controlled such that ions are rapidly moved through the quad lenses into the extraction region of the TOF-MS. Typical ion migration times through the IMS tube are on the order of 10 to 100 ms, while the migration times in the MS are on the order of 10 to 100 μs. Thus, IMS drift time measurement errors due to the flight time of the MS are less than 1%. These slight errors can be corrected by using IMS standards whose Ω and K_o values are accurately known.

Figure 9.10 shows an example of a metabolomics sample separated by ambient pressure IMS coupled with TOF-MS. In this 2D spectrum, mass is plotted along the x-axis, and mobility data are plotted along the y-axis. A hot methanol extract of *E. coli* cells that had been separated from the extracellular fluid was used for this analysis. The hot methanol lysed the cells and produced an extract of the intercellular metabolome. This methanol extract was then diluted with a water solution of acetic acid to produce the final electrospray solution that was sprayed into the IMS for separation and detection of the metabolites by IMS-TOF-MS.

The major feature of the 2D spectrum in Figure 9.10 is a broad band of "features" tracing from the lower left-hand corner of the plot toward the upper right-hand corner. This trend band corresponds roughly to the mobility–mass correlation of the metabolites. Each of the "features" in the spectrum represents an ion obtained from the ESI of the sample. Some of these ions are background ions and were present when a pure methanol/water/acetic acid solvent was sprayed into the IMS. Most of the ions, however, are the result of ions produced from metabolites in the solution. Ions produced in this manner are mostly protonated $(M + H)^+$; the proton attaches to an electropositive region of the metabolite. For example, primary amines form an RNH_3^+ ion. Compounds containing carbonyl structures, however, form adducts with sodium ions to produce the $(M + Na)^+$ ion. Sugars, for example, form the $C_6H_6O_6Na^+$ at a mass of 197 Da.

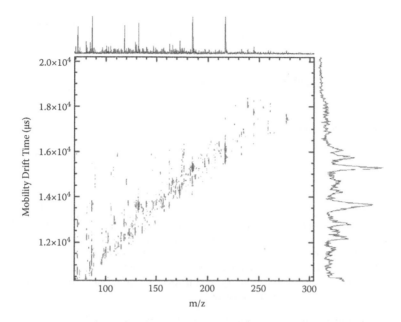

FIGURE 9.10 Two-dimensional direct infusion ESI ion mobility MS spectrum of *E. coli* culture producing over 1,000 metabolite peaks.[40] (Reprinted from Dwivedi et al., Metabolic profiling by ion mobility–mass spectrometry (IMMS), *Metabolomics* 2007, 21, 1115–1122. With permission.)

Other ions in the spectrum are the result of ion fragmentation at the IMS–MS interface. The ion fragments are easy to recognize in the spectrum because they produce a mass spectrum at a single mobility. Within the metabolite trend band, individual metabolite ions are positioned above or below the average trend line as a result of differences in their SCS (Ω/m). Thus, isomers and isobars are separated within the metabolite trend band. Hundreds of specific metabolites can be identified and quantified from a single ion mobility MS run in a matter of seconds.

9.3.3 DTIMS Ion Trap Mass Spectrometry

The exception to the rule "IMS must be interfaced to MSs through a pinhole leak" is when they are interfaced to some type of ion-trapping MS. Because ion traps collect ions for several seconds, IMS mobility resolution is lost. Thus, ion traps such as Paul traps, linear traps, Orbitraps, and FTICR instruments require a mobility-filtering type of IMS. When drift tube IMS instruments are interfaced to trapping MSs, a two ion gate system must be employed to select specific mobilities to trap in the MS.

Figure 9.11 shows an ambient pressure drift tube interfaced to a Paul trap. In this figure, the ions travel from right to left. As with other ESI-IMS instruments described in this chapter, ions are electrosprayed into a desolvation region where bare molecular ions are produced before introduction to the drift region of the spectrometer. As before, the ions are pulsed through an ion gate into the drift region, where they are separated in time by their time of drift. In this case, however, the end for the drift space is not

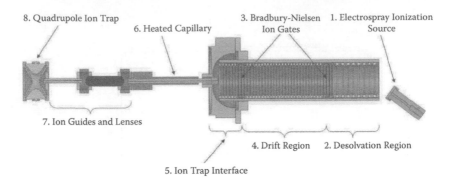

FIGURE 9.11   Schematic of the electrospray ionization, ambient pressure, dual-gate ion mobility, quadrupole ion trap mass spectrometer. This instrument consisted of six primary units: an electrospray ionization source, an ion mobility spectrometer, a vacuum interface, ion guides and lenses, a quadrupole ion trap, and a PC-based data acquisition system (not shown).[41] (Reprinted from Clowers and Hill, Mass analysis of mobility-selected ion populations using dual gate, ion mobility, quadrupole ion-trap mass spectrometry, *Anal. Chem.* 2005, 77, 5877–5885. With permission.)

defined by a Faraday plate or a pinhole leak into the MS but rather by a second ion gate. When most of the ions arrive at the second gate, the gate is closed, and those ions are not allowed to traverse into the MS. When an ion swarm of a selected mobility arrives at the second ion gate, the gate is opened, and that ion swarm is allowed to migrate through the gate and into the ion trap interface region of the spectrometer.

In the interface region, linearity of the electric field is no longer required, and the ion can be focused onto the entrance orifice of a heated capillary similar to the typed used for interfacing MSs with electrospray ionizers. The mobility-selected ions are then transferred through a capillary interface hydrodynamically into the ion guides and to the ion trap, where they can be concentrated and stored for MS^n analysis. This two-gate IMS approach for mobility selecting an ion with a high-resolution DTIMS for MS^n analysis has also been demonstrated with FTICR and with the Waters G2 instrument.

An example of IMS-TrapMS analysis is shown in Figure 9.12. Spectrum 1 illustrates the ion mobility separation of three isomers (hesperidin, neohesperidin, and rutin) adducted with silver. Spectrum 2 shows the overlaid ion mobility spectra of the respective standards. Through the use of single-mobility monitoring, the ions contained in the drift time windows (a), (b), and (c) were fragmented to produce the mass spectra shown in 1(a), 1(b), and 1(c), respectively. Shown in bold text in 1(a) and 1(b), the ions 409 and 411 may be used to confirm the presence of either hesperidin or neohesperidin. However, the IMS separation prior to mass analysis is necessary to distinguish conclusively among the three isomers.

9.4   DIFFERENTIAL MOBILITY SPECTROMETRY–MASS SPECTROMETRY

Differential mobility spectrometry (DMS) and FAIMS find their niche as mobility filters for MSs.[27] They were first interfaced to MS by Guevermont et al.; they demonstrated an

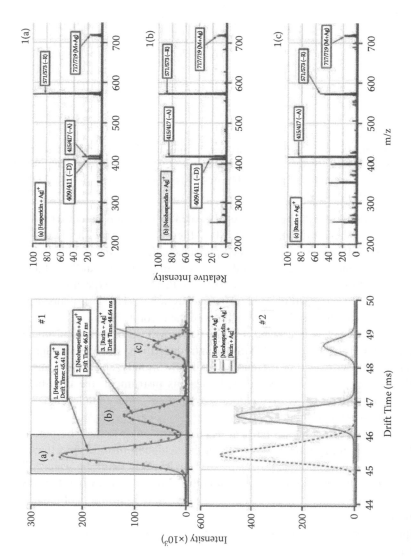

FIGURE 9.12 Drift time ion mobility (ion trap) mass spectrometric data illustrating the necessity of isomer separation prior to MS analysis.[42] (From Clowers and Hill, Influence of cation adduction on the separation characteristics of flavonoid diglycosides isomers using dual-gate ion mobility-quadrupole-ion trap mass spectrometry, *J. Mass Spectrom.* 2006, 41, 339–351. With permission.)

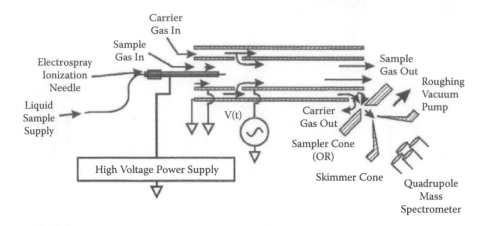

FIGURE 9.13 Schematic of ESI-FAIMS instrument interfaced to quadrupole mass spectrometer (QMS). (From Purves and Guevremont, Electrospray ionization high-field asymmetric waveform ion mobility spectrometry–mass spectrometry, *Anal. Chem.* 1999, 71(13): 2346–2357. With permission.)

improved signal-to-noise ratio due to the removal of the low-mass solvent cluster ions, reducing the background chemical noise of the MS.[28] Figure 9.13 provides a schematic of this early FAIMS-MS design. As described previously, the FAIMS ion filter was composed of two concentric cylinder electrodes about 5 mm apart with an RF voltage of 2 kHz and a peak-to-peak voltage from 0 to 5 kV. For these initial experiments, the dispersion voltage (DV) was –3300 V, and the frequency was 2 kHz. The compensation voltage (CV) was applied to the inner cylinder of the mobility filter.

The FAIMS analyzer was interfaced to a PE Sciex API 300 triple-quadrupole MS. Ions were filtered through the FAIMS and introduced into the MS through a "sampler cone" that was placed at the end of the FAIMS analyzer at a 45° angle relative to the axis of the FAIMS cylinders. Using this instrument, the high-field mobility of conformer ions, ions of multiple charge states, and complex ions of leucine enkephalin were investigated.[29]

Perhaps the most spectacular of the early results that demonstrated the potential of FAIMS as a mobility filter for MS was in the detection of nine chlorinated and brominated haloacetic acids at the part-per-trillion levels in drinking water.[30,31] The selectivity and efficiency of ion transmission through the FAIMS into the MS improved the detection limits of these compounds by three to four orders of magnitude over conventional ESI-MS methods.

After improvements made to the concentric design to enhance ion transmission to the MS, Kapron et al. demonstrated the use of FAIMS as an interface between liquid chromatography (LC) and MS.[27,32] The online FAIMS device improved the relative accuracy and precision of drug analysis by removing metabolite interferences before entrance to the MS.

Figure 9.14 shows the results of a commercial instrument using FAIMS as an ion filter between the LC and MS.[33] In this figure, the LC on the right was an assay

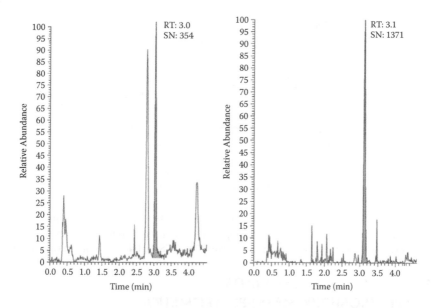

FIGURE 9.14 Mass-selected liquid chromatogram of norverapamil in urine. The chromatogram on the left is without FAIMS selectivity, and the one on the right shows significant improvement of signal-to-noise ratio when FAIMS is selected for norverapamil. Shaded peak is that of nornerapamil. (From Kapron and Barnett, Selectivity improvement for drug urinalysis using FAIMS and H-SRM on the TSQ Quantum Ultra, *Thermo Sci. Appl. Notes* 2006, 362, 1–4. With permission.)

for norverapamil in human urine using LC-MS^2. Only the reaction ion was monitored, so the detection was in the selective reaction monitoring (SRM) mode. The chromatographic peak for norverapamil was shaded in the two chromatograms. When the FAIMS was not used, the chromatogram on the left, several interfering compounds could be seen, and the signal-to-noise level was 354. When FAIMS was added between the LC and the MS, the results (on the right) showed that with the FAIMS online, the interfering peaks were eliminated from the chromatogram, and the signal-to-noise ratio was increased by almost a factor of four to 1,371.

Recently, the linear DMS has been interfaced to a MS^2-type instrument and is also commercially available for use as an ion filter between LC and ESI-MS^2. The primary benefit is the elimination of solvent ions that interfere with detection. As with the FAIMS interface, the DMS interface is expected to increase the signal-to-noise ratio of MS^2 detection

In an alternative approach to the use of DMS as a filter for ESI before a high-end MS, a low-resolution QMS was fitted with a DMS for real-time chemical analysis in the field.[34] The instrument had a mass resolution of 140 with two stages of differential pumping and an electrodynamic ion funnel to transport the ion beam from ambient pressure to the MS. This prototype DMS–MS detected approximately 1 ppb of dimethyl methyl phosphonate (DMMP) as a simulent for chemical warfare agents.

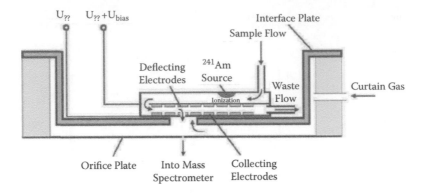

FIGURE 9.15 Schematic diagram of aspiration-type IMS interface to a quadrupole mass spectrometer. (From Adamov et al., Interfacing an aspiration ion mobility spectrometer to triple quadropole mass spectrometer, *Rev. Sci. Instrum.* 2007, 78(4): 044101. With permission.)

9.5 ASPIRATION ION MOBILITY SPECTROMETRY–MASS SPECTROMETRY

As described, the simplest of the IMSs is the aIMS. This light handheld device produces low-resolution patterns to predict the presence of toxic agents. The major investigation combining aIMS with MS uses a triple quadrupole interfaced to an IMCell.[35] This prototype IMCell contained eight electrodes connected to a measuring channel. To transfer ions from the cell to the MS, a hole was drilled into the third electrode of the IMCell as shown in Figure 9.15. The IMCell was then positioned such that the third electrode was aligned with the entrance orifice of the atmospheric interface to the MS. The bias voltage between the deflecting electrodes and the collecting electrode was scanned between 0 and 5 V to focus onto the third electrode and into the MS. While this is not the normal operating procedure for the IMCell, voltage scanning similar to that used in the DMA was required to introduce ions into the MS. The primary use for coupling MS to aIMS was to gather fundamental information on the mechanism of ion separation by aIMS.

Unlike aIMS, which is normally operated as a handheld device without MS, DMA is normally coupled to a MS. Developed from the concepts of a gas phase electrophoretic mobility molecular analyzer (GEMMA),[36] DMA was constructed around a parallel plate platform.[37] Figure 9.16 provides a schematic diagram of a DMA that can be fitted to a MS. In this design, ions are created using an ESI source and sprayed in the direction of the electric field and orthogonally into a laminar flow of buffer gas flowing through an electric field. The electric field moves the ions toward the exit orifice on the opposite plane electrode, while the buffer gas flow moves the ions in the direction of the flow. The complex relation among the strength of the electric field, the flow of the buffer gas, and the mobility of the ion in the buffer gas determines if an ion will reach the orifice to the MS. Ions are brought into focus on the orifice by adjusting the electric field. When operated in the scan mode, the electric field is changed linearly to bring different ion swarms into focus on the MS entrance.

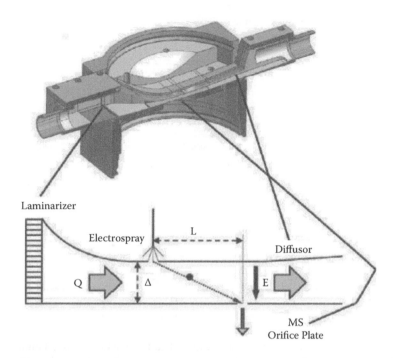

Laminarizer

Electrospray

L

Diffusor

Q

Δ

E

MS
Orifice Plate

FIGURE 9.16 Sketch of a high-transmission ESI–DMA–MS interface. (From Fernandez de le Mora; Ude, S.; Thomson, B.A., The potential of differential mobility analysis coupled to MS for the study of very large singly and multiply charged proteins and protein complexes in the gas phase, *Biotechnol. J.* 2006, 1, 988–997. With permission.)

When operated in the filter mode, the electric field is constant and adjusted to the field strength that will allow the target ion swarm to enter the MS.

Figure 9.17 provides a DMA–MS spectrum of a peptide digest of bovine serum albumin (BSA).[38] The top graph is the 2D plot of mass as a function of the voltage applied to the DMA. The long black lines represent the arrival of individual ions at specific m/z and mobility/z, where z is the number of charges on the ion. In this case, as with the tryptic digest discussed previously, only two charge states were observed. They are indicated by the trend lines in the 2D spectrum. Note that the mobility resolving power for this spectrum is about 2,800/300 = 9.3. The mass spectrum of the window shown in the 2D figure is plotted as the mass spectrum on the bottom of the figure, demonstrating clearly that the DMA was able to separate the doubly charged ions from the singly charged ions.

The primary objective of the DMA–MS is to provide an ion filter much like that of FAIMS and DMS. Tremendous improvements in DMA designs have been made over the past decade, but DMA has so far not demonstrated the resolving power or the selectivity that is available with FAIMS/DMS-type instruments. Nevertheless, its future potential is promising, especially for high molecular weight compounds. It may be that DMA evolves into an ion filter for high molecular weight compounds, while FAIMS and DMS will be used to transport selected ions of low molecular weight.

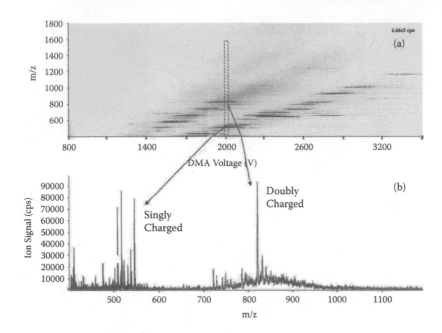

FIGURE 9.17 DMA–MS spectrum for a tryptic digest of bovine serum albumin (BSA). The bottom mass spectrum is for the selected mobility window indicated on the upper figure. Doubly charged peptides are separated by mobility from singly charge peptides. Mobility resolving power of this apparatus is about 9.3. (From Fernandez de le Mora.; Ude, S.; Thomson, The potential of differential mobility analysis coupled to MS for the study of very large singly and multiply charged proteins and protein complexes in the gas phase, *Biotechnol. J.* 2006, 1, 988–997. With permission.)

9.6 ION MOBILITY MS AND THE FUTURE

Ion mobility and MS complement each other so well they almost seem to be a single technique (ion mobility MS) rather than a hyphenated method (IMS–MS) as they are normally described. This chapter has demonstrated many types of IMSs interfaced with many types of MSs. This combinatorial approach to IMS–MS has led to the production of a variety of ion mobility MS instruments with multiple applications. The qualitative and quantitative analytical information gained when IMS separations are added to MS separations provides a significant leap in our ability to improve national security, environmental protection, industrial safety, reaction monitoring, biological assays, systems biology research, disease detection, health monitoring, pharmaceutical analysis, and forensic investigation. Advantages of the addition of IMS to MS include the following:

1. rapid separation prior to mass analysis;
2. increased sample throughput of complex samples;
3. separation of isomers and isobars;
4. rapid separation of enantiomers;
5. reduction in chemical noise;

6. measurement of ion size as collision cross section Ω;
7. correlation of mobility with mass in the form of trend bands or lines;
8. separation of multiply charged ions into families of charge states;
9. ion identification using Ω/m;
10. alternate separation patterns with multiple buffer gases; and
11. protection of MS from contamination from complex samples.

These are some (but perhaps not all) of the benefits that come for the combination of these two powerful analytical methods. Advancements in ion mobility coupled to MS are happening at such a rapid rate that this chapter certainly will be out of date as soon as it is finished, but one thing is clear: The value-added information that comes from measuring mobility and mass simultaneously over that gained from their individual application, coupled with the ease with which mobility devices can be interfaced to MSs, lead to the conclusion that almost all future MSs will be fitted with some type of mobility-discriminating device.

REFERENCES

1. Karasek, F.W.; Kim, S.H.; Rokushika, S., Plasma chromatography of alkyl amines, *Anal. Chem.* 1978, 50(14), 2013–2016.
2. Karpas, Z., Ion mobility spectrometry of aliphatic and aromatic amines, *Anal. Chem.* 1989, 61(7), 684–689.
3. Woods, A.S.; Ugarov, M.; Egan, T.; Koomen, J.; Gillig, K.J.; Fuhrer, K.; Gonin, M.; Schultz, J.A., Lipid/peptide/nucleotide separation with MALDI-ion mobility-TOF MS, *Anal. Chem.* 2004, 76(8), 2187–2195.
4. McLean, J.A.; Ruotolo, B.T.; Gillig, K.J.; Russell, D.H., Ion mobility–mass spectrometry: a new paradigm for proteomics, *Int. J. Mass Spectrom.* 2005, 240, 301–315.
5. Ruotolo, B.T.; Verbeck, G.F.; Thomson, L.M.; Woods, A.S.; Gillig, K.J.; Russell, D.H., Distinguishing between phosphorylated and nonphosphorylated peptides with ion mobility-mass spectrometry, *J. Proteome Res.* 2002, 1, 303–306.
6. Counterman, A.E.; Clemmer, D.E., Gas phase polyalanine: assessment of $i \rightarrow i + 3$ and $i \rightarrow i + 4$ helical turns in [Ala$_n$ + 4H]$^{4+}$ (n = 29–49) ion, *J. Phys. Chem.* 2002, 106, 12045–12051.
7. McDaniel, E.W.; Martin, D.W.; Barnes, W.S., Drift-tube mass spectrometer for studies of low-energy ion-molecule reactions, *Rev. Sci. Instrum.* 1962, 33, 2–7.
8. McAfee, K.B., Jr.; Sipler, D.; Edelson, D., Mobilities and reactions of ions in argon, *Phys. Rev,* 1967, 160, 130.
9. Edelson, D.; Morrison, J.A.; McKnight, L.G.; Sipler, D.P., Interpretation of ion-mobility experiments in reacting systems, *Phys. Rev.* 1967, 164, 71.
10. Young, C.E.; Edelson, D.; Falconer, W., Water cluster ions: rates of formation and decomposition of hydrates of the hydronium ion, *J. Chem. Phys.* 1970, 53, 4295.
11. Albritton, D.L.; Miller, T.M.; Martin, D.W.; McDaniel, E.W., Mobilities of mass-identified ions in hydrogen, *Phys. Rev.* 1968, 171, 94.
12. McDaniel, E.W., Possible sources of large error in determinations of ion-molecule reactions rates with drift tube-mass spectrometers, *J. Chem. Phys.* 1970, 52, 3931.
13. Kuk, Y.; Jarrold, M.F.; Silverman, P.J.; Bower, J.E.; Brown, W.L., Preparation and observations of Si10 clusters on a Au(001)-(5 × 20) surface, *Phys. Rev. B Condens. Matter* 1989, 39, 11168.

14. Bowers, M.T.; Kemper, P.R.; von Helden, G.; van Koppen, P.A.M., Gas-phase ion chromatography: transition metal state selection and carbon cluster formation, *Science* 1993, 260 (June 4), 1446–1451.

15. Bohringer, H.; Arnold, F., Temperature dependence of three-body association reactions from 45 to 400 K. The reactions $N_2^+ + 2N_2 \rightarrow N_4^+ + N_2$ and $O_2^+ + 2O_2 \rightarrow O_4^+ + O_2$, *J. Chem. Phys.* 1982, 77(11), 5534–5541.

16. Henderson, S.C.; Valentine, S.J.; Counterman, A.E.; Clemmer, D.E., ESI/ion trap/ion mobility/time-of-flight mass spectrometry for rapid and sensitive analysis of biomolecular mixtures, *Anal. Chem.* 1999, 71, 291.

17. Hoaglund, C.S.; Valentine, S.J.; Clemmer, D.E., An ion trap interface for ESI-ion mobility experiments, *Anal. Chem.* 1997, 69(20), 4156–4161.

18. Tang, K.; Shvartsburg, A.A.; Lee, H.-N.; Prior, D.C.; Buschbach, M.A.; Li, F.; Tolmachev, A.V.; Anderson, G.A.; Smith, R.D., High-sensitivity ion mobility spectrometry/mass spectrometry using electrodynamic ion funnel interfaces, *Anal. Chem.* 2005, 77, 3330–3339.

19. Gillig, K.J.; Ruotolo, B.; Stone, E.G.; Russell, D.H.; Fuhrer, K.; Gonin, M.; Schultz, A.J., Coupling high-pressure MALDI with ion mobility/orthogonal time-of flight mass spectrometry, *Anal. Chem.* 2000, 72(17), 3965–3971.

20. Dempster, A.J., On the mobility of ions in air at high pressures, *Phys. Rev.* 1912, 84(1), 53–57.

21. Franck, J.; Pohl, R., *Verh. der. Deutsch. Phys. Ges.* 1907, 69.

22. Cohen, M.J.; Karasek, F.W., Plasma chromatography—a new dimension for gas chromatography and mass spectrometry, *J. Chromatogr. Sci.* 1970, 8, 330–337.

23. Cohen, M.J., *J. Chromatogr. Sci.* 1970, 8, 330.

24. Karasek, F.W.; Kim, S.H.; Hill, H.H., Mass identified mobility spectra of p-nitrophenol and reactant ions in plasma chromatography, *Anal. Chem.* 1976, 48(8), 1133–1137.

25. Wu, C.; Siems, W.F.; Asbury, G.R.; Hill, H.H., Electrospray ionization high-resolution ion mobility spectrometry-mass spectrometry, *Anal. Chem.* 1998, 70(23), 4929–4938.

26. Spangler, G., The pinhole interface for IMS/MS, *NASA Conf. Pub.* 1995, 3301(3), 115–133.

27. Kapron, J.; Wu, J.; Mauriala, T.; Clark, P.; Purves, R.W.; Bateman, K.P., Simultaneous analysis of prostanoids using liquid chromatography/high-field asymmetric waveform ion mobility spectrometry/tandem mass spectrometry, *Rapid Commun. Mass Spectrom.* 2006, 20, 1504–1510.

28. Purves, R.W.; Guevremont, R., Electrospray ionization high-field asymmetric waveform ion mobility spectrometry-mass spectrometry, *Anal. Chem.* 1999, 71, 2346–2357.

29. Guevremont, R.; Purves, R.W., High field asymmetric waveform ion mobility spectrometry-mass spectrometry: an investigation of leucine enkephalin ions produce by electrospray ionization, *Am. Soc. Mass Spectrom.* 1999, 10, 492–501.

30. Ells, B.; Barnett, D.A.; Froese, K.; Purves, R.W.; Hrudey, S.; Guevremont, R., Detection of chlorinated and brominated byproducts of drinking water disinfection using electrospray ionization-high-field asymmetric waveform ion mobility spectrometry-mass spectrometry, *Anal. Chem.* 1999, 71(20), 4747–4752.

31. Ells, B.; Barnett, D.A.; Purves, R.W.; Guevremont, R., Detection of nine chlorinated and brominated haloacetic acids at part-per-trillion levels using ESI-FAMIMS-MS, *Anal. Chem.* 2000, 72, 4555–4559.

32. Kapron, J.T.; Jemal, M.; Duncan, G.; Kolakowski, B.; Purves, R.W., Removal of metabolite interference during liquid chromatography/tandem mass spectrometry using high-field asymmetric waveform ion mobility spectrometry, *Rapid Commun. Mass Spectrom.* 2005, 19, 1979–1983.

33. Kapron, J.; Barnett, D.A., Selectivity improvement for drug urinalysis using FAIMS and H-SRM on the TSQ Quantum Ultra, *Thermo Sci. Appl. Notes* 2006(362, 1–4.

34. Manard, M.J.; Trainham, R.; Weeks, S.; Coy, S.L.; Krylov, E.V.; Nazarov, E.G., Differential mobility spectrometry/mass spectrometry: the design of a new mass spectrometer for real-time chemical analysis in the field, *Int. J. Mass Spectrom.* 2010, 295, 138–144.

35. Adamov, A.; Viidanoja, J.; Karpanoja, E.; Paakkanen, H.; Ketola, R.A.; Kostiainen, R.; Sysoev, A.; Kotiaho, T., Interfacing an aspiration ion mobility spectrometer to a triple quadrupole mass spectrometer, *Rev. Sci. Instrum.* 2007, 78, 044101.

36. Kaufman, S.L.; Skogen, J.W.; Dorman, F.D.; Zarrin, F.; Lewis, K.C., Macromolecule determination based on electrophoretic mobility in air globular proteins, *Anal. Chem.* 1996, 68, 1895–1904.

37. Ude, S.; Fernandez de le Mora, J.; Thomson, B.A., Charge-induced unfolding of multiply charged polyethylene glycol ions, *J. Am. Chem. Soc.* 2004, 126, 12184–12190.

38. Fernandez de le Mora, J.; Ude, S.; Thomson, B.A., The potential of differential mobility analysis coupled to MS for the study of very large singly and multiply charged proteins and protein complexes in the gas phase, *Biotechnol. J.* 2006, 1, 988–997.

39. Pringle S. D.; Giles, K.; Wildgoose J. L.; Williams, J. P.; Slade S. E.; Thalassions, K.; Bateman, R. H.; Bowers, M. T. and Scrivens, J. H., An investigation of the mobility separation of some peptide and protein ions using a new hybrid quadrupole/travelling wave IMS/oa-ToF instrument, *Int. J. Mass Spectrom.* 2007, 261, 1–12.

40. Dwivedi, P.; Wu, P.; Klopsch, S.J.; Puzon, G.J.; Xun, L.; Hill, H.H., Jr., Metabolic profiling by ion mobility-mass spectrometry (IMMS), *Metabolomics* 2007, 21, 1115–1122.

41. Clowers, B.H.; Hill, H.H., Jr., Mass analysis of mobility-selected ion populations using dual gate, ion mobility, quadrupole ion-trap mass spectrometry, *Anal. Chem.* 2005, 77, 5877–5885.

42. Clowers, B.H.; Hill, H.H., Jr., Influence of cation adduction on the separation characteristics of flavonoid diglycosides isomers using dual-gate ion mobility-quadrupole-ion trap mass spectrometry, *J. Mass Spectrom.* 2006, 41, 339–351.

10 Ion Characterization and Separation: Mobility of Gas Phase Ions in Electric Fields

10.1 INTRODUCTION

The first part of this chapter is a brief presentation of the interactions encountered by an ion that is moving through a neutral gas under the influence of a weak electric field. This simplified treatment pertains mainly to the classic form of linear ion mobility spectrometers (IMSs) and aspiration IMS devices. The motion of ions in other ion mobility devices, like the differential mobility spectrometer (DMS) and traveling wave (TW) IMS is also discussed. In the second part of the chapter, the implications on ion behavior in these embodiments of IMSs are discussed. The effects of the experimental parameters temperature, drift gas composition, moisture level of the supporting atmosphere, and concentration of the analytes are described in Chapter 11.

The theory underlying IMS describes the motion of slow ions in gases. As the ion moves through a neutral gas (the supporting atmosphere) under the influence of an external electric field, different forces act on it. On the one hand, there are forces due to the resistance encountered by the ion from the gas molecules. These are electrostatic forces as well as forces arising from the geometry (size and structure) of the ion and molecules. On the other hand, a diffusive force, arising from a concentration gradient of the ions and the influence of the electric field, acts to enhance ion motion. Therefore, analysis of ion mobility must take into account the diffusive and nonelectrostatic interactions between the ion and gas molecules, the electrostatic interactions between the ion and the dipole moment or induced dipole moment of polarizable gas molecules, and the effect of the electric field on ion motion.

The foundations for IMS were laid close to commencement of the 20th century by Langevin[1] and Townsend[2] and were refined by several other investigators. Especially noteworthy are the books written by McDaniel and Mason[3] and by Mason and McDaniel,[4] both of which deal extensively with the phenomenon of ion transport in gases from a theoretical point of view as well as from experimental and practical considerations. A more detailed treatment of ion motion in the drift tube of an IMS was presented by Mason[5] and by Revercomb and Mason[6] and can also be found in other works, especially those by Mason, McDaniel, Viehland, and their coworkers.[7–9]

It is customary to start discussions of the motion of ions in gases with the theory of diffusion, then to introduce the electric field and its effect on ion motion, and finally to observe the combined effect of the two forces. This approach is taken in the following material, and the connection between theory and observations with IMS is emphasized. Most of this chapter presents an analysis of the more popular and simple theoretical models that deal with the motion of ions in a weak electric field, mainly by making assumptions on the type of potential that represents the interaction between the ion and gas molecules of the supporting atmosphere. The correlation between experimentally measured mobilities and those calculated according to the different models is discussed. The capabilities and limitations of these models to predict the mobility of ions of different masses in several drift gases at different temperatures are also discussed. Finally, a separate section is dedicated to the treatment of ion motion under high-field conditions[10–18] and the theory underlying devices that rely on a radio frequency (RF) or combination of direct current (DC) and alternating current (AC) electric fields for measurement of ion mobility. This is presented as the DMS under conditions of asymmetric electric fields.

10.2   MOTION OF SLOW IONS IN GASES

In this section, the approach of Mason and McDaniel[3–6] and their colleagues has been adopted, and emphasis is given to analysis of the motion of relatively large polyatomic ions in a buffer gas at atmospheric pressure in a weak electric field.

10.2.1   DIFFUSION OF GASEOUS IONS

Dispersion will occur through normal diffusion for a bundle or swarm of ions of a single type, with a density of n ions per unit volume, in a gas of neutral molecules or in the supporting atmosphere. This will be the only active process if all of the following conditions are met: no temperature gradient, no electric or magnetic fields, and density of ions low enough that coulomb repulsion may be neglected. The dispersion of ions will create a concentration gradient ∇_n, and ions will flow from regions of higher concentration toward regions of lower concentration at a rate that is proportional to the magnitude of the concentration gradient and according to Fick's law, as shown in Equation 10.1:

$$J = -D \, \nabla_n \tag{10.1}$$

where J is the number of ions flowing through a unit area normal to the direction of gas flow in a unit of time, and D is the proportionality constant. The diffusive force depends on the nature of the ions and the gas and is typical for a given combination of ions and neutral molecules. The vector J may also be written as the product of the velocity of the diffusive flow v and the number of ions per unit volume n as shown in Equation 10.2 and represents the total charge or electric current carried by the ions:

$$J = vn \tag{10.2}$$

Fick's law may be rewritten in the form of Equation 10.3 after Equation 10.1 and Equation 10.2 are combined and rearranged:

$$v = -(D/n)\nabla_n \tag{10.3}$$

This diffusive flow of ions continues until all the ions are uniformly dispersed in the neutral gas, and the concentration gradient becomes zero.

10.2.2 EFFECT OF ELECTRIC FIELD ON ION MOTION

When an electric field is imposed on the ion swarm in the supporting atmosphere, ion motion will be influenced by this field. In contrast, neutral molecules will barely be affected, if at all, by the electric field, and any effect will depend on the dipole or quadrupole moments of the gases. An additional factor is the electrostatic interaction between the ion and gas molecules; the ion may attract gas molecules that have permanent dipole, quadrupole, or higher moments. The electrostatic forces would also lead to ion-induced dipole interactions with the gas molecules, the magnitude of which depends on the polarizability of the gas. The interaction potential used to represent these forces is discussed in Section 10.3.

In this section, only the effects of the imposed electric field on ion motion are considered. If the electric field is weak and uniform, the ion swarm will flow along the field lines so that ion motion will be superimposed on the diffusive motion described previously. The drift velocity of the ions v_d will be proportional to the magnitude of the electric field E as given in Equation 10.4:

$$v_d = KE \tag{10.4}$$

The term K, called the mobility coefficient, like the diffusion coefficient D, is unique at a fixed temperature for a given combination of an ion and neutral gas molecules of the supporting atmosphere. The relation between the diffusion coefficient and weak-field ion mobility shown in Equation 10.5, known as the Einstein equation, and is sometimes called the Nernst–Townsend relationship:

$$K = (eD/kT) \tag{10.5}$$

where e is the ion charge, k is the Boltzmann constant, and T is the gas temperature. The mobility coefficient is directly proportional to the diffusion coefficient because both express the resistance of ion motion through the gas atmosphere. If the appropriate units (K in cm^2 V^{-1}s^{-1}, D in cm^2/s, and T in kelvin) for a singly charged ion are substituted in Equation 10.5, the mobility coefficient will be directly proportional to the diffusion coefficient and inversely proportional to gas temperature as shown in Equation 10.5a:

$$K = 11605(D/T) \tag{10.5a}$$

This expression is valid only when the electric field does not cause heating of the ions, that is, when the ions are no longer thermalized and retain energy acquired from the field. This condition is normally used in IMS drift tubes in which only a few collisions are needed for the ion to reach thermal equilibrium with the neutral molecules.

When the electric field intensity is increased at fixed pressure, the ions may acquire an average energy well above their thermal energy.[10-18] The ions are no longer thermalized, and the mobility coefficient becomes dependent on the ratio E/N (N is the density of the neutral molecules) and represents the condition in which the ion gains energy from the field in excess of the thermal energy. The diffusive forces are no longer spherically symmetrical, and the Einstein equation no longer holds. However, this limitation should not affect the understanding of ion motion in the conventional or linear IMS drift tube in which thermalized conditions apply to all standard applications.

10.2.3 EFFECT OF GAS DENSITY

Until now, the effect of the density of the drift gas molecules N on ion motion has not been considered. The motion of ions in an electric field can be regarded as a kind of spasmodic motion in which the ion is accelerated by the electric field until it collides with a gas molecule and loses, on collision, part of the acquired momentum. This process is repeated throughout the transit of the ion swarm through the drift gas under influence of the electric field (see Section 10.4). Therefore, an increase in the electric field strength will increase the drift velocity as per Equation 10.4, while an increase in the neutral gas density will directly diminish this effect with proportional increases in collision frequency and losses in kinetic energy. Thus, the motion of ions in an electric field is governed by E/N in combination rather than as separate terms. The mobility coefficient will be independent of E/N only if the energy acquired by the ion from the electric field is negligible compared with the thermal energy,[4] according to Equation 10.6:

$$\left(m/M + M/m\right)eE\lambda \ll kT \tag{10.6}$$

where m is the mass of the ion, M is the mass of the neutral gas molecules, and λ is the mean free path between collisions. Thus, $eE\lambda$ is the energy gained by an ion with charge e moving in an electric field E over a distance λ, and $(m/M + M/m)$ describes the efficiency of the elastic energy transfer from the ion to the gas molecules. If the gas is assumed to be ideal, then the mean free path is equal to the inverse of the product of the gas density and the collision cross section Q; that is, $\lambda = 1/NQ$, so that Equation 10.6 can be rewritten as

$$\left(m/M + M/m\right)eE \ll kTNQ \tag{10.7}$$

which, after rearrangement, becomes Equation 10.8:

$$E/N \ll kTQ/\left[e\left(m/M + M/m\right)\right] \tag{10.8}$$

The units of E/N are given in townsends, defined as 1 Td $= 10^{-17}$ V cm^2.

A so-called low-field condition in analytical IMS holds as long as E/N is below approximately 2 Td, and the previous discussion is a valid, although approximate, representation of the effect an external electric field has on the transport of ions in gases.

The relevance of this discussion to IMS is that the ion swarm is moved in the direction of the electric field, but the drift velocities are comparatively small due to the high frequency of collisions and small mean free path (see Frame 10.1). Furthermore, normal processes of diffusion within the ion swarm will cause broadening of the peak width while the ions are moved through the drift tube. These are opposing phenomena that limit the resolution obtainable with a drift tube and act as practical boundaries for the dimensions of IMS drift tubes.

10.3 MODELS FOR ION-NEUTRAL INTERACTIONS

In addition to the diffusive forces and effects of an external electric field mentioned, ion motion in an IMS drift tube is affected by the electrostatic interactions between the ion and the gas molecules of the supporting atmosphere. The electron cloud surrounding the neutral gas molecule is polarized by the ion, thus inducing a dipole moment in the neutral molecule. This results in an electrostatic interaction between the ion and the neutral molecule: the ion-induced dipole effect. Furthermore, molecules that have permanent dipole or quadrupole moments will be attracted to the ion through ion–dipole or ion–quadrupole interactions.

Several models have been proposed to account for the overall effect of these three forces on the motion of the ion, and some of the classical models are discussed here in brief, and their usefulness in predicting the mobility of polyatomic ions in different drift gases is examined. Two simple models are considered first: the rigid sphere model and the polarization limit model. Next, a more refined yet relatively simple-to-use model is described in which a 12,4 hard-core potential represents the ion-neutral interaction. The more complex three-temperature model[19–21] is not discussed because ions in linear IMS are traditionally regarded as thermalized. This is the one-temperature assumption, in which ion temperature is assumed to be equal to the temperature of the drift gas.

The critical test of any model is the capability to reproduce experimentally observed correlations between the mass of an ion and its mobility and between temperature and mobility, as well as to predict the mobility coefficients of ions in different gas atmospheres accurately. Normally, the success of a model is examined with relatively simple systems such as a homologous series of ions. For example, one such series is that in which the only changes arise from the addition of methylene groups in the ion with minimal changes in ion size and internal charge distribution. A complication in this discussion is the limited number of such tests with ions and gases representative of analytical IMS conditions.

Most of the early published investigations on models of ion mobility were made by physicists in relatively simple systems, mainly those in which monoatomic ions drifted through an inert monoatomic or diatomic neutral gas.[15–17,22–24] This is evident in Table 1 given in Appendix 1 of Reference 4. As this chapter is concerned with polyatomic ions drifting through polar or polarizable gases, especially air, there are not many detailed experimental and theoretical studies that can be cited as relevant.[25–37]

FRAME 10.1   SOME EXAMPLES FOR NEUTRAL–NEUTRAL COLLISION THEORY*

The following hold for calculation of molecular velocity, mean free path, mean time between collisions, and collision frequency for molecules with a diameter of $3*10^{-10}$ m, of mass 28 Da, at pressure of 760 torr, and temperature of 60°C:

Molecular speed: 501.9 m/s
Mean free path: $\lambda = 1.14*10^{-7}$ m
Mean time between collisions: $2.26*10^{-10}$ s
Collision frequency: $0.44*10^{10}$ s^{-1}

Changing the temperature to 150°C, while keeping everything else the same, yields the following:

Molecular speed: 565.7 m/s
Mean free path: $\lambda = 1.44*10^{-7}$ m
Mean time between collisions: $2.55*10^{-10}$ s
Collision frequency: $0.39*10^{10}$ s^{-1}

Calculation for a molecule with a diameter of $10*10^{-10}$ m, of mass 88 Da, at pressure of 760 torr and temperature of 150°C provides

Molecular speed: 319.1 m/s
Mean free path: $\lambda = 1.30*10^{-8}$ m
Mean time between collisions: $0.407*10^{-10}$ s
Collision frequency: $2.46*10^{10}$ s^{-1}

If we look at the conditions pertaining to a linear IMS, ignoring the effects of the electric field and the ion charge, the residence time of ions would be around 5 ms, so at atmospheric pressure and a temperature of 60°C, an ion of mass 28 in nitrogen (MW = 28 amu) would undergo $5*10^{-3}*0.44*10^{10} = 2.2*10^{7}$ collisions. Thus, it would be fully thermalized. The average drift velocity of this ion packet in the drift tube is about 10 m/s compared with the calculated velocity of a molecule of mass 28 amu, which is 501.9 m/s. Thus, for each millimeter of advance the ions have to perform 50 steps of 1 mm.

---

* From Frequency of Molecular Collisions, http://hyperphysics.phy-astr.gsu.edu/hbase/kinetic/frecol.html.

10.3.1 MOBILITY EQUATIONS

One equation for the mobility coefficient of thermalized ions drifting through a gas atmosphere in an electric field was given by Mason and coworkers,[3–6] as shown in Equation 10.9:

$$K = \left(3e/16N\right)*\left(2\pi/\mu kT_{eff}\right)^{1/2}*\left[\left(1+\alpha\right)/\Omega_D\left(T_{eff}\right)\right] \qquad (10.9)$$

where e, k, and N are as defined previously, and μ is the reduced mass of the ion-neutral collision pair defined as $\mu = mM/(m + M)$. The term T_{eff} is the effective temperature of the ions, and if the one-temperature approximation is regarded as valid, T_{eff} is equal to the temperature of the drift gas. Alpha (α) is a correction factor, generally less than 0.02 for $m > M$ (see Reference 5, p. 50), and $\Omega_D(T_{eff})$ is the collision cross section. The term $\Omega_D(T_{eff})$ is similar to Q in Equations 10.7 and 10.8, but we chose to retain the same terminology of the authors who developed the theory (for some further thoughts, see Frame 10.2).

Mason and coworkers[3-6] described the collision cross section using Equation 10.10:

$$\Omega_D = \pi r_m^2 \Omega(1,1) * \left(T^*\right) \tag{10.10}$$

where $\Omega(1,1)*(T^*)$ is a good approximation of the dimensionless collision integral that depends on the ion-neutral interaction potential and is a function of the dimensionless temperature $T^* = kT/\varepsilon_0$. The term ε_0 is the depth of the minimum in the potential surface (Figure 10.1), and r_m is the position of this minimum as shown in Equation 10.11:

$$\varepsilon_0 = e^2 \alpha_p \Big/ \left[3 r_m^4 \left(1 - a^*\right)^4 \right] \tag{10.11}$$

where α_p is the polarizability of the drift gas molecules (not to be confused with α from Equation 10.9). The term $a^* = a/r_m$ represents the separation between the center of charge and the center of mass of the ion and is not negligible for polyatomic ions of complex structure.

10.3.2 Brief Description of the Models

10.3.2.1 The Rigid Sphere Model

A rigorous treatment of the rigid sphere model was presented by Mason and McDaniel;[4] here, we give an abbreviated version. According to this model, the collision between an ion and a gas molecule is treated as that of rigid spheres in which the ion is equally likely to be scattered in any direction (in the center-of-mass coordinate system). Mason and McDaniel[4] described the mean ion energy ($\frac{1}{2}\overline{mv^2}$) in Equation 10.12:

$$\tfrac{1}{2}\overline{mv}^2 = \tfrac{1}{2}\overline{MV}^2 + \tfrac{1}{2}\overline{mv_d^2} + \tfrac{1}{2}\overline{Mv_d^2} \tag{10.12}$$

where v and V are the velocities of the ion and neutral molecules, respectively; and v_d, m, and M are as defined previously. The underscore represents the average kinetic energy. The term $\frac{1}{2}\overline{MV}^2$ is simply equal to the thermal energy acquired by the ion ($3kT/2$). The term $\frac{1}{2}\overline{mv_d^2}$ is the energy gained by the ion from the external electric field, and $\frac{1}{2}\overline{Mv_d^2}$ represents the random motion of the neutral gas molecules gained from the field. Where a heavy ion drifts through a light gas as the supporting atmosphere (as is common in IMS), this last term may be neglected. The mean relative energy $\hat{\varepsilon}$ may also be taken as the effective temperature and represents the total

FRAME 10.2   ON MASON'S FORMULAS[*]

The following is the Mason–Schamp equation:

$$K = \frac{3}{16} \frac{q}{N} \left(\frac{2\pi}{\mu k T_{eff}} \right)^{\frac{1}{2}} \frac{1}{\Omega}$$

The equation is in centimeter-gram-second (cgs) units (Mason):

$$q = 4.803 \times 10^{-10} \text{ esu;}$$

$$k = 1.381 \times 10^{-16} \text{ erg K}^{-1};$$

$$1 \text{ amu} = 1/N_A = 1.6605 \times 10^{-24} \text{ g}$$

A required conversion factor for K is 1 statvolt (stV) = 299.792 V.

Consider N_2^+ in N_2 at 300 K and 1 atm and suppose that $K = 2.0 \text{ cm}^2 \text{ V}^{-1} \text{ s}^{-1}$. Calculate the cross-section Ω with the assumption that at low field T_{eff} = the ambient temperature.

$$\mu = 14 \times 1.6605 \times 10^{-24} \text{ g}$$

$$K = 2.0 \times 299.792 = 599.6 \text{ cm}^2 \text{ stV}^{-1} \text{ s}^{-1}$$

$$599.6 = \left(\frac{3}{16} \right) \left(\frac{4.803 \times 10^{-10}}{2.45 \times 10^{19}} \right) \left(\frac{2\pi}{14 \times 1.66 \times 10^{-24} \times 1.381 \times 10^{-16} \times 300} \right)^{\frac{1}{2}} \left(\frac{1}{\Omega} \right)$$

Then, $\Omega = 1.56 \times 10^{-14} \text{ cm}^2 = 156 \text{ Å}^2$

The calculated Langevin cross section is about 80 Å^2, so the result appears to be in the correct ballpark.

If we consider a more realistic value for the mobility, say $K = 2.0 \text{ cm}^2 \text{ V}^{-1} \text{ s}^{-1}$, then we obtain $\Omega = 1.03 \times 10^{-14} \text{ cm}^2 = 103 \text{ Å}^2$, even closer to the calculated cross section.

---

[*] Compliments of Prof. J. A. Stone.

random energy of the ions. Thus, contributions arise from the thermal motion and the effect of the external field as shown in Equation 10.13:

$$3kT_{eff}/2 = 3kT/2 + \frac{1}{2}\overline{MV}^2 \tag{10.13}$$

The expression for the mobility coefficient in Equation 10.14 can be obtained from Equation 10.4 and Equation 10.9:

$$K = v_d/E = (3e/16N)(2\pi/\mu k T_{eff})^{1/2} [1/Q_D] \tag{10.14}$$

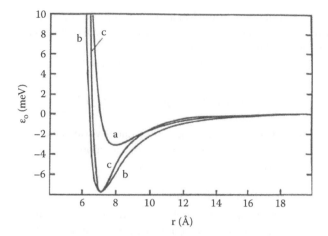

FIGURE 10.1 The calculated potential energy surface for protonated octylamine ions (130 amu) drifting through air at 250°C. Curve a was calculated taking $a* = 0.1$, $r_m = 7.988$ Å, and $\varepsilon_0 = 3.11$ meV; curve b was calculated using $a* = 0.2$, $r_m = 7.143$ Å, and $\varepsilon_0 = 7.79$ meV; and curve c with $a* = 0.4$, $r_m = 7.143$ Å, and $\varepsilon_0 = 7.79$ meV. (From Berant et al., The effects of temperature and clustering on mobility of ions in CO_2, *J. Phys. Chem.* 1989, 93, 7529–7532. With permission; Berant et al., Correlation between measured and calculated mobilities of ions, *J. Phys. Chem.* 1991, 95, 7534–7538. With permission.)

where Q_D is the collision cross section (same as Q and Ω_D). For an ion in a neutral gas, Equations 10.13 and 10.14 represent the complete momentum transfer theory for mobility.

The preceding treatment was based on considerations of the conservation of momentum and energy, as well as a few approximations, such as the one-temperature assumption. The rigid sphere model qualitatively describes some aspects of observed mobility measurements: For a given effective temperature, the mobility coefficient is inversely proportional to the neutral gas density, and the drift velocity of the ions v_d depends on E/N.

From this, we can define that "low-field" conditions are not valid when the ion's drift velocity is no longer proportional to E/N as described in Equations 10.7 and 10.8. This will occur when the energy gained by the ion from the field is no longer small compared to the thermal energy, as shown in Equation 10.15 from Mason and McDaniel[4]:

$$E/N \ll 0.78 \left[m/(m+M) \right]^{1/2} Q_D \qquad (10.15)$$

where E/N is given in townsend units, and Q_D is given in units of Å² (10^{-16} cm²).

From Equation 10.15, we can conclude that what may be low-field conditions for a heavy ion may still be high-field conditions for a light ion (or an electron) in a heavy gas.

According to the rigid sphere model, the collision cross section given in Equation 10.10 may be written per Equation 10.16 as

$$\Omega_D = \pi d^2 \qquad (10.16)$$

where d is the sum of the radii of the ion and neutral molecule and can be estimated, at least crudely, from the ion size. The mobility coefficient of an ion is approximately given by Equation 10.17 because when $m > M$, the correction term α in Equation 10.9 is less than 0.02 (Reference 4, p. 50) and can be neglected:

$$K = \left(3e/16N\right)\left[2\pi/\left(\mu kT_{eff}\right)\right]^{1/2}\left[1/\left(\pi d^2\right)\right] \tag{10.17}$$

According to Equation 10.17, the mobility coefficient varies inversely as the square root of the temperature and reduced mass. This is contradictory to experimental results under conditions used in analytical IMS.

10.3.2.2 The Polarization Limit Model

According to the polarization limit model, the polarization is added to the interaction between the ion and the drift gas molecule. If the neutral molecule does not have a permanent dipole or quadrupole moment and if there are no ion-neutral repulsive forces, then the interaction between the ion and the neutral molecule is due solely to the ion-induced dipole interaction. This interaction is a function of the polarizability of the neutral molecule α_p. The interaction potential V_{pol} varies as a function of the distance r between the ion and the neutral molecule (this r is not to be confused with r_m from Equations 10.10 and 10.11), according to Equation 10.18:

$$V_{pol}\left(r\right) = \varepsilon^2\alpha_p/\left(2r^4\right) \tag{10.18}$$

The collision cross section is proportional to the expression $[\varepsilon^2\alpha_p/(kT)]^{1/2}$, and all mobility coefficients approach a common limit as the temperature approaches 0 K. This is dependent on the polarizability and is therefore called the polarization limit K_{pol} per Equation 10.19:

$$K_{pol} = K\left(T \rightarrow \text{zero}\right) = \left[13.853/\alpha_p^{1/2}\right]\left[1/\mu\right]^{1/2} \tag{10.19}$$

The number 13.853 is obtained for K_{pol} when α_p is in units of Å^3, m and M are in daltons, and K has units of square centimeters per volt per second at 273 K and 760 torr. When the mass of the ion is much larger than the mass of the neutral molecule, $1/m$ in the reduced mass term is negligible compared to $1/M$, so that the mobility is essentially independent of the ion mass, and the reduced mass simply becomes the mass of the drift gas. This contradicts physical intuition as well as experimental observations. In summary, the polarization limit model provides a poor description of several empirical observations in IMS.

10.3.2.3 The 12,4 Hard-Core Potential Model

The 12,4 hard-core potential model takes into account repulsive (power of 12 dependence on the distance of approach) and attractive (power of 4 dependence) potentials that arise when the ion and neutral molecule approach each other at short ranges (see Equation 10.20). When a short-range repulsive term is added to the polarization limit potential of Equation 10.18, the interaction potential is modified to a $(n,4)$

potential. Choosing different values for n will change the steepness of the repulsion as well as the width and depth of the potential well. An example of this was given by Viehland[36] for $n = 8$, 12, and 16 (Reference 4, p. 248). Naturally, attractive forces will increase the ion-neutral interaction and the resistance of the neutral gas to ion motion, resulting in a decrease in ion mobility. Repulsive forces will lead to the opposite result.

The type of interaction potential used in most cases is the 12,4 or the so-called hard-core potential as given in Equation 10.20:

$$V(r) = (\varepsilon_0/2)\left\{\left[(r_m - a)/(r - a)\right]^{1/2} - 3\left[(r_m - a)/(r - a)\right]^4\right\} \qquad (10.20)$$

where r is the distance between the ion and neutral drift gas molecule (Equation 10.18), and all other parameters are as defined previously. As mentioned in the preceding text, in large polyatomic ions, the center of mass and center of charge do not necessarily coincide; therefore, a in Equation 10.20 cannot be neglected.

A slightly more sophisticated model includes a third interaction potential and is called the 12,6,4 hard-core potential model. This is formulated by adding a further term to the 12,4 potential to include some attractive energy in the form of an r^{-6} term as shown in Equation 10.21:

$$V(r) = \left\{n\varepsilon_0/\left[n(3+\gamma) - 12(1+\gamma)\right]\right\}\left[(12/n)(1+\gamma)(r_m/r)^{12}\right.$$

$$\left. -4\gamma(r_m/r)^6 - 3(1-\gamma)(r_m/r)^4\right] \qquad (10.21)$$

where γ (gamma) is a dimensionless fourth parameter that measures the relative strength of the r^{-6} and r^{-4} attraction energies. However, as shown in the following section, in most practical cases the 12,4 hard-core potential gives a sufficiently good agreement with experimental data, and these last refinements are often unnecessary.

10.4 LINEAR ION MOBILITY SPECTROMETERS: MODELS AND EXPERIMENTAL EVIDENCE

The validity of the models described can be tested by comparing experimentally measured reduced mobilities of several ions in the linear IMS with the predicted coefficients calculated according to the three models. The main features of interest were the correlations of mass with mobility and temperature with mobility; another interesting feature is the effect of the drift gas on mobility coefficients (the last two are discussed in Chapter 11). Six parameters are needed in the modeling: a^*, r_0, z, polarizability, reduced mass, and temperature. The last three arise from direct physical measurements, while the other parameters (a^*, r_0, z) are optimized by a fitting procedure to minimize the deviation between calculated and measured mobility constants.[37] The values of T^* and r_m were calculated from a^*, r_0, and z, and the dimensionless collision cross section Ω^* was taken from Table 1 in Reference 9. In practice, a discrete value of a^* was chosen, and initial values for r_0 and z were estimated. The parameters r_0 and z were then optimized to obtain a good fit with experimental data points by minimizing the squared sum of deviations X^2 between theory and experiment. Special attention was given to the fact that a broad range of ion masses had to be tested because a reasonable fit could be obtained with most models over a narrow mass range. Amines were chosen as the probe compounds to test the models because of the simple ionization chemistry[38] in all the drift gases used in these studies, namely, helium, argon, nitrogen, air, carbon dioxide, and sulfur hexafluoride.[39]

10.4.1 Ion Radii and the Mass-Mobility Correlation in Homologous Series

One test of the three models is to compare the experimentally measured reduced mobility values and cross sections of ions in a linear IMS within a homologous series with the predicted values. In such a series, the ion radius strongly affects the distance of approach to the neutral gas molecule and, correspondingly, the interaction parameters. In such a homologous series of ions, the ion radius is assumed to vary approximately as the cube root of its mass.

In the rigid sphere model, the sum of the radii of the ion and the neutral molecule d will increase slightly as the chain length and ion mass in the homologous series increase. In the polarization limit model, the ion size is totally neglected, whereas in the hard-core potential model, r_m (the minimum in the interaction potential) depends on the ion mass, as shown in Equation 10.22:

$$r_m = r_0 \left[1 + b \left(m/M \right)^{1/3} \right] \tag{10.22}$$

where b is a constant representing the relative density of the ion and the neutral molecule, generally taken as unity, and r_0 is a constant representing the ion radius for a given homologous series. Because the reduced mass term, $\mu = (1/m + 1/M)$, is insensitive to small changes in the mass of heavy ions drifting through a light-neutral gas, the ion radius and calculated mobility will not quantitatively follow the experimentally determined values.[37] A correction for the effect of changes in ion mass for large ions can be added with a semiempirical correction term, mz. This greatly improves agreement between the calculated and measured values over a broad range of ion masses in a variety of drift gases.[37–41] The physical significance of this correction term is the underlying assumption that a better representation of the position of the minimum of the interaction potential is obtained by adding a term that depends on the ion mass and accounts for the compressibility of the collision pair. Thus, the dependence of r_m on the ion mass is not just through the cube root of the ratio between the ion mass and neutral gas mass (Equation 10.22) but is also affected by the z factor mentioned. A modified expression for r_m may thus be obtained in Equation 10.23 as

$$r_m = \left(r_0 + mz \right) \left[1 + \left(m/M \right)^{1/3} \right] \tag{10.23}$$

Another factor that affects the ion radius and the ion mass arises from the observation that ions drifting in polarizable gases, especially at low temperatures, tend to form clusters with the drift gas molecules.[39–41] Thus, the mass of the ion may be incremented from its original mass m by clustering with n neutral molecules, each having a mass of M, such that its effective mass m_{eff} is altered per Equation 10.24:

$$m_{eff} = m + M_n \qquad (10.24)$$

It should be noted that n here is the average of the number of drift gas molecules clustered on the core ion drifting through the buffer gas and is not necessarily a natural number. This mass increment will also affect r_m obtained in Equation 10.23 when the effective mass is substituted in the equation, giving Equation 10.25:

$$r_m = (r_0 + mz)\left[1 + (m_{eff}/M)^{1/3}\right] \qquad (10.25)$$

Several years ago, Griffin et al.[42] demonstrated that the masses of ions in a homologous series may be determined from mobility coefficients in a linear IMS. The reduced mobility values of a series of protonated primary amines and a series of tertiary aliphatic amines measured at 250°C in air and in helium are shown in Table 10.1.[37-41] In Figure 10.2, a difference can be seen in the behavior of the two homologous series in air and in helium. Whereas in air the mobility of protonated tertiary amines is significantly higher than that of their respective isomeric primary amines, in helium there is little or no difference between isomeric ions, thus confirming some of the assumptions made previously regarding the ion-neutral interaction.

TABLE 10.1

Experimentally Measured Reduced Mobilities of Protonated Normal Primary Amines and Tertiary Amines in Air and in Helium at 250°C

	Primary Amines $CH_3(CH_2)_nNH_2$		Tertiary Amines $[CH_3(CH_2)_n]_3N$	
Now Carbon Atoms	Air	Helium	Air	Helium
1	2.65	—	—	—
2	2.38	12.1	—	—
3	2.20	10.1	2.36	10.3
4	1.98	9.1	—	—
5	1.85	7.8	—	—
6	1.72	7.2	1.95	7.4
7	1.61	7.1	—	—
8	1.50	6.4	—	—
9	1.42	5.9	1.62	5.7
10	1.35	5.6	—	—
11	1.28	—	—	—
12	1.23	5.1	1.35	4.8
14		4.1	1.19	3.9
15		—	—	—
16		3.8	—	—
21			0.95	3.0
24			0.85	2.7
36			0.64	2.0

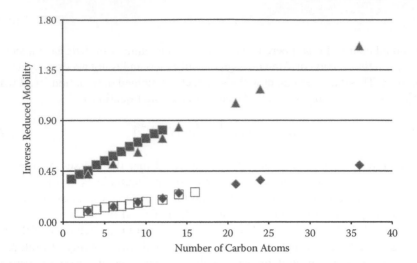

FIGURE 10.2 The inverse reduced mobility values of a homologous series of normal primary amines in air (solid squares) and in helium (hollow squares) and of normal tertiary amines in air (triangles) and in helium (diamonds).

In helium, in which this interaction is weak due to its low polarizability, the protonated primary and tertiary ions are affected similarly (and weakly) by the drift gas molecules. On the other hand, in air, the neutral molecules, especially nitrogen, cluster preferentially with the localized charge on the primary amines, leading to larger electrostatic interactions, rather than with the delocalized and shielded charge of the tertiary amine ions. Thus, a difference in the mobility in air of isomeric ions should arise, as is, indeed, observed.

To summarize this section, the radius of ions in a homologous series is assumed to increase as the cubed root of the ion mass. The effect of the ion radius on the interaction potential according to the hard sphere model is through the sum of the radii of the ion and drift gas molecules (d in Equation 10.17), changing the ion radius (or mass for a heavy ion in a light drift gas) does not affect the interaction potential in the polarization limit model (Equation 10.19), and the hard-core model predicts that the closest approach r_m of the ion–molecule collision pair will change according to Equation 10.22. However, experimental results showed that the dependence of the interaction potential, expressed in the cross section and mobility, has a stronger dependence on ion mass, and two correction terms are added as shown in Equation 10.25: One is an empirical term z that multiplies the ion mass, and the second takes into account the fact that the true ion mass m_{eff} is larger than that of the core ion due to clustering with molecules in the drift gas.

To study mass–mobility correlations, Equation 10.9 is rewritten by substitution of the appropriate constants and units, and uses the inverse reduced mobility,[3–6] as shown in Equation 10.26:

$$K_o^{-1} = 1.697 \times 10^{-4} \left(\mu_{eff} T_{eff} \right)^{1/2} r_{m2} \left[\Omega(1,1) * \left(T^* \right) \right]$$ (10.26)

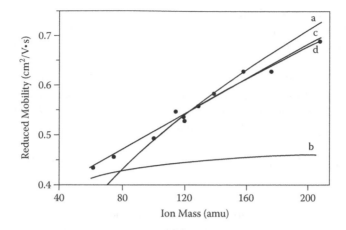

FIGURE 10.3 The measured inverse mobility of protonated acetyl compounds in air at 200°C as a function of ion mass. Curve a was calculated according to the rigid sphere model with $r_0 = 2.60$ Å; curve b according to the polarization limit model; curve c according to the hard-core model with $a^* = 0.2$, $z = 0$ Å/amu, and $r_0 = 2.40$ Å; curve d with $a^* = 0.2$, $z = 0.0013$ Å/amu, and $r_0 = 2.20$ Å. (From Berant and Karpas, Mass–mobility correlation of ions in view of new mobility data, *J. Am. Chem. Soc.* 1989, 111, 3819–3824. With permission.)

The factor 1.697×10^{-4} arises from substituting the appropriate units of K_o cm^2 V^{-1} s^{-1} when the masses are in daltons, r_m in angstroms, and T in degrees kelvin.

An example of the fitting procedure for a series of protonated acetyl compounds can be seen in Figure 10.3 and in Appendix A of Chapter 2 in the second edition of *Ion Mobility Spectrometry*.[13] In this example, the rigid sphere model can reproduce the experimental mobility results only over a very limited mass range even with the appropriate adjustment of r_0. The polarization limit model dramatically over-estimates the reduced mobility over the entire range of ion masses and totally fails to reproduce the dependence of mobility on the ion mass. However, by choosing the best-fit parameters for a^*, r_0, and z, an excellent fit with experimental results was obtained over the entire mass range, in no case deviating by more than 3%, with the model employing the 12,4 hard-core potential. This is even more impressive keeping in mind that the acetyl compounds were not a homologous series.[37]

In conclusion, the capability, or inability, of each of the three models to account for the correlation between ion mass and reduced mobility has been demonstrated previously. The capacity of each model to account for the dependence of the reduced mobility on the drift gas composition, temperature, and moisture level is further discussed in Chapter 11.

10.5 DIFFERENTIAL MOBILITY SPECTROMETER AND THE DEPENDENCE OF ION MOBILITY ON THE ELECTRIC FIELD STRENGTH

Most applications of analytical IMS include the understanding, or implied condition, that the mobility coefficient of an ion is independent of the applied electric field, such that $v_d = K \times E$ and $K \neq f(E/N)$.[44] However, Townsend and others recognized in the early 1900s that K is dependent on the energy obtained by ions between collisions from the applied electric field. When E/N is small, energy acquired by the ion from the electric field is considered negligible because collisions with molecules of the supporting gas atmosphere will dissipate any field-acquired energy (i.e., K will not be influenced by the applied electric field at a given pressure). The mobility coefficient will then become dependent on the electric field (i.e., $K = f(E/N)$) with increasing values of E/N as shown in Equation 10.27[15–17,45]:

$$K(E/N) = K(0)\left[1 + \alpha_2 (E/N)^2 + \alpha^4 (E/N)^4 + \ldots \alpha_{2n} (E/N)^{2n}\right] \quad (10.27)$$

where the term $K(0)$ is the mobility coefficient under zero field conditions; α_2, α_4, ... , α_{2n} are the specific coefficients of even powers of the electric field; and E/N is the electric field normalized for pressure (neutral gas density). In Equation 10.27, an even power series in E/N is used due to symmetry considerations (i.e., the absolute value for ion velocity is independent of electric field direction).

Ions exhibit patterns of K versus E/N due to characteristic values of α_{2n}, and Equation 10.27 can be simplified as an α function[46] to describe the electric field dependence of the coefficient of mobility as per Equation 10.28:

$$K(E/N) = K(0)\left[1 + \alpha(E/N)\right] \quad (10.28)$$

where $\alpha(E/N) = \alpha_2 (E/N)^2 + \alpha_4 (E/N)^4 + \ldots \alpha_{2n} (E/N)^{2n}$. This function describes the nonlinear electric field dependence of mobility for an ion. The characterization of mobility dependence has been studied in detail for simple ions and can be found in references such as the *Atomic Data of Nuclear Data Tables*[15–17,45] However, these measurements were typically made at reduced pressure and often in inert gases. Only a few studies for the measurement of $K(E/N)$ at ambient pressure have been made owing to the technical difficulties associated with creating high fields with conventional time-of-flight drift tubes. For example, an electric field of 21,360 V/cm would be needed to attain values of $E/N \sim 80$ Td at ambient pressure; this would require a power supply of 106.8 kV for studies with a 5-cm long drift tube. An alternative approach in IMS technology has enabled studies of field dependence using a field asymmetric ion mobility spectrometer (FAIMS) or DMS. The method of high-field asymmetric IMS for ion separations is based on a nonlinear, high-field dependence of mobility coefficients. In 1983, the method was proposed,[47] and in 1993 it was experimentally demonstrated.[46] A detailed presentation of this method can be found in Chapter 6, and the following discussion is restricted to the dependence of ion mobility on the electric field. Field dependence studies of K at ambient pressure have been reported for amines,[48] chlorides,[49] positive and negative amino acid ions,[50] organophosphates,[50] and ketones.[51]

Results from studies of the field dependence of the ions from ketones are shown in Figure 10.4 as plots of the function α (in Equation 10.28) versus E/N (in townsends) between 0 and 80 Td. The protonated monomers exhibit a positive dependence of K on E/N (Figure 10.4a) in which the ion at low values for E/N had substantially lower

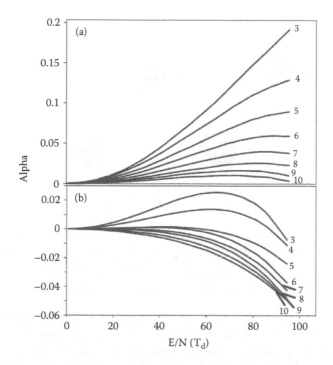

FIGURE 10.4 Results from a differential mobility spectrometer (DMS) show the change in alpha value as a function of the electric field strength E/N for a series of ketones with different molecular weights. (A) Protonated monomer showed that alpha increases with the field strength, and that the larger the effect, the lower the molecular weight of the ketone. (B) Proton-bound dimers generally showed a decrease in alpha with increasing electric field, but low molecular weight ketones initially showed an increase. The increase in alpha was due to declustering of the ion with increasing field, that led to a reduction in the effective mass of the ion. The effect was more pronounced for protonated monomers due their higher degree of clustering and larger relative change in the effective mass. The numbers to the right of each plot are carbon numbers for each ketone. (From Krylov et al., Field dependence on mobilities for gas phase protonated monomers and proton bound dimers of ketones by planar field asymmetric waveform ion mobility spectrometer (PFAIMS), *J. Phys. Chem. A* 2002, 106, 5437–5444. With permission.)

mobility coefficients than that shown by the ion under high E/N values. The core ion was unchanged in these experiments, and the trend is contrary to the expected influence of temperature, as in Equation 10.9. Rather, the dependence must be associated with ion size as governed by the formation of clusters between the core ion and small neutrals in the supporting atmosphere. The cross section will be governed by the degree of solvation of the protonated molecule by neutral molecules such as water. At high field, weakly bound molecules will be removed as the effective temperature of the cluster ion increases, while at low field, solvation will increase as the effective temperature is lowered. The same desolvation and solvation processes will occur in each field cycle. Cluster ions such as $MH^+(H_2O)_n$ will, under the low-field portion of the duty cycle, favor clustering and maximum values for n under the experimental conditions of moisture and gas temperature. However, when subjected to the

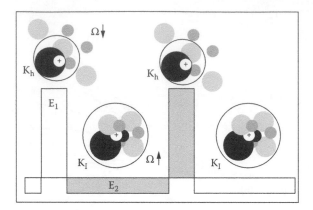

FIGURE 10.5 A simplified model for positive change in alpha with increasing electric field. During the high-field part of the duty cycle (E1), ions are declustered, leading to a decrease in Ω (represented by the faint circle around the ion cluster) and an increase in K. At the low-field part of the duty cycle (E2), ions are reclustered, leading to an increase in Ω and a decrease in K. Consequently, a positive ΔK may be expected with the trends shown in this model.

high-field portion of the waveform and electric fields of 20,000 V/cm or greater, ion heating should arise, and the ion will experience dissociation of hydrates with increase in ion temperature.[52] Thus, the basis of α in Equation 10.28 is understood to arise from the declustering and clustering of the core ion as schematically shown in Figure 10.5. In the high-field portion of the waveform, the cross section of the ion is reduced due to declustering. In the low-field portion of the waveform, the ion mobility is decreased as a result of formation of clusters. Under this condition, the declustered ion will have a smaller size and lower mass than the clustered ion, and correspondingly, the mobility coefficient will be higher than that of the clustered ion. The processes of cluster formation and declustering are dynamic and continuous throughout ion transport in the drift region, and the overall effect is a weighted average of the lifetimes and compositions of the clustered and declustered forms of the ion. Thus, the α function is gradual and lacks any apparent steps or discontinuous stages. This gradual change in mobility coefficient with E/N has been studied with a large number of organic compounds and observed uniformly regardless of functional group (Figure 10.5).

Returning to the measurement of ketones,[51] as the molecular weight of the ketone increases, effects on the changes in mobility from declustering by ion heating become less pronounced, and the $\alpha(E/N)$ function for ketones with large molar masses nearly flattens (Figure 10.4a). The changes in α function with changes in the molecular weight of the ketone are also gradual, with the greatest effect on field dependence observed with the ketone of the lowest mass. The gradual change is also supportive of a general phenomenon rather than a structure- or size-specific process. Nonetheless, the positive α function is small for large ketones, and this observation is supported by the model for clustering–declustering. An ion of high mass will slightly increase in size and weight with the addition of adducts of water or other small neutrals in the clustered stage; therefore, there will be a small difference between the clustered and declustered ions. In contrast, an ion of low mass will undergo proportionally significant changes in size and mass with the addition of a few water molecules to the core ion. Between these extremes, intermediate effects should be expected, and these are seen in Figure 10.4a.

The dependence of mobility coefficients on electric fields for proton-bound dimers is given in Figure 10.4b and shows a negative α function that changes with the molar mass of the ketone. As E/N is increased, the mobility of the proton-bound dimer $M_2H^+(H_2O)_n$ decreases, indicating that the ion in motion is slower at high electric fields than at low electric fields. This is consistent with decreases in mobility under high-field conditions or increased temperature, according to Equations 10.17 and 10.26, and presumably occurs through increases in collision frequency with increased temperature. The negative α functions for the proton-bound dimer changed gradually as the mass of the ion was increased, suggesting a weighted average of mobility under conditions of moisture and gas temperature.

The results for ketones shown were obtained with air containing 0.1 ppm moisture as the supporting atmosphere. Temperature and pressure were near ambient, and no changes were made in temperature or moisture. However, the cluster model used to explain the positive α dependence could be tested with changes in either moisture or temperature. In another study, the effects of water content on α functions with a series of organophosphorous compounds were explored.[50] These compounds were chosen based on high thermal stability and lack of chemical reactivity toward water. Moreover, organophosphorous compounds with the general form $RR_2P = O$ offered a large selection of structures in which R and R_2 were alkyl groups (organophosphates) or in which R was an alkoxy group organophosphonates).

The results of this study are summarized in Figure 10.6, where α values are shown for protonated monomers at two E/N levels of 80 Td (bottom) and 20 Td (top) for moisture levels in air between 0.1 and 10,000 ppm. In these plots, a dependence of α on moisture can be divided into two regions, with an apparent break in behavior at about 50 ppm moisture in air. Below this value, $\alpha(E/N)$ was not dependent on moisture, although absolute values for α were different for each of the protonated phosphates (see previous discussion of ketones). However, above about 50 ppm, $\alpha(E/N)$ increased monotonically with moisture for ions of each chemical. This same behavior was observed at all values for E/N, although only two settings (20 and 80 Td) are shown in Figure 10.6.

The model mentioned for clustering and declustering and the influence of moisture on α may be valid only if there is sufficient time during the low-field period for solvent molecules to encounter the ion. At 50 ppm water in air at ambient temperature and pressure, there are 1.3×10^{15} molecules cm^{-3}. If we take a typical collision rate constant of about 1×10^{-9} cm^3 molecule^{-1} s^{-1} for ion-neutral encounters, and knowing that the concentration of ions is far less than that of water, the time between collisions is approximately $1/k[H_2O] = 0.8$ μs. In the low-field period of about 1.6 μs, an ion will undergo approximately two collisions with water molecules, a sufficient number if the association reaction is efficient, to change the mass and the cross section of the ion. At elevated water concentrations, the number of collisions in the low-field period will increase proportionately and reach 400 at 10,000 ppm. With an increase of water concentration, the potential cluster size will increase, but the strength of the

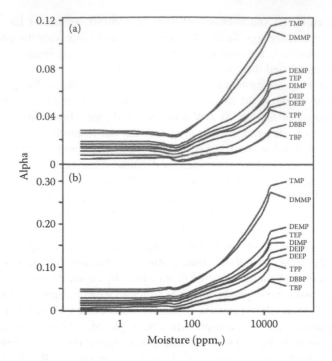

FIGURE 10.6 The dependence of the alpha function on moisture for a series of organo-phosphorous compounds at two different field strengths of 20 Td (a) and 80 Td (b). Up to a level of 50 ppm in the drift gas, there is hardly any change in alpha at both field strengths. The changes in mobility between high and low fields for an ion cluster cannot be associated with hydration reactions. In addition, weakly bound nitrogen molecules are thought to contribute to the effective collision cross section of an ion at low field. This is considered likely owing to the large number of ion-nitrogen collisions in air at ambient pressure. (From Krylova et al., Effect of moisture on high field dependence of mobility for gas phase ions at atmospheric pressure: organophosphorus compounds, *J. Phys. Chem.* 2003, 107, 3648–3648. With permission.)

attachment of each successive water molecule will decrease. The larger clusters will be more susceptible to size diminution in the same high field than are the smaller ones, and hence the value of α will increase with increasing water concentration.

The effect of mass of the organophosphorous compounds is also consistent with this model of α in which the lower-mass ions show the greatest changes in the values of α as the water concentration is increased. For the ions in this study, lower mass also implies a smaller ion whose cross section will change to a greater extent than will that of a larger ion by the addition of the same number of water molecules. Below about 50 ppm water, an α dependence exists for each ion, and if this change is also due to a change in ion solvation between high and low fields, then the solvating molecule must be either an impurity in the air drift gas or the air drift gas itself. This last conjecture is consistent with a recent study of the effect of carrier gas type on the high- to low-field mobility of several types of ions. The nitrogen molecule with no dipole moment and a very low polarizability is a very poor solvent for gas phase ions compared with water. Because an ion will experience a collision with a nitrogen

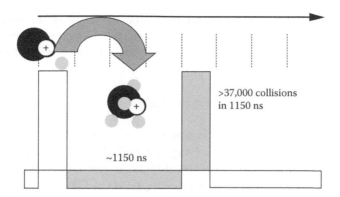

FIGURE 10.7   A schematic of the processes that take place in a DMS at high moisture levels.

molecule of the drift gas roughly once every 40 ps, an equilibrium solvation can occur in the low-field period (Figure 10.7). Although the associations with nitrogen molecules should be weak at about 19 kJ/mol, the cumulative effect of over 37,000 collisions during the low-field portion of the waveform may account for an increase in effective cross section.

To summarize this section, the basis for separation of ions in DMS is the different effect low electric fields and high electric fields have on the mobility of ions. This in turn strongly depends on the clustering of neutral molecules that occurs when the ions are in the low electric field and the declustering that is induced by the high electric field. The clustering depends mainly on the moisture level, the temperature, and the analyte concentration (dimer formation) in the cell.

10.6   TRAVELING WAVE IMS

The description of TW-IMS,[53–55] for which the drift gas is usually nitrogen or helium and the drift gas pressure is only a few millibars, is presented in Chapter 6. A detailed theory of TWs is beyond the scope of this monograph; however, a comprehensive analysis of the forces that govern the mobility of ions in a TW-IMS device was published by Shvartsburg and Smith using derivations and ion dynamics simulations.[55] According to that analysis, the motion of the ion depends on the velocity of the ion in relation to the propagating wave front **c**. In analogy with a cork floating in the sea, the ions may behave like a cork bobbing slightly on fast waves (**c** << 1), but slow waves (**c** > 1) will push the cork with them. The cork will behave like a surfer in the intermediate case **c** ≈ 1, that is, following the wave but occasionally falling behind.

$$\mathbf{c} = KE_{\max}/s \qquad\qquad (10.29)$$

where K is the ion mobility, E_{max} is the electric field strength, and s is the wave velocity. In their analytical model, Shvartsburg and Smith treated ion diffusion as independent from drift,[55] so that the separation parameters depended only on the drift, while diffusion determined the ion packet width and thus the resolution.

The transit time t_t needed for the ion to traverse the length of the drift tube L in the TW-IMS when the ratio c is low is proportional to the squares of mobility K and electric field intensity E, as opposed to linear scaling of the mobility with the electric field in linear IMS and DMS.

$$t_t = Ls/(KE)^2 \qquad (10.30)$$

As the propagating wave speed decreases, the ion transit velocity asymptotically approaches the wave speed. The resolving power of TW-IMS depends on mobility, scaling as $K^{1/2}$ in the low-c limit and less at higher c. This nonlinear dependence of the transit time on mobility affects the resolving power of TW-IMS, and it is claimed that near-optimum resolution is achievable over a approximately 300–400% range of mobilities.[55]

Bearing in mind that TW-IMS is usually operated at low drift gas pressure P, the mobility can be expressed in the low-field regime as

$$K = K_0\left(N_0/N\right) = K\left(P_0/P\right)\left(T/T_0\right) \qquad (10.31)$$

where N is the number density of gas molecules; N_0 and P_0 are the value of N at standard temperature ($T_0 = 273$ K) and pressure ($P_0 = 1$ atm), respectively; and K_0 is the reduced mobility under those conditions. Thus, if the amplitude of the electric field U and pressure are changed simultaneously, and the U/P ratio is retained, the value of KE and the mean ion velocity will remain the same with any waveform. This is analogous to linear IMS, where keeping V/P (or E/N) constant will not affect the drift time.

In summary, the main difference between TW-IMS with linear IMS with regard to the effect of the electric field on mobility is that the dependence of the ion velocity on K and E is linear in linear IMS but nonlinear in TW-IMS. In TW-IMS at the limit of low KE (KE is much smaller than the wave velocity s), the dependence is quadratic but becomes progressively stronger with increasing KE. In any case, the drift velocity cannot exceed the wave velocity like a cork that cannot travel faster than the wave that carries it. Like the high-field declustering effect in DMS, the low gas pressure and high electric fields in TW-IMS will cause substantial heating of ions, which may lead to fragmentation of macromolecular structures, complicating spectral interpretation.

10.7   SUMMARY

The motion of ions in a buffer gas is governed by diffusive forces, the external electric field and the electrostatic interactions between the ions and neutral gas molecules. Ion–dipole or ion–quadrupole interactions, as well as ion-induced dipole interactions, can lead to attractive forces that will slow the ion movement, mainly due to clustering effects. The interaction potential can be calculated according to different theories, and three such approaches—the hard-sphere model, the polarization limit model, and the 12,4 hard-core potential model—were introduced here. Under the influence of a weak electric field, the ions drift along the field lines at a velocity that is proportional to the field strength and affected by clustering that increases the drag forces. This approximation is relevant in the case of the linear IMS drift tube. Strong electric fields, or elevated temperatures, will lead to breakdown of the clusters and in some cases even to fragmentation of the core ions. This occurs in the DMS and in the TW-IMS and is the basis of separation of ions according to their behavior in the strong electric field.

An attempt to comprehensively, yet concisely, present an overview of the theory underlying the mobility of ions with regard to the linear IMS and DMS was made by Spangler.[56] However, here we present the fundamental aspects of the forces that govern the motion of an ion under conditions that pertain to the practical IMSs.

REFERENCES

1. Langevin, P., Une formule fondamentale de théorie cinétique, *Ann. Chim. Phys.* 1905, 5, 245–288. English translation in *Collision Phenomena in Ionized Gases* (Appendix II), Editor McDaniel, E.W., Wiley, New York, 1984.
2. Townsend, J.S., On the diffusion of ions, *Philos. Trans. R. Soc. London A* 1899, 193, 129–158.
3. McDaniel, E.W.; Mason, E.A., *The Mobility and Diffusion of Ions in Gases*, Wiley, New York, 1973.
4. Mason, E.A.; McDaniel, E.W., *Transport Properties of Ions in Gases*, Wiley, New York, 1987.
5. Mason, E.A., Ion mobility: its role in plasma chromatography, in *Plasma Chromatography*, Editor Carr, T.W., Plenum Press, New York, 1984, pp. 43–93.
6. Revercomb, H.E.; Mason, E.A., Theory of plasma chromatography gaseous electrophoresis—a review, *Anal. Chem.* 1975, 47, 970–983.
7. Hahn, H.; Mason, E.A., Field dependence of gaseous-ion mobility: theoretical tests of approximate formulas, *Phys. Rev. A* 1972, 6, 1573–1577.
8. McDaniel, E.W.; Viehland, L.A., The transport of slow ions in gases: experiment, theory and applications, *Phys. Rep.* 1984, 110, 333–367.
9. Mason, E.A.; O'Hara, H.; Smith, F.J., Mobilities of polyatomic ions in gases: core model, *J. Phys. B At. Molec. Phys. B* 1972, 5, 169–176.
10. (a) Wannier, G.H., On the motion of gaseous ions in a strong electric field. I, *Phys. Rev.* 1951, 83, 281–289; (b) Wannier, G.H., Motion of gaseous ions in a strong electric field. II, *Phys. Rev.* 1952, 87, 795–798; (c) Wannier, G. H., Motion of gaseous ions in strong electric fields, *Bell Syst. Tech. J.* 1953, 32, 170–254.
11. (a) Viehland, L.A.; Kumar, K., Transport coefficient for atomic ions in atomic and diatomic neutral gases, *Chem. Phys.* 1989, 131, 295–313; (b) Viehland, L.A.; Robson, R.E., Mean energies of ion swarms drifting and diffusing through neutral gases, *Int. J. Mass Spectrom. Ion Proc.* 1989, 90, 167–186.
12. Ness, K.F.; Viehland, L.A., Distribution functions and transport coefficients for atomic ions in dilute gases, *Chem. Phys.* 1990, 148, 255–275.
13. Skullerud, H.R., On the relation between the diffusion and mobility of gaseous ions moving in strong electric fields, *J. Phys. B* 1976, 9, 535–546.
14. Viehland, L.A.; Mason, E.A., Gaseous ion mobility and diffusion in electric fields of arbitrary strength, *Ann. Phys. (N.Y.)* 1978, 110, 287–328.
15. Ellis, H.W.; Pai, R.Y.; McDaniel, E.W.; Mason, E.A.; Viehland, L.A., Transport properties of gaseous ions over a wide energy range, *At. Nucl. Data Tables* 1976, 17, 177–210.

16. Ellis, H.W.; McDaniel, E.W.; Albritton, D.L.; Lin, S.L.; Viehland, L.A.; Mason, E.A., Transport properties of gaseous ions over a wide energy range—part II, *At. Nucl. Data Tables* 1978, 22, 179–217.

17. Ellis, E.W.; Thackston, M.G.; McDaniel, E.W.; Mason, E.A., Transport properties of gaseous ions over a wide energy range. Part III, *At. Nucl. Data Tables* 1984, 31, 113–131.

18. Paranjape, B.V., Field dependence of mobility in gases, *Phys. Rev.* 1980, A21, 405–407.

19. Viehland, L.A.; Lin, S.L., Application of the three-temperature theory of gaseous ion transport, *Chem. Phys.* 1979, 43, 135–144.

20. Lin, S.L.; Viehland, L.A.; Mason, E.A., Three-temperature theory of gaseous ion transport, *Chem. Phys.* 1979, 37, 411–424.

21. Waldman, M.; Mason, E.A., Generalized Einstein relations from a three temperature theory of gaseous ion transport, *Chem. Phys.* 1981, 58, 121–144.

22. Robson, R.E.; Kumar, K., Mobility and diffusion. II. Dependence on experimental variables and interaction potential for alkali ions in rare gases, *Aust. J. Phys.* 1973, 26, 187–201.

23. Skullerud, H.R., Mobility, diffusion and interaction potential for potassium ions in argon, *J. Phys. B* 1973, 6, 918–928.

24. Helm, H., The mobilities of $Kr^+(2P_{3/2})$ and $Kr^+(^2P_{1/2})$ in krypton at 295 K, *Chem. Phys. Lett.* 1975, 36, 97–99.

25. Viehland, L.A.; Fahey, D.W., The mobilities of NO_3^-, NO_2^-, NO^+ and Cl^- in N_2. A measure of inelastic energy loss, *J. Chem. Phys.* 1983, 78, 435–441.

26. Mason, E.A., Higher approximations for the transport properties of binary gas mixtures, I. General formulas, *J. Chem. Phys.* 1957, 27, 75–84.

27. Mason, E.A., Higher approximations for the transport properties of binary gas mixtures, II. Applications, *J. Chem. Phys.* 1957, 27, 782–790.

28. Mason, E.A.; Hahn, H., Ion drift velocities in gaseous mixtures at arbitrary field strengths, *Phys. Rev.* 1972, A5, 438–441.

29. Mason, E.A.; Schamp, H.W., Jr., Mobility of gaseous ions in weak electric fields, *Ann. Phys. (N.Y.)* 1958, 4, 233–270.

30. Robson, R.E., Mobility of ions in gas mixtures, *Aust. J. Phys.* 1973, 26, 203–206.

31. Whealton, J.H.; Mason, E.A.; Robson, R.E., Composition dependence of ion transport coefficients in gas mixtures, *Phys. Rev.* 1974, A9, 1017–1020.

32. Eisele, F.L.; Perkins, M.D.; McDaniel, E.W.; Mobilities of NO_2^-, NO_3^- and CO_3^- in N_2 over the temperature range 217–675 K, *J. Chem. Phys.* 1980, 73, 2517–2518.

33. Eisele, F.L., Perkins, M.D.; McDaniel, E.W., Measurement of the mobilities of Cl^-, NO_2^- H_2O^-, $NO_3^-H_2O^-$, $CO_3^-H_2O^-$ and $CO_4^-H_2O^-$ in N_2 as a function of temperature, *J. Chem. Phys.* 1981, 75, 2473–2475.

34. Holstein, T., Mobilities of positive ions in their parent gases, *J. Phys. Chem.* 1952, 56, 832–836.

35. Patterson, P.L., Mobilities of negative ions in SF_6, *J. Chem. Phys.* 1970, 53, 696–704.

36. (a) Viehland, L.A.; Mason, E.A., Gaseous ion mobility in electric fields of arbitrary strength, *Ann. Phys. N.Y.* 1975, 91, 499; (b) Viehland, L.A., Lin, S.L.; Mason, E.A., Kinetic theory of drift-tube experiments with polyatomic species, *Chem. Phys.* 1981, 54, 341–364.

37. Berant, Z.; Karpas, Z., Mass-mobility correlation of ions in view of new mobility data, *J. Am. Chem. Soc.* 1989, 111, 3819–3824.

38. Karpas, Z., Ion mobility spectrometry of aliphatic and aromatic amines, *Anal. Chem.* 1989, 61, 684–689.

39. Karpas, Z.; Berant, Z., The effect of the drift gas on the mobility of ions, *J. Phys. Chem.* 1989, 93, 3021–3025.

40. Karpas, Z.; Berant, Z.; Shahal, O., The effect of temperature on the mobility of ions, *J. Am. Chem. Soc.* 1989, 111, 6015–6018.

41. (a) Berant, Z.; Karpas, Z.; Shahal, O., The effects of temperature and clustering on mobility of ions in CO_2, *J. Phys. Chem.* 1989, 93, 7529–7532; (b) Berant, Z., Shahal, O., Karpas, Z. Correlation between measured and calculated mobilities of ions: sensitivity analysis of the fitting procedure, *J. Phys. Chem.* 1991, 95, 7534–7538.

42. Griffin, G.W., Dzidic, I., Carroll, D. I., Stillwell, R. N., Horning, E. C., Ion mass assignments based on mobility measurements, *Anal. Chem.* 1973, 45, 1204–1209.

43. Eiceman, G.A.; Karpas, Z., *Ion Mobility Spectrometry*, CRC Press, Boca Raton, FL, 1993.

44. Purves, R.W.; Guevremont, R., Electrospray ionization high-field asymmetric waveform ion mobility spectrometry-mass spectrometry, *Anal. Chem.* 1999, 71, 2346–2357.

45. Viehland, L.A.; Mason, E.A., Transport properties of gaseous ion over a wide energy range, *At. Data Nucl. Data Tables* 1995, 60, 37–95.

46. Buryakov, I.A.; Krylov, E.V.; Nazarov, E.G.; Rasulev, U.Kh., A new method of separation of multi-atomic ions by mobility at atmospheric pressure using a high-frequency amplitude-asymmetric strong electric field, *Int. J. Mass Spectrom.* 1993, 128, 143–148.

47. Gorshkov, M.P., Invention Certificate No. 9666583, Russia, G01N27/62, 1983.

48. Viehland, L.A.; Guevremont, R.; Purves, R.W.; Barnett, D.A., Comparison of high-field ion mobility obtained from drift tubes and a FAIMS apparatus, *Int. J. Mass Spectrom.* 2000, 197, 123–130.

49. Guevremont, R.; Barnett, D.A.; Purves, R.W.; Viehland, L.A., Calculation of ion mobilities from electrospray ionization high-field asymmetric waveform ion mobility spectrometry mass spectrometry, *J. Chem. Phys.* 2001, 114, 10270–10277.

50. Krylova, N.; Krylov, E.; Eiceman, G.A.; Stone, J.A., Effect of moisture on high field dependence of mobility for gas phase ions at atmospheric pressure: organophosphorus compounds, *J. Phys. Chem.* 2003, 107, 3648–3648.

51. Krylov, E.; Nazarov, E.G.; Miller, R.A.; Tadjikov, B.; Eiceman, G.A., Field dependence on mobilities for gas phase protonated monomers and proton bound dimers of ketones by planar field asymmetric waveform ion mobility spectrometer (PFAIMS), *J. Phys. Chem. A* 2002, 106, 5437–5444.

52. Kim, S.H.; Betty, K.R.; Karasek, F.W., Mobility behavior and composition of hydrated positive reactant ions in plasma chromatography with nitrogen carrier gas, *Anal. Chem.* 1978, 50, 2006–2012.

53. Giles, K.; Pringle, S.D.; Worthington, K.R.; Little, D.; Wildgoose, J.L.; Bateman, R.H., Applications of a travelling wave-based radio-frequency-only stacked ring ion guide, *Rapid Commun. Mass Spectrom.* 2004, 18, 2401–2414.

54. Pringle, S.D.; Giles, K.; Wildgoose, J.L.; Williams, J.P.; Slade, S.E.; Thalassinos, K.; Bateman, R.H.; Bowers, M.T.; Scrivens, J.H., An investigation of the mobility separation of some peptide and protein ions using a new hybrid quadrupole/travelling wave IMS/oa-ToF instrument, *Int. J. Mass Spectrom.* 2007, 261, 1–12

55. Shvartsburg, A.A.; Smith, R.D., Fundamentals of traveling wave ion mobility spectrometry, *Anal. Chem.* 2008, 80, 9689–9699.

56. Spangler, G.E., New developments in ion mobility spectrometry, *J. Process Anal. Chem.* 2009, 6, 88–93.

11 Control and Effects of Experimental Parameters

11.1 INTRODUCTION

The performance of an ion mobility spectrometer and consequently the results from a measurement depend on choices of design and fabrication of each component in the instrument, including the sample inlet, ion source, ion injector, mobility method for the drift tube, dimensions of the drift tube, detector characteristics, and speed of electronics, including parameters of signal processing. While these are controllable with research-grade instruments, they are commonly pre-determined with commercial instruments and are not easily altered by the operator. Measurements with all ion mobility instruments are affected significantly by experimental parameters, including the chemical composition of the drift gas, levels of moisture of the supporting atmosphere inside the drift tube, temperature of this same gas, and any intentional or unintentional change in the identity of reactant ions. All of these are controllable, in principle, with laboratory- and research-grade instruments, and some, such as temperature and drift gas moisture, may be controlled with commercial instruments. As a rule, experimental parameters for handheld mobility analyzers are inflexible apart from moisture and gas purity, which are controlled by onboard and exchangeable gas purification filters.

11.2 CHEMICAL COMPOSITION OF THE SUPPORT GAS ATMOSPHERE

11.2.1 EFFECTS ON MOBILITY COEFFICIENTS

Intentional changes in the composition of the drift gas in ion mobility spectrometry (IMS) have been compared functionally to chromatographic methods for which relative retention and separation factors can be controlled with changes in mobile or stationary phase.[1] While this analogy can only be used loosely, the influence and value of controlling the drift gas composition on mobility of ion swarms was recognized and investigated almost from the advent of IMS.[2] Although nitrogen, ambient air, and purified air have been used commonly as drift gases and for introducing sample to the drift tube, mobility coefficients in several other drift gases (helium, argon, and neon, to a lesser degree) have been examined. Others include low molecular weight substances such as ammonia and methane and relatively heavy gases, such as carbon tetrafluoride and sulfur hexafluoride.[3–9] Apart from the role of the molecular weight of the drift gas in K_o values, two other characteristics of gas atmospheres have strong influences: the dipole moment and polarizability. Although collision frequencies of an ion traveling in a gas atmosphere can be calculated from the theories for ion–molecule collisions (see Chapter 10), sophisticated models to predict mobility coefficients for ions in a range of gases are not available, particularly for polar drift gases. In contrast, existing elementary models of ion mobility (Chapter 10) are satisfactory in predicting the mobility coefficients within a homologous series of ions in a given drift gas at a given temperature or, more precisely, when interpolating and in some cases extrapolating "missing" members of the series. Nevertheless, the theory and tools to relate the ion-neutral interactions in the gas phase to the measurements for analytical ion mobility methods are lacking, and experimental knowledge of ion mobility in one drift gas is not directly useful in predicting the mobility coefficients in a gas of another chemical composition.[1]

The models for relating ion structures and experimental collision cross sections were originally developed by physicists for monoatomic ions with helium as the neutral drift gas. Due to the low polarizability of helium, and therefore its weak electrostatic interactions with ions, the theoretical descriptions for interactions between an ion and the He drift gas are in relatively good agreement with experimental results.[10–12] In one example with a series of primary and tertiary aliphatic amines, mobility values could be predicted for several different drift gases from theory combined with empirical results.[13] Later, the ionic radii for several compounds in different drift gases could be estimated using a less-rigorous model.[14] This model facilitated an empirical linear correlation between the drift gas polarizability and calculated ion radii, showing that ions with equivalent masses but different structures could still be separated in certain drift gases. Today, systematic experimental studies of the mobilities of polyatomic ions in various gases are needed to refine and validate predictive models.

The formation of ion–molecule clusters is strongly influenced by the gas temperature and by the polarizability of the drift gas; these combined can govern the effective mass of the ion, and its collision cross section, and thus affect mobility as shown in Figure 11.1. In Chapter 10, it was shown that helium does not cluster with ions at ambient pressure; therefore, the agreement between the theoretical mobility of ions drifting through helium atmosphere and the measured mobility values is good. In contrast, substances with high polarizability and mass will associate with ions, forming clusters and changing the effective mass and mobility of the core ion. This same behavior can be observed for impurity molecules when present in a drift gas intentionally or inadvertently; examples include water molecules as moisture or substances such as impurities from material off-gassing. These effects are more pronounced when the gas temperature in the drift tube is low or in the low electric field part of the cycle with differential mobility spectrometry (DMS). A more detailed discussion of the mass–mobility correlation and the changes in the dimensionless collision cross section $\Omega^*(T^*)$ for each ion in each drift gas at any given temperature is found in Chapter 10. In brief, for a given ion, $\Omega^*(T^*)$ is much smaller in helium than in any other drift gas, and the value of the product $\Omega_D T^{1/2}$ increases with ion mass in all drift gases. This product also increases with the molecular weight of the drift gas for a given ion. Finally, the reasons were given in Chapter 10 for why $\Omega_D T^{1/2}$ increases with temperature in helium and air, is practically unchanged in CO_2, and actually decreases with temperature in SF_6.

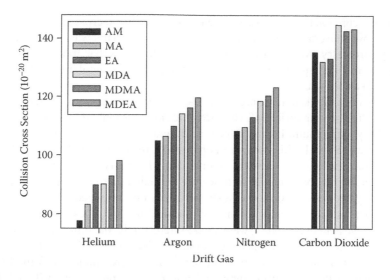

FIGURE 11.1 Effect of supporting gas atmosphere on collision cross section of six amphetamines in four gases. Although the core ion (MH⁺) is the same in each gas, adducts and associations are influenced distinctly by each gas. Differences in cross section will lead to differences in mobility. The substances are as follows: AM, amphetamine; MA, methamphetamine; EA, ethylamphetamine; MDA, 3,4-methylenedioxy amphetamine; MDMA, 3,4-methylenedioxy methamphetamine; and MDEA, 3,4-methylenedioxy ethylamphetamine. (From Matz et al., Investigation of drift gas selectivity in high resolution ion mobility spectrometry with mass spectrometry detection, *J. Am. Soc. Mass Spectrom.* 2002, 300–307. With permission.)

The most commonly used drift gas in IMS is air, usually ambient air that is passed over a molecular sieve filter to remove moisture or reduce its content as well as that of other vapor impurities that may be present. Helium is the drift gas of choice for some applications, particularly when a mobility spectrometer serves for preseparation, or as a filter, for study of macromolecules in ion mobility–mass spectrometry (IM–MS) instruments (see Chapter 9). Argon has also been used in some studies, but it should be noted that when inert gases like He or Ar are used, the electric field gradient should be kept low (usually less than 100 V/cm) to prevent arcing, which could cause serious signal-to-noise issues or even damage the device. Heavier drift gases such as carbon dioxide and sulfur hexafluoride have also been used in some studies and can be deployed in special cases either in pure form or as mixtures with lighter gases to improve separation of isomers.

In principle, almost any permanent gas can be used in the drift region of a mobility spectrometer, and some may even yield interesting and unexpected results. Initially, the approach of researchers was to do their utmost that no reactions should take place in the drift region of the mobility spectrometer so that any ion entering the drift region should traverse the distance between the shutter and the detector plate intact without any change in its composition. In more recent understandings, the ion is viewed as a core ion that is in dynamic equilibrium with drift gas molecules and water molecules, and this core ion picks up and loses attached molecules while moving through the drift region. Thus, an ion moving through a drift gas can be viewed as a core ion that has an effective mass and collision cross section affected or modified by an average number of clustered molecules. This was established when mass spectrometry (MS) measurements revealed that ions that appeared as a single peak in the mobility spectrum actually were composed of several different molecules attached to a core ion.

Effects on analytical response from the selection of gas atmosphere are also seen with field asymmetric IMS (FAIMS) methods for ions of cesium, gramicidin S, tetrahexylammonium, heptadecanoic acid, and aspartic acid in gases of N_2, O_2, CO_2, N_2O, and SF_6.[15] In this early work on FAIMS, the change in the ratio of high- to low-field ion mobility was shown to be invariant with gas type provided the depth of the well for interaction potential significantly exceeded thermal energy (i.e., a stable ion mobility adduct was formed or a strong interaction occurred). This was seen to be mass dependent, and interactions for gramicidin S, tetrahexylammonium, and heptadecanoic acid in N_2 and O_2 were weak. Interestingly, the influence of the gas on changing the alpha function was not governed by the polarizability of the gas alone.

In a next step, blends or mixtures of these gases were employed in the FAIMS method, and results were surprisingly different from the prediction of Blanc's law.[16] In Blanc's law, the mobility of an ion in a binary mixture of the purified gases is a linear combination of fractional composition times the mobility coefficients in individual gases. This deviation was exploited with planar FAIMS separation of biological molecules.[17] Helium-rich gas mixtures with up to 74% He in nitrogen permitted the separation of phosphopeptides with variant modification sites, and the results were superior in resolution possible with pure gases alone.

11.2.2 INFLUENCE ON ION FORMATION

The composition of a supporting gas atmosphere not only may alter the reagent ion as described in Section 11.2.6 but also may lead to characteristic ionization patterns without any addition of reagents. One well-known example is the ionization of vapors of trinitrotoluene (TNT) that in air produces mainly $(TNT-H)^-$, which arises from the formation initially of a $TNT \cdot O_2^-$ ion. The $(TNT-H)^-$ is formed through a subsequent proton abstraction reaction; a somewhat acidic proton from the TNT is removed by the adducted O_2^-. In contrast, the ion of highest abundance is (TNT) when the $[O_2]$ is 2% or less in nitrogen[18] and is formed through charge transfer reactions (Chapter 12). The negative ion formation from a radioactive (^{63}Ni) ion source in ambient air produces O_2^-, CO_2^-, and some other ionic species; in pure nitrogen, only electrons are observed. In general, the role of moisture and temperature are pronounced in affecting ionization processes and are treated in separately in Section 11.3.

11.2.3 SEPARATION OF ISOMERS

The mobility coefficient is dependent on shape and charge distribution and not only on mass; thus, structural and geometric isomers that are isobaric (and unresolved in most MSs) may be separated by IMS. This was demonstrated early in development of IMS and seen then as an attraction of mobility-based measurements.[19–21] This feature of isomer separation can be controlled by the composition of the supporting gas atmosphere. Isomers of primary and tertiary amines that have very similar K_o values in helium[5] are easily distinguishable in air (see Table 10.1, Chapter 10). Another example is the similarity in the reduced mobility values of the isomers trimethylamine (TMA) and isopropylamine in helium, CO_2, and SF_6, while in air they are readily separated.

The mobilities of 18 amines from three types of structural classes (primary, secondary, and tertiary) were measured by IMS using four types of drift gases (helium, argon, nitrogen, and CO_2). The predictions of mobility values based on the commonly used theoretical models were compared with the experimental data. The model that appeared to have the best fit was the model that considered the displacement of the center of mass from the center of charge for the polyatomic ions.[22] However, the main challenge is to separate isomeric ions that overlap in the mobility spectrum in air, the most commonly used drift gas.

11.2.4 SEPARATION OF OVERLAPPING PEAKS

Ion mobility spectrometry is generally considered as a method with low resolving power, and the overlap of peaks from different compounds in the mobility spectrum is a common occurrence with measurements in purified air as the drift gas. One method to control peak position and thus resolution is to change the drift gas by introducing a component and creating a gas mixture. Since this can be beneficial for resolving peaks of isomers, there may be benefits for separating compounds that differ in molecular weights. Indeed, this has been demonstrated on numerous occasions for IMS and DMS measurements.

One example of such control of peak separation is that of fluoroaniline and aminobenzonitrile, which are almost fully convolved in nitrogen and argon, even when measured with an ultrahigh-resolution IMS.[23] When He is used as the drift gas, the two compounds are partially separated with a larger K_o for fluoroaniline than aminobenzonitrile, and the ion peaks are completely resolved in a CO_2 atmosphere with a larger K_o for aminobenzonitrile than fluoroaniline. These changes in relative drift time are attributed to changes in the interactions between the ion and molecules of the drift gas. Another example is separation of 19 compounds belonging to different chemical groups or classes (cocaine and metabolites, amphetamines, benzodiazepines, and small peptides) by control of four gases (helium, argon, nitrogen, and CO_2) with high-resolution IMS.[1] Peaks due to diazepam and lorazepam were not separated in nitrogen and strongly overlapped in argon but were well separated in helium and CO_2.

11.2.5 ADDITION OF REAGENT GASES, DOPANTS, AND SHIFT REAGENTS

Substances can be introduced into the drift gas, or throughout the supporting gas atmosphere of a drift tube, to control mobility coefficients through means not possible with simple changes in composition of gases as described in previous sections. Such substances also may influence the formation of ions with a degree of flexibility not possible with simple gas mixtures. These substances are termed reagent gases, dopants, or shift reagents, and they may affect ionization processes, ion mobility, or both. Regardless, all may be considered additives rather than major constituents in the gas atmosphere since the concentration levels for the substances range usually from a low of 1 ppm to a maximum, rarely used, of 5,000 ppm.

Reagent gases or dopants may control response by moderating the formation of ions in a mobility spectrometer,[24] and this was utilized in an early military application for which suppression of response from sample matrices was necessary with battlefield-deployed technology. The chemical agent monitor (CAM) of Graseby Dynamics was originally equipped with a permeation source containing acetone (Ac), which replaced the hydrated proton of a purified air atmosphere with a proton-bound dimer $(Ac)_2H^+$. Response by CAM to airborne vapors then was restricted largely to molecules with dipoles, proton affinities, and concentrations greater than acetone, effectively washing out response to a large number of potential interferences. The presence of a large amount of acetone throughout the entire drift tube did affect ion swarms with product ions that were formed by displacement of Ac by the analyte molecule in the proton-bound dimer (see also Chapter 13).[25] This concept of alternative reagent gases was eventually patented.[26]

An example of the control of ionization over a large range of compounds of several chemical families was made with three reagent gases: water, acetone, and dimethyl sulfoxide (DMSO). In studies with nearly 120 volatile organic compounds from a range of chemical families (i.e., association energies to a proton), each reagent gas provided a level of selectivity in response through chemistry of ion formation.[27] Response with hydrated protons provides little selectivity, and response to a complex mixture, even with a gas chromatograph (GC) to prefractionate sample, provided little possibility of determining organophosphorous compounds in the mixture. When acetone was introduced as a reagent gas at about 1 ppm in the reaction region, some chemicals were no longer ionized, providing enhanced selectivity through ion molecule reaction chemistry. Response to virtually all volatile organic compounds from alcohols, aldehydes, ketones, and esters was eliminated or strongly suppressed. When DMSO was used as the reagent gas, response was limited to amines, organophosphorous compounds, and a few chemicals from the other families. Mobility spectra for the organophosphorous compounds, in the presence of DMSO as the reagent gas, were nonetheless distinctive and characteristic of each chemical.

A recent systematic discussion of reagent gases or dopants in IMS has been provided and includes discussion for several of the most commonly used reagent gases, including acetone, ammonia, and nicotinamide.[28] In addition, discussion was given for some of the less-well-distributed chemical dopants, and presentation clearly distinguished between reagents intended largely to affect ionization chemistry and those intended to control ion mobility once an ion was formed. A distinct advantage of mobility measurements at ambient pressure rather than 1 to 4 torr with He or N_2 drift gas, as practiced in most electrospray ionization (ESI) IM–MS experiments with biological molecules, is the additional control of peak resolution using reagents that displace or shift ion peak drift times. One example is the addition of methylsalicylate to the drift tube gas atmosphere to cluster with the reagent ion in negative polarity in air and O_2^- and displace its peak well away from the peaks of the small product ions. Thus, an obvious and uncluttered response to F^- was observed in monitoring ambient air for the presence of HF.[29]

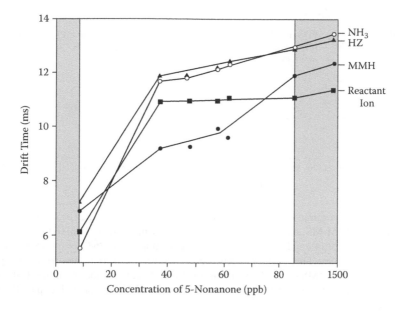

FIGURE 11.2 Plots of drift times for ammonia, hydrazine (HZ), and monomethylhydrazine (MMH) with vapor levels of 5-nonanone in the drift gas ranging from trace levels (10 ppb or less) to 1.5 ppm. Although all ions undergo increases due to the formation of ion-neutral adducts, the ordering of drift time is by ion mass in relatively unmodified gas, and when vapor levels of 5-nonanone are large, drift times are altered and ordered by the number of N–H bonds in an ion to adduct a ketone molecule. (Drawn from data in Eiceman et al., Ion mobility spectrometry hydrazine, monomethyhydrazine, and ammonia in air with 5-nonanone reagent gas, *Anal. Chem.* 1993, 65, 1696–1702.)

A similar example, except one for which the product ions were directly affected by the dopant, was the simultaneous separation of peaks for ammonia, hydrazine, methylhydrazine, and dimethylhydrazine using nonanone.[30] The ketone formed cluster ions with each of the available hydrogen atoms, and thus four ketone molecules were clustered on protonated hydrazine, three formed clusters with methylhydrazine, and just two formed them with dimethylhydrazine. The increase in mass and cross section to the core ion from the adducted ketones caused shifts in the drift time of the product ion peaks, so that clustered hydrazine had the longest drift time and dimethylhydrazine the shortest.[31] This change in drift times is shown in Figure 11.2 with nonanone vapors near 1 ppm in the gas atmosphere throughout the drift tube.

Pharmaceutical and biochemical applications of shift reagents have also occurred, for example, with pharmaceutical ingredients,[32] which form noncovalent complexes with polyethylene glycol (PEG) and lamivudine, used for treatment of chronic hepatitis B and HIV. After separation in a drift tube (Figure 11.3), the ion complexes were dissociated using collision-induced dissociation (CID) to recover the protonated lamivudine. This made the measurement free from interfering matrix ions and with a drift time associated with the precursor complex.[32] Other shift reagents for biological molecules have included crown ethers[33] and metal ions, specifically lanthanum and lutetium.[34]

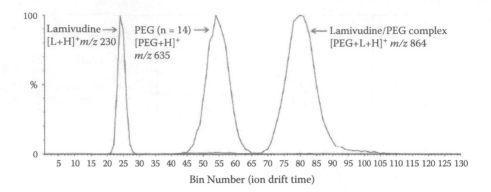

FIGURE 11.3 An example of the formation of adducts in drift tubes can be seen in mobility spectra from ESI-IM-MS analysis of a solution of a pharmaceutical, lamivudine (*m/z* 230), of PEG (*m/z* 635, *n* = 14), and for a lamivudine/PEG complex (*m/z* 864, *n* = 14) overlaid with normalized intensities. This example occurred with special efforts to modify the supporting atmosphere of an IMS drift tube and was attributed to clusters formed in the ESI spray. In this instance, the ion clusters were strongly held and had lifetimes enough to pass through the IMS drift region.

More demanding separations in ion mobility, than those noted directly preceding this paragraph, can be found in the resolution of peaks for enantiomers. In these methods, a chiral dopant is added to control drift times based on stereospecific interactions with the selected enantiomers. In the mobility separation of optically active amino acids, an asymmetric volatile chiral reagent [in this instance, (S)-(+)-2-butanol] permitted enantiomeric separations of atenolol, serine, methionine, threonine, methyl r-glucopyranoside, glucose, penicillamine, valinol, phenylalanine, and tryptophan from their respective racemic mixtures.[35] A similar approach was used to separate valinol from serine by adding 2-butanol to form sterically hindered clusters with serine, reducing its mobility by 13.6%.[36] When the temperature was increased from 100°C to 250°C, the two peaks were no longer separated, as could be expected from declustering at this elevated temperature.

11.2.6 Multiple Reagent Gases to Simplify or Clarify Response

Since drift tubes for mobility spectrometers can be comparatively inexpensive, several analyzers (CAMs) were used in parallel, each with a particular dopant, and operated continuously for ambient air monitoring.[37] The dopants were water, acetone, and DMSO as used with GC-IMS analysis of a complex mixture of volatile organic compounds.[27] The CAMs were not modified otherwise and were operated with positive ion polarity, in parallel, and without any chromatographic prefractionation of sample. Ambient air inside a university work site, the stockroom of the Department of Chemistry and Biochemistry at New Mexico State University, was sampled continuously for several hours before and during laboratory classes for undergraduate students when activity increases in the stockroom with dispensing materiel and chemicals. Each analyzer provided response from general to specific, based on selectivity of ionization with each of the reagent gases. Thus, the ionization chemistry alone can bracket a chemical family; for example, alcohols should be detectable in the CAM with water as the reagent gas but not in the CAMs containing acetone or DMSO. Similarly, ketones and esters should be observable in the CAMs with water and acetone but not the CAM with DMSO. In practice, the reactant ion peak intensity was recorded against clock time, although the mobility spectra provided additional information on the K_o values and concentrations for individual analytes. This concept may be revitalized with the development of miniature drift tubes and the small DMS instruments, which can be manufactured inexpensively.

11.2.7 MODIFIERS IN GAS ATMOSPHERES OF FAIMS AND DMS

The addition of reagents to supporting gas atmospheres is now understood as essential to the analytical value of FAIMS or DMS. Early studies on the effects of moisture on ion behavior in planar small DMS analyzers demonstrated that clusters between ion and neutrals significantly and favorably affected the location of ion peaks on the compensation voltage (CV) scale, opening analytical space over measurements with lesser levels of moisture.[38] Alpha functions increase hugely above 1,000 ppm moisture, although eventually levels of moisture began to interfere with formation of reactant and product ions. This suggested that other substances could be used to control peak location and separation in DMS without deleterious effects of moisture on ionization chemistry, and methylene chloride was successful in distributing ion peaks from explosives over an expanded range of CV values,[39] in contrast to a comparatively narrow range of CV values in purified air alone at 100°C. Peaks were confined to a narrow range of CVs between −1 and +3 V, which arose through a low dependence of mobility for the ions in electric fields at E/N values between 0 and 120 Td. When methylene chloride was added at 1,000 ppm in the drift gas, the scale of CVs expanded between +3 and +21 V for peaks. This improved separation of peaks was attributed to modeling of ions clustering with vapor neutrals during the low-field portion of the separation field waveform and ion declustering of the same core ion when heated and during the high-field portion of waveform. This concept of modification of the gas atmosphere has been confirmed for explosives and extended for negative ions of phthalate acids.[40]

The extension of DMS with an ESI source for measurements of biomolecules, here peptides, was first reported in 2006 with a DMS-MS instrument.[41] The separation of peptides was poor and attributed to the formation of higher-order peptide aggregate ions (ion complexes) from the ESI source. Drift gas modifiers were used in this instance to reduce aggregate ion size and improve ion separation in the CV scale (Figure 11.4). This was extended and expanded in a following work,[42] for which results were consistent with a cluster formation model, also referred to as the dynamic cluster–decluster model. The uniqueness of chemical interactions that occur between an ion and cluster-forming neutrals increases the selectivity of the separation, and the depression of low-field mobility relative to high-field mobility increases the compensation voltage and peak capacity.

The effect on peak capacity in DMS or DMS–MS with polar modifiers was examined with favorable results over broad range of chemicals.

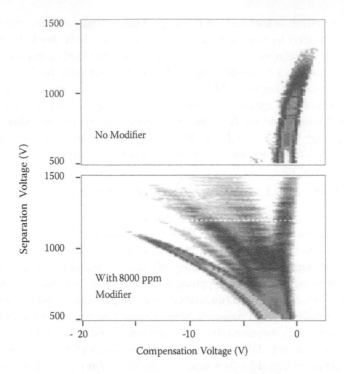

FIGURE 11.4 The influence of a modifier in the supporting atmosphere of a DMS measure. The dispersion plots is a graph of ion intensity, separating voltage (*y*-axis) and compensation voltage (*x*-axis). Shown in the top plot are ions of an angiotensin fragment and neurotensin. In the bottom plot, the same sample is characterized in 8,000 ppm of 2-butanol. (Modified from Levin et al., Rapid separation and quantitative analysis of peptides using a new nanoelectrospray-differential mobility spectrometer–mass spectrometer system, *Anal. Chem.* 2006, 78, 5443–5452.)

11.3   MOISTURE AND TEMPERATURE OF THE SUPPORTING GAS ATMOSPHERE

The importance of moisture and temperature in overall performance of ion mobility methods can hardly be overstated and is certainly the most significant of parameters that can be controlled on most analyzers. Operating mobility analyzers without knowledge of these two parameters is equivalent to making measurements in MS without knowledge or control of vacuum or, in the instance of electron impact ion sources, the electron energy. Remarkably, mobility spectrometers function well across a wide range of moisture levels and temperatures, yet without control of these parameters, response and reproducibility will be difficult to understand, and comparisons between laboratories may be difficult to achieve with a high level of confidence. Although fundamental measurements of association energies of gas phase protons with water molecules were made in the 1960s and early 1970s,[43] the importance on response has been experimentally explored only in the past two decades for atmospheric pressure ionization (API) MS[44,45] and only recently for IMS.[46]

11.3.1 Effects in Ion Mobility Spectrometry

The influence of moisture in an IMS measurement begins with the formation and composition of reactant ions available in the reaction region of the drift tube. While the formation of hydrated protons can be assumed when moisture is 1 ppm and above, at levels of moisture of 100 ppb and below, the lifetime of ions such as H_2O^+, N_4^+, and N_2^+ may be sufficient for reactions with analyte.[43] At practical extreme moisture levels, 10 ppb and below, metastable ions may exist in helium.[47] Consequently, a continuum of ion populations with moisture can be anticipated in both the reactant ion, and the associated reactions depend on moisture.

In purified air or nitrogen at ambient pressure and temperature, the principal reactant ions in positive polarity will be hydrated protons, and the product ions will be formed through displacement-type reactions of Equation 11.1:

$$M \;+\; H^+(H_2O)_n \;\leftrightarrow\; MH^+(H_2O)_n^* \;\leftrightarrow\; MH^+(H_2O)_{n-x} \;+\; xH_2O \qquad (11.1)$$

| Sample neutral | + | Reactant ion | Cluster Ion | Product ion Protonated monomer | + | Water |

Generally, only molecules with dipole moments and proton affinities greater than that of water will be able to displace the water molecules and remain intact at moisture levels greater than 10 to 100 ppm and thus provide response in the instrument. This excludes alkanes, alkenes, aromatic hydrocarbons, and perhaps some alcohols. As moisture levels reach 0.1 to 0.5 ppm, charge exchange reactions may be observed as shown in Equation 11.2:

$$M \;+\; H_2O^+ \;\leftrightarrow\; M^+ \;+\; H_2O \qquad (11.2)$$

| Sample neutral | + | Reactant ion | Product ion | + | Water |

In charge exchange reactions, the main consideration is ionization energy of the analyte, and that of H_2O is 12.62 eV. Thus, product ions should be formed from a range of molecules, including alkenes and some alkanes and most of all other organic compound classes.[48] The analytical consequences of moisture then include changes in the relative sensitivities of response to analytes, in the formation of ions characteristic of the reactant ions, and finally some fragmentation in air at ambient pressure.[49] Fragment ions have been found for nearly all chemical families even at low temperatures when moisture levels are below 0.5 ppm.[50–52] These are found near the reactant ion peak and were discovered through neural network analysis of regions of spectra where ions of low intensity and peaks clustered near the reactant ion peak could be assessed.

Recently, the mobility spectra for TMA were obtained over a range of moistures, and emphasis was given to quantitative characteristics and how moisture inside the drift tube affected detection.[53] As a molecule with strong proton affinity, TMA was ionized over a range of moistures. When response was based on the monomer ion peak or the sum of peaks generated by the analyte, the sensitivity of detection was not strongly dependent on humidity. The proton-bound dimer was quantitatively useful only when moisture was low. Moisture in the drift region altered drift times of ionic species, and this was reasonably attributed to the average degree of solvation, which increased with elevated levels of moisture. Changes in drift time for a given increase in moisture levels are more pronounced at low levels. The drift time of proton-bound dimers of TMA did not change with increased levels of moisture, and this was understood as poor, if any, hydration of the proton-bound dimer ion.

Temperature of the gas atmosphere in an IMS drift tube, and hence the ion that is thermalized under ambient pressure and ordinary fields of 200 to 400 V/cm, affects mobility coefficients, ion identities, and finally ion stabilities. In a nonclustering gas, increase in temperature (i.e., T_{eff}) largely follows an inverse while causing a decrease in the reduced mobility coefficients K_o as discussed in Chapter 10. The least-developed aspect of predictive and interpretative capabilities in formulas for mobility coefficients is the relationship between ion temperature T_{eff} and the collision cross section $\Omega(T_{eff})$. Quantitatively unclear, in a polarizable gas, is the dependence of $\Omega(T_{eff})$ on ion temperature. Qualitatively, increases in temperature cause declustering of ions in polarizable atmospheres with decreases in $\Omega(T_{eff})$ and increases in K_o. Simultaneously, the effective mass of the core ion is reduced when water molecules are stripped from the hydration shell. This occurs for all ions, including product ions and reactant ions, which to some measure can be stripped of waters of hydration, improving response factors, as demonstrated with API MS but not yet with IMS. In an early study of effect of temperature on mobility, all reactant ion peaks in positive polarity underwent steady increases in mobility with temperature (Figure 11.5a), and this can be associated with the ion distributions in a swarm moving through the drift tube (Figure 11.5b).

In given concentrations of sample vapor neutrals, spectra for product ions can be altered by control of temperature, and this was seen at cryogenic temperature, at which ion clusters not usually observed in mobility spectra were formed and appeared in mobility spectra. For example, proton-bound trimers of alcohols were observed when temperatures were decreased to −20°C and dissociated at temperatures from −20°C to +10°C.[54] Increases in temperatures will lead to dissociation of complex ions, such as proton-bound trimers and proton-bound dimers. As temperature is increased, the intensity of peaks for protonated monomer increase, and the peak abundance of proton-bound dimers decreases. This has been developed and explored for dimethyl methyl phosphonate (DMMP),[54] amines,[55] and ketones.[56,57] For example, proton-bound dimers of alkyl amines underwent dissociation above −30°C on a 2- to 20-ms time scale, which is within the range of drift times for these ions. Consequently, the dissociation pathway can be observed as a distortion in the peak shape and baseline of a mobility spectrum since an ion entering the drift region as a proton-bound dimer dissociates to a protonated monomer before arriving at the detector. These studies permitted the determination of kinetics of dissociation for thermalized ions and illustrated that the appearance of an ion in a mobility spectrum is governed by ion lifetimes in comparison to ion residence times in drift tubes, and ion lifetimes are controlled by temperature.

Increases in temperature can also affect the stability of ions, resulting in fragmentation in ways reminiscent of electron impact ionization in MS. The earliest systematic report on such fragmentations was for butyl acetates; thermalized ions were heated to induce decomposition, forming carboxylic acid fragments at 100–125°C.[58] There was

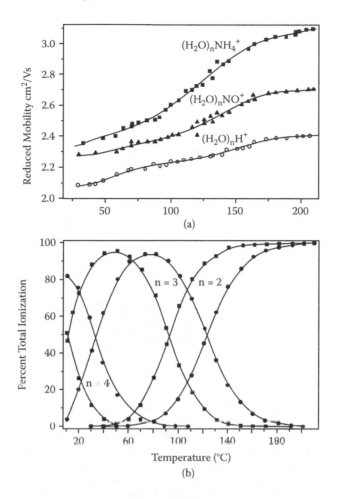

FIGURE 11.5 Effects of temperature on the reduced mobility of reactant ions in IMS. (a) The reduced mobility for $H^+(H_2O)_n$ gradually increased as temperature increased due to decreasing extent of hydration of the ion. The resultant mobility is the weighted average of all hydrated protons. (b) The distribution of ions according to n is shown. (Adapted from Kim et al., Mobility behavior and composition of hydrated positive reactant ions in plasma chromatography with nitrogen carrier gas, *Anal. Chem.* 1978, 50, 2006–2016.)

evidence for fragmentation directly of a proton-bound dimer; alternatively, dissociation to a protonated monomer preceded fragment ion formation, and the pathway was unclear. Increases in temperature of ions of alcohols resulted in decreased intensity of the protonated monomers and the appearance of fragment ions for alcohols from 125°C to 150°C.[59] Other chemical classes exhibited similar behavior, although the temperature at which fragmentation of the ions occurred was class dependent. For example, proton-bound dimers of ketones underwent dissociation only above 175°C, and fragmentation was not observed until the temperature exceeded 225°C.

Temperature control factored into a low-pressure, variable-temperature (80 to 400 K) IM–MS with an electron ionization source. This instrument with an

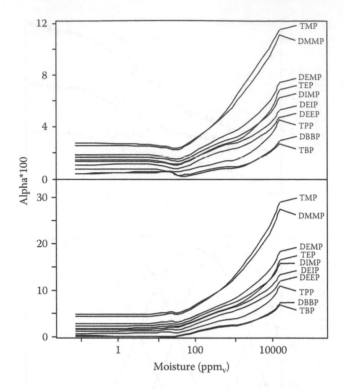

FIGURE 11.6 Plot of alpha values for protonated monomers of organophosphate compounds obtained at two fields of 80 (bottom) and 140 (top) Td as a function of moisture. DMMP, dimethyl-methylphosphonate; TMP, trimethyl-phosphate; DEMP, diethyl-methylphosphonate; DEEP, diethyl-ethylphosphonate; DIMP, di-iso-prophyl-methylphosphonate; DEIP, diethyl-iso-prophylphosphonate; TEP, triethyl phosphate; TPP, tripropyl phosphate; DBBP, dibutyl-butylphosphonate; TBP, tri-n-butylphosphate. (From Krylova et al., Effect of moisture on high field dependence of mobility for gas phase ions at atmospheric pressure: organophosphorus compounds, *J. Phys. Chem.* 2003. With permission.)

orthogonal time-of-flight MS was used to mass identify mobility separated ions, specifically long-lived electronic states of Ti[+].[60]

11.3.2 Effects in Differential Mobility Spectrometry

The effects of variations in moisture with differential mobility methods in ion formation are identical to those observed in low-field methods, although the consequences for ion mobility are distinctive to this method. Effects were studied in detail using a set of organophosphorus compounds whose mobility dependence on field, or alpha functions, were determined at E/N values between 0 and 140 Td at ambient pressure in air with moisture levels between 0.1 and 15,000 ppm.[38] At moisture levels of 0.1 to 10 ppm, the alpha function for protonated monomers was unchanged (Figure 11.6). At 50 ppm, there was an onset of change in the alpha function that increased twofold when moisture level was raised from 100 to 1,000 ppm at all E/N values. Changes of the alpha function with moisture continued with another twofold increase with moisture from 1,000 to 10,000 ppm. These measurements were used to establish a model for field dependence for mobility through changes in collision cross sections that are governed by the degree of solvation of the protonated molecule by neutral molecules. The model was supported by calculations of collision rates for ions and neutrals and the duty cycle of the waveform applied to the drift tube.

Any discussion of the effects of temperature in a differential mobility measurement is more complicated than in an IMS measurement since ions are not thermalized. In these measurements with asymmetric waveforms with field extremes, at ambient pressure, of −1,000 to 30,000 V/cm or greater, ions are near thermalized in the low portion of the waveform (−1000 V/cm); however, ions are heated by the electric field during the high portion of the waveform when the energy from the field is large compared to thermal energy. This causes an increase in ion temperature in a cooler surrounding gas atmosphere. While this topic might be discussed in the following section on effects of electric fields, descriptions of the effects are included here since temperature, including ion temperature, is the theme of this section.

Increases in ion temperature by field heating is known to have several effects, including

a. declustering of proton-bound dimers and other clusters
b. production of reactions within an ion cluster, such as hydrogen abstraction
c. fragmentation of ions through the breaking of covalent bonds

In nearly all of these, the best graphic representation of the role of electric field on the reactions is a contour plot of ion intensity, separation voltage, and CV, which is known as a dispersion plot. A dispersion plot is shown in Figure 11.7 for the fragmentation of esters. The first link between ion energy gain from fields and subsequent reactions was the decomposition of aromatic hydrocarbons with increased field strength.[61] Extensive fragmentations were also observed with esters, leading to the production of carboxylic acid fragments.[62] A particular discovery was that energy flow into the ion was mass dependent, with small ions undergoing greater heating than larger ions. This seemed to provide evidence that the protonated monomer in the butyl acetate studies with IMS, described previously, was rapidly heated and fragmented, and a rate-limiting step was the slower heating of proton-bound dimer to protonated monomer. The proton-bound dimer was too large to be heated directly to fragment ions.[63] The minimum field for first observation of the product ion decreased by 0.68 Td for every 1°C increase in instrument temperature; that is, 1 Td of high-field heating raised the effective temperature of the decomposing ion by 1.5°C. The combination of thermal energy and energy derived from collisional heating by acceleration in the asymmetric electric field caused ion decomposition at an effective temperature T_{eff} higher than ambient. Proton-bound dimers of ketones[57] were also examined by field heating, as was DMMP, a well-studied substance in IMS.

Ion decomposition by field heating is becoming a strategy for chemical identification with the ultraFAIMS technology of Owlstone Nanotech; strong electric fields (60,000 V/cm) and high-frequency waveforms (26 MHz) promote ion decomposition. Dispersion plots in the region of ion dissociation or fragmentation are proposed to contribute analytical information that supports chemical identifications.[64]

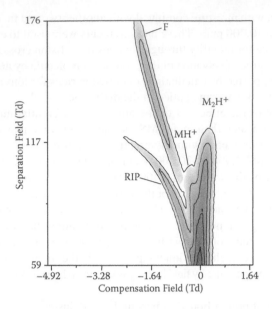

FIGURE 11.7 Dispersion plot for propyl acetate in a DMS analyzer at a temperature of 100°C, from 59 Td (500 V) to 176 Td (1,500 V). The *CV* range is −5 (−43 V) to 1.8 Td (15 V). The signal intensity is both shaded and contour coded, and the trajectory of RIP, M_2H^+, MH^+, fragment (protonated acetic acid) are labeled as RIP, M_2H^+, MH^+, and F, respectively. (From An et al., Gas phase fragmentation of protonated esters in air at ambient pressure through ion heating by electric field in differential mobility spectrometry and by thermal bath in ion mobility spectrometry, *Int. J. Mass Spectrom.* 2011. With permission.)

11.4 EFFECTS OF PRESSURE

11.4.1 Ion Mobility Spectrometry

Two regimes of pressure have been favored in IMS, pressures of 1 to 3 torr and ambient pressure, with virtually no experience or practice between these. Recently, efforts have been made with pressures much greater than ambient pressure, although this is an isolated and developing concept.[65] Historically, there is an understanding that ion losses occur as pressure is reduced, and those instruments operating at 1 to 3 torr in helium or nitrogen have low tolerance for even slight increases in pressures. The only systematic study of effects of pressure in IMS was released in 2006; spectra were obtained at pressures ranging from 29 torr to atmospheric pressure.[66] Results showed that drift times increased with pressure in a "perfectly linear" manner, so that separation factors were unaffected by pressure. Pressure was reported to affect strongly resolving power and resolution. Ion shutter widths ranged from 50 to 225 μs, and reducing the pressure at a constant pulse width decreased the resolving power and resolution. This was compensated by shortening the ion pulse width. In a prior study,[67] the same team explored the role of temperature and pressure in IMS reaching 15 torr and concluded that pressure affects the clustering reactions linearly, but temperature affects it exponentially. This is a somewhat underdeveloped topic in IMS despite the decades of experience with analytical applications of IMS.

11.4.2 DIFFERENTIAL MOBILITY SPECTROMETRY

In contrast to the development of understandings of pressure in IMS only recently, pressure was systematically examined in a planar differential IMS operated in a supporting atmosphere of purified air at pressures from 0.4 to 1.55 atm.[68] The CVs for ion peaks were found to vary with pressure and could be simplified by expressing both compensation and separation fields as ratios of E/N in townsend units. The greatest CV and the greatest resolution were expected at E/N values that are unreachable at elevated pressure because of electrical breakdown. Nonetheless, breakdown voltage for air near 1 atm is nonlinear, and higher E/N values could be reached at reduced pressure, resulting in improved resolving power. An increase of approximately 15% in E/N can be achieved by reducing the operating pressure from 1 to 0.5 atm, and separation voltages can be reduced by a factor of 2. Operating the DMS instrument at pressures below 1 atm provided reduced clustering, and improved performance in DMS was observed; however, decreases in pressure also led to ion losses, and limits to reducing pressure exist if ion intensities are impractically low.

11.5 EFFECTS OF FIELD STRENGTH AND ION RESIDENCE TIME

11.5.1 ION MOBILITY SPECTROMETRY

The influence of electric field strength in IMS has been explored for resolving power, and it can be said that increasing field strength improves resolving power.[69] Resolving power goes up with E, but this trend is ion shutter limited. That is, increasing V for a fixed length increases the relative contribution of shutter pulse width to peak width since t_d decreases linearly with V. This means that, for a drift tube of given length, each value for shutter pulse width will have optimum drift voltage V_{opt} at which R is maximized. This was reported as

$$V_{opt} = \left(\frac{\alpha T L^2}{2 \left(\gamma + \beta t_g^2 \right)^{1/2} K^2} \right)^{1/3}$$

where other terms are: α, β, and γ, parameters from fitting peak widths; T, temperature in K; L, length of drift region; K, mobility of ion; and t_g, time of shutter pulse.

Regarding examples of strong fields and resolving power, Leonhardt et al. reported drift fields up to 700 V/cm to reach R values as high as 120 for singly charged species,[70] and Dugrourd et al.[71] reported a low-pressure, high-resolution IMS instrument for which a potential of 14,000 V across a 63-cm drift tube at 500 torr produced a resolving power of 172.

Ion residence inside an analyzer is perhaps the most overlooked parameter in all of IMS methods and fundamentally governs the appearance of a mobility spectrum. The principle is reasonably simple: Response to a substance will be favorable if the lifetime of an ion is greater than the residence time of the same ion in a mobility analyzer. In contrast, response will be poor or nil when the lifetime of an ion is shorter than the residence time; the ion will decompose before reaching the detector. This concept is significant not only in the drift region of an analyzer but also governs the very first step in the formation of ions in an ion source. Ions that are long lived due to strong associations to the reactant charge should be seen in a mobility spectrum, providing decomposition does not occur in a timeframe faster than drift time.

Mobility spectra will exhibit protonated monomers for most polar or strongly polarizable molecules when the reactant ions are hydrated protons and vapor levels of analyte are more than 10 to 100 ppb, the detection limits for most such compounds. Mobility spectra may contain a proton-bound dimer when vapor levels are increased to 0.5 to 1 ppm, yet a proton-bound trimer or tetramer is never observed, even if vapor concentrations exceed those needed to form these higher cluster ions according to equilibrium calculations. In mobility spectrometers today, ions are formed and then drawn into purified air or gases excluding neutrals of sample. Thus, equilibrium does not exist in analytical mobility spectrometers, and ion passage through purified gas should be seen as a kinetic experiment.

Ion lifetimes for protonated monomers are commonly much greater than 20 ms at temperatures up to 100°C and beyond. In contrast, some proton-bound dimers have lifetimes under a few milliseconds and thus are not observed unless temperatures are comparatively low (e.g., –20°C); molecules with strong dipoles have proton-bound dimers that are comparatively long lived and can drift for 20 ms or more. In contrast, the lifetimes for proton-bound trimers in a purified gas atmosphere at ambient pressure and temperatures are under 1 to 5 ms and undergo rapid decomposition; these are never seen in analytical IMS drift tubes unless control of sample vapors is lost and ion-neutral reactions in the drift region occur. Higher clustered ions have even shorter lifetimes than proton-bound trimers and thus are not ever observed in mobility spectra.[54]

11.5.2 Differential Mobility Spectrometry

The immediate effects of electric field in DMS or FAIMS are seen in the dependence of the mobilities of gas phase ions on field strength at two extremes of the asymmetric waveform. The waveform is designed in field strength and duty cycle so that an ion without any dependence of mobility on field passes through the analyzer, center line of sight, carried by a flow of gas. Ions that do have a dependence of mobility on field (K_o at high field strength is not identical to K_o at low field strength) will undergo with each complete cycle of the waveform successive net displacement from the central axis of ion flow. Eventually, the ion swarm will collide with a plate that defines the analyzer volume and be annihilated or discharged and removed from the measurement. When a direct current (DC) potential is applied to the plates, the effects of the electric field with K_o dependence are compensated, and ion motion can be restored to the center of the analyzer. This is a central feature of all mobility-dependent analyzers regardless of the name applied to the method.

A large record exists on the mobility of small ions often in nonclustering atmospheres,[72] and in analytical DMS or FAIMS, the emphasis has been on molecules of industrial, environmental, or medical importance, which tend to be large, complex structures of organic molecules with dipoles, and measurements are made in purified

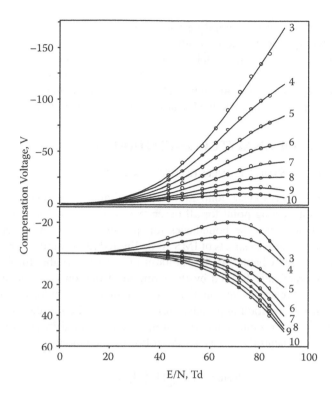

FIGURE 11.8 Plots of compensation voltage versus separation field for a homologous series of ketones from acetone (carbon number 3) to decanone (carbon number 10) for protonated monomer (top) and proton-bound dimers (bottom). These plots show field dependence of mobility for small molecules; nonetheless, a mass relationship to the field-dependent behavior is clearly shown. (From Krylov et al., Field dependence on mobilities for gas phase protonated monomers and proton bound dimers of ketones by planar field asymmetric waveform ion mobility spectrometer (PFAIMS). *J. Phys. Chem.* 2002. With permission.)

air. A systematic study of field dependences under such conditions was made in 2002; a homologous series of ketones was used to determine quantitatively the effect of field strength on mobility from 0 to 90 Td at ambient pressure.[73] Dependence on field for protonated monomers and proton-bound dimers, from acetone to decanone, has been described as a normalized function of mobility $\alpha(E/N)$ versus E/N, which was increased monotonically from 0 to 90 Td for acetone, butanone, and pentanone (Figure 11.8). Functions for hexanone to octanone exhibited plateaus at high fields, and those for nonanone and decanone showed inversion of slopes above 70 Td. Proton-bound dimers for ketones with carbon numbers greater than five exhibited slopes that decreased continuously with increasing E/N.

The formation of positive alpha functions was described in a model that involved ion declustering during the high-voltage portion of the waveform where ions are heated with strong fields (up to 60,000 V/cm in the ultraFAIMS) and then clustered during the weak-field region (which at −1,000 V/cm with small planar DMS analyzers still could heat some ions above thermal energies). The negative dependence of ions is explained by invoking ion drag and collision frequencies, and these effects are shallow functions compared to positive alpha functions. Ion elongation is also a possible contribution to negative alpha functions; increased electric fields result in increased ion cross sections by elongation and thus reduced mobility coefficients with strong electric fields. The secondary effects of electric field on ion stability and decomposition of ions by electric field heating were described in previous sections.

11.6   EFFECTS OF ANALYTE CONCENTRATION

11.6.1   Effect of Sample Concentration on Response

The first step in IMS response, ionization of neutral analyte molecules, fundamentally establishes a starting point for all other events that occur subsequently in the drift tube and determines the boundaries for quantitative response. Central to all discussions of quantitative aspects of IMS is the limited reservoir of charge available for ion formation (see Chapter 4). The amount of charge available for formation of product ions is practically limited by the strength of the ionization source, by the overall sequence of reactions leading to the formation of reactant ions, and finally by the reaction rate of formation of product ions. Thus, when the source is not saturated with analyte vapors, reactant ions $[H^+(H_2O)_n]$ are consumed in proportion to the density of the vapor neutrals $[M]$ as per the following equation:

$$\text{Rate} = k[M]\left[H^+(H_2O)_n\right] \tag{11.3}$$

and the rate constant k in Equation 11.3, which is a correction on the collision-rate constant and controls the rate of formation of product ions. The density of product ions, for a sample at a constant concentration, is dependent on both the residence time of the sample in the source and its concentration. Depletion of the reactant ions is rooted in the kinetics of Equation 11.3 and parallels inversely the formation of product ions. The relationship is virtually stoichiometric. For a fixed time of residence of sample molecules in the reaction region, the number of product ions and their peak intensity in the mobility spectrum will be quantitatively proportional to the concentration of sample molecules. Also, increases in residence time of M in the source region will lead to an increase in the number of products ions when the source is not saturated with sample vapor. Once the reservoir of reactant ion charge is totally consumed, no additional increase in product ion intensity will be seen even with additional increase in the concentration of sample molecules. In practice, this limits the upper end of the linear range associated with ion sources at ambient pressure.

An equally critical component to understanding analytical response curves in IMS is the principles and parameters that establish the limit of detection or profile in the lower concentration range of the response curve. When product ion is formed from molecules of strong dipole or attachment energy to a gas phase proton in positive polarity, the rate of formation is thought to be collision rate based, namely, every collision results in a product ion. The limitation in product ion formation and thus limit of detection is governed fundamentally by the statistics of a collision occurring in the time during which the sample vapors and ions of the source are together in the source region, which is dynamically swept with gas flows (other practical details will affect limit of detection, including signal noise, shutter width, and drift tube designs). In practice with conventional ^{63}Ni sources, this amounts to 1 ms or a few milliseconds, and if, at a given concentration of ions and sample vapor neutrals both diluted in the supporting gas atmosphere, enough collisions can occur to produce enough ions to elevate the signal above background noise, detection will be observed. In most mobility spectrometers, this is in the range of 10^5 to 10^6 ions/s.

One of the best-recognized quantitative patterns in mobility spectra is the relationship between protonated monomers and proton-bound dimers, common to several types of compounds, including esters, ketones, alcohols, amines, and organophosphorus compounds. As the vapor concentration of the analyte M is increased in an ion source, a protonated monomer peak first appears with a corresponding decrease in the reactant ion peak intensity. The reaction yields an energetically excited intermediate adduct ion or transition state ($[MH^+(H_2O)_n]^*$) that may dissociate back to the reactants or may form product ions through other reaction pathways. Such reaction pathways involve a third body Z to stabilize the products, as shown in Equation 11.4:

$$M \quad + \quad H^+(H_2O)_n \quad \leftrightarrow \quad MH^+(H_2O)_n^* \quad \leftrightarrow \quad MH^+(H_2O)_{n-x} \quad + \quad xH_2O \quad (11.4)$$

| Sample neutral | + | Reactant ion | Cluster Ion | Product ion Protonated monomer | + | Water |

A second peak appears corresponding to the proton-bound dimer $M_2H^+-(H_2O)_{n-x}$ with a further increase in analyte concentration; as this occurs, intensity declines for peaks of both the reactant ion and the protonated monomer (Equation 11.5). When the analyte vapors are removed from the ion source and the analyte concentration decreases, these patterns are reversed: Intensity declines for the proton-bound dimer, and it increases for the protonated monomer peak.

$$MH^+(H_2O)_n \quad + \quad M \quad \leftrightarrow \quad M_2H^+(H_2O)_{n-x} \quad + \quad xH_2O \qquad (11.5)$$

| Protonated monomer | + | Sample | Proton-bound dimer | + | Water |

Eventually, the protonated monomer peak intensity declines, and the reactant ion peak is restored to the original intensity. Although this pattern is commonly associated with positive-polarity IMS, similar patterns may be observed with negative ions, such as MCl^- and M_2Cl^-, as observed with volatile halogenated anesthetics[74] and other chemicals like trichloroanisole[75] that form long-lived negative-ion clusters.

When the reactant ion population is completely exhausted, the proportional relationship between product ion intensity and neutral vapor density no longer holds, and this negates conventional calibrations. Thus, the quantitative value of the response of the analyzer is lost with existing methods and interpretations. Another troubling phenomenon is the diffusion of analyte molecules from the source region into the drift region when concentrations of the sample are increased in the source or reaction region. In the drift region, neutral adducts of the sample may associate with product ions of the same chemical, leading to the formation of cluster ions of the type $M_nH^+(H_2O)_x$, in which **n** can have a value equal to or greater than three. Cluster ions in the ion swarm will rapidly undergo loss, followed by the addition of neutral adducts while the swarm traverses the drift region. The observed reduced mobilities for such ions are the weighted averages of the reduced mobilities of all the individual or cluster ions that participate in the "localized" equilibrium.[25] A practical implication of this is that drift times of the product ions can vary over a large range, determined by the concentration of sample neutrals in the drift region. Because distribution of sample neutrals in the drift region will be irregular and dependent on sample concentration, attempts to assign identity to ions using K_o values becomes difficult or impossible, and even peak detection could be difficult due to band broadening. Thus, complete depletion of reactant ions should be avoided, and a residual level of peak intensity for the reactant ions should be maintained to ensure that the reaction region is not saturated, which would lead to quantitatively unclear behavior of and concentration-dependent drift times for product ions. The most modern IMS drift tubes are designed to avoid this condition, but user carelessness can circumvent good designs.

11.6.2 Analytical Facets of Gas Phase Ion Reactions

Ion mobility spectrometers have performed successfully in several applications in which only a threshold response is necessary or important, for example, the detection of explosives and screening of suspect object for drugs of abuse. In these applications, the detection of even trace amounts of the target analyte, above a predetermined or preset level, will trigger an alarm, and no further quantitative information is necessary. In other instances, knowledge of both the presence of a substance and the quantitative amounts is required. Examples of such a quantitative use of an IMS instrument include the determination of chemical warfare agents and the screening of air for volatile organic compounds. In both these examples, human mortality or vitality is associated with dose–response relationships, and the level of dose is essential information.

Improvements in threshold applications of IMS have arisen from improved sampling methods, more sensitive and rugged instrumentation, and better data-processing techniques. These are seen with advances in the detection of explosives with handheld as well as with fixed or stationary analyzers. Parallel advances have occurred with instrumentation and procedures requiring quantitative determinations. These have occurred with the increased availability of drift tubes that exhibit fast response, low memory effects, and reproducible delivery of sample to the ion source. Together, these applications have resulted in enhanced precision, sensitivity, and linear range compared with previous generations of analyzers.

11.7 SUMMARY

Measurements with ion mobility methods depend on the identity of the gas, the ion, and the interactions of this ion with the supporting gas atmosphere. When collision frequencies are large and ions are often found at thermal energies, gaseous ions are influenced fundamentally and strongly by a range of parameters. Virtually every parameter of the mobility measurement, when not carefully controlled, will affect ion identities and thus the appearance of spectra and K_o values. Mobility measurements or K_o values are exquisitely sensitive to changes in ion structure, including waters of hydration, adducts with other polar neutrals, and indeed the clustering with gases considered "unreactive." These effects and consequences are pronounced with mobility methods operated at ambient pressure in clustering atmospheres and fade in importance with low-pressure mobility methods in nonclustering gases. Drift tubes operated at low temperatures are particularly sensitive to the effects of clustering, as are ions of low mass (see Chapter 10). This would be even more severe if a polar drift gas with high molecular weight is used.

At a fundamental or instrument development level, all the "knobs" possible with parameters of these instruments at ambient pressure provide a high level of flexibility and opportunity for mobility measurements. Control of temperature, moisture, and composition of the supporting atmosphere offers flexibilities that extend the value and use of mobility spectrometers from the simple one-temperature, purified nitrogen, fixed-parameter experiment. As a cautionary note here, models and quantitative supporting understanding for most of these parameters are lacking direct support with mobility spectrometers and at present may have one or a few journal articles to support understandings. While ion lifetimes and residence times underlie this method, there has been to date little benefit derived from controlling mobility spectrometers through varying residence times in analyzers.

At a level of practical or routine use, mobility spectrometers provide high value in analytical chemistry with control of experimental parameters made through good engineering and measurement procedures. This performance is good enough that lives of soldiers depend on the reliability of in-field findings with handheld version of ion mobility instruments, and desktop instruments are central to international commercial aviation security.

REFERENCES

1. Matz, L.M.; Hill, H.H.; Beegle, L.W.; Kanik, I., Investigation of drift gas selectivity in high resolution ion mobility spectrometry with mass spectrometry detection, *J. Am. Soc. Mass Spectrom.* 2002, 13, 300–307.
2. Revercomb, H.E.; Mason, E.A., Theory of plasma chromatography/gaseous electrophoresis, *Rev. Anal. Chem.* 1975, 47(7), 970–983.
3. Eisele, F.L.; Thackston, M.G.; Pope, W.M.; Ellis, H.W.; McDaniel, E.W., Experimental test of the generalized Einstein relation for Cs + ions in molecular gases: H_2, N_2, O_2, CO and CO_2, *Chem. Phys.* 1977, 67, 1271–1279.
4. Sennhauser, E.S.; Armstrong, D.A., Ion mobilities in gaseous ammonia, *Can. J. Chem.* 1978, 56, 2337–2341.
5. Carr, T.W., Comparison of the negative reactant ions formed in the plasma chromatograph by nitrogen, air, and sulfur hexafluoride as the drift gas with air as the carrier gas, *Anal. Chem.* 1979, 51, 705–711.
6. Sennhauser, E.S.; Armstrong, D.A., Ion mobilities and collision frequencies in gaseous CH_3CI, HCI, HBr, H_2S, NO, and SF: effects of polarity of the gas molecules, *Can. J. Chem.* 1980, 58, 231–237.
7. Rokushika, S.; Hatano, H.; Hill, H.H., Jr., Ion mobility spectrometry in carbon dioxide, *Anal. Chem.* 1986, 58, 361–365.

8. Berant, Z.; Karpas, Z.; Shahal, O., The effects of temperature and clustering on mobility of ions in CO_2, *J. Phys. Chem.* 1989, 93, 7529–7532.

9. Yamashita, T.; Kobayashi, H.; Konaka, A.; Kurashige, H.; Miyake, K.; Morii, M.M.; Nakamura, T.T.; Nomura, T.; Sasao, N.; Fukushima, Y.; Nomachi, M.; Sasaki, O.; Suekane, F.; Taniguchi, T., Measurements of the electron drift velocity and positive-ion mobility for gases containing CF_4, *Nucl. Instrum. Methods Phys. Res. Sect. A* 1989, A 283(3), 709–715.

10. Shvartsburg, A.A.; Jarrold, M.F., An exact hard-spheres scattering model for the mobilities of polyatomic ion, *Chem. Phys. Lett.* 1996, 261, 86–91.

11. Mesleh, M.F.; Hunter, J.M.; Shvartsburg, A.A.; Schatz, G.C.; Jarrold, M.F., Structural information from ion mobility measurements: effects of the long-range potential, *J. Phys. Chem.* 1996, 100, 16082–16086.

12. Shvartsburg, A.A.; Mashkevich, S.V.; Siu, M.K.W., Incorporation of thermal rotation of drifting ions into mobility calculations: drastic effect for heavier buffer gases, *J. Phys. Chem. A* 2000, 104, 9448–9453.

13. Karpas, Z.; Barant, Z., The effect of the drift gas on the mobility of ions, *J. Phys. Chem.* 1989, 93, 3021–3025.

14. Asbury, G.R.; Hill, H.H., Jr., Using different drift gases to change separation factors (alpha) in ion mobility spectrometry, *Anal. Chem.* 2000, 72, 580–584.

15. Barnett, D.A; Ellis, B.; Guevremont, R.; Purves, R.W.; Viehland, L.A., Evaluation of carrier gases for use in high-field asymmetric waveform ion mobility spectrometry, *J. Am. Soc. Mass Spectrom.* 2000, 11(12), 1125–1133.

16. Shvartsburg, A.A.; Tang, K.; Smith, R.D., Understanding and designing field asymmetric waveform ion mobility spectrometry separations in gas mixtures, *Anal. Chem.* 2004, 76(24), 7366–7374.

17. Shvartsburg, A.A.; Creese, A.J.; Smith, R.D.; Cooper, H.J., Separation of peptide isomers with variant modified sites by high-resolution differential ion mobility spectrometry, *Anal. Chem.* 2010, 82(19), 8327–8334.

18. Spangler, G.E.; Lawless, P.A., Ionization of nitrotoluene compounds in negative ion plasma chromatography, *Anal. Chem.* 1978, 50, 884–892.

19. Carr, T.W., Plasma chromatography off isomeric dihalogenated benzene, *J. Chrom. Sci.* 1977, 15(2), 85–88.

20. Hagen, D.F., Characterization of isomeric compounds by gas and plasma chromatography, *Anal. Chem.* 1979, 51, 872–874.

21. Karasek, F.W.; D.M. Kane, Plasma chromatography of isomeric halogenated nitrobenzenes, *Anal. Chem.*, 1974, 46(6),780–782.

22. Steiner, W.E.; English, W.A.; Hill, H.H., Jr., Ion-neutral potential models in atmospheric pressure ion mobility time-of-flight mass spectrometry IM(tof)MS, *J. Phys. Chem. A* 2006, 110, 1836–1844.

23. Asbury G.R.; Hill, H.H., Jr., Using different drift cases to change separation factors (alpha) in ion mobility spectrometry, *Anal. Chem.* 2000, 72(3), 580–584.

24. Proctor, C.J.; Todd, J.F.J., Alternative reagent ions for plasma chromatography, *Anal. Chem.* 1984, 56(11), 1794–1797.

25. Preston J.M.; Rajadhyax, L., Effect of ion–molecule reactions on ion mobilities, *Anal. Chem.* 1988, 60, 31–34.

26. Cox, S.J., Ion mobility spectrometer system with improved specificity, Patent number 4551624; filing date September 23, 1983.

27. Eiceman, G.A.; Harden, C.S.; Wang, Y.F.; Garcia-Gonzalez, L.; Schoff, D.B., Enhanced selectivity in ion mobility spectrometry analysis of complex mixtures by alternate reagent gas chemistry, *Anal. Chim. Acta* 1995, 306, 21–33.

28. Puton, J.; Nousiainen, M.; Sillanpää, M., Ion mobility spectrometers with doped gases, *Talanta* 2008, 76, 971–987.

29. Spangler G.E.; Epstein, J., Detection of HF using atmospheric pressure ionization (API) and ion mobility spectrometry (IMS), paper presented at the 38th ASMS Conference on Mass Spectrometry and Allied Topics, Tucson, AZ, June 1990.

30. Eiceman, G.A.; Salazar, M.R.; Rodriguez, M.R.; Limero, T.F.; Beck, S.W.; Cross, J.H.; Young, R.; James, J.T., Ion mobility spectrometry of hydrazine, monomethylhydrazine, and ammonia in air with 5-nonanone reagent gas, *Anal. Chem.*1993, 65, 1696–1702.

31. Bollan, H.R.; Stone, J.A.; Brokenshire, J.L.; Rodriguez, J.E.; Eiceman, G.A., Mobility resolution and mass analysis of ions from ammonia and hydrazine complexes with ketones formed in air at ambient pressure, *J. Am. Soc. Mass Spec.* 2007, 18(5), 940–951.

32. Howdle, M.D.; Eckers, C.; Laures, A.M.F.; Creaser, C.S., The use of shift reagents in ion mobility-mass spectrometry: studies on the complexation of an active pharmaceutical ingredient with polyethylene glycol excipients, *J. Am. Soc. Mass Spectrom.* 2009, 20(1),1–9.

33. Hilderbrand, A.E.; Myung, S.; Clemmer, D.E., Exploring crown ethers as shift reagents for ion mobility spectrometry, *Anal. Chem.* 2006,78(19), 6792–6800.

34. Kerr, T.J., Development of novel ion mobility-mass spectrometry shift reagents for proteomic application, PhD dissertation, Vanderbilt University, Nashville, TN, May 2011.

35. Dwivedi, P.; Wu, C.; Matz, L.M.; Clowers, B.H.; Seims, W.F.; Hill, H.H., Jr., Gas-phase chiral separations by ion mobility spectrometry, *Anal. Chem.* 2006, 78, 8200–8206.

36. Maestre, R.F.; Wu, C.; Hill, H.H., Jr., Using a buffer gas modifier to change separation selectivity in ion mobility spectrometry, *Int. J. Mass Spectrom.* 2010, 298, 2–9.

37. Meng, Q.; Karpas, Z.; Eiceman, G.A., Monitoring indoor ambient atmospheres for VOCs using an ion mobility analyzer array with selective chemical ionization, *Int. J. Environ. Anal. Chem.* 1995, 61, 81–94.

38. Krylova, N.; Krylov, E.; Eiceman, G.A., Effect of moisture on high field dependence of mobility for gas phase ions at atmospheric pressure: organophosphorus compounds, *J. Phys. Chem.* 2003, 107(19), 3648–3654.

39. Eiceman, G.A.; Krylov, E.V.; Krylova, N.S.; Nazarov, E.G.; Miller, R.A., Separation of ions from explosives in differential mobility spectrometry by vapor-modified drift gas, *Anal. Chem.* 2004, 76(17), 4937–4944.

40. Rorrer, L.C., III, Yost, R.A., Solvent vapor effects on planar high-field asymmetric waveform ion mobility spectrometry, *Int. J. Mass Spectrom.* 2011, 300, 173–181.

41. Levin, D.S.; Miller, R.A.; Nazarov, E.G.; Vouros, P., Rapid separation and quantitative analysis of peptides using a new nanoelectrospray-differential mobility spectrometer-mass spectrometer system, *Anal. Chem.* 2006, 78(15), 5443–5452.

42. Schneider, B.B.; Covey, T.R.; Coy, S.L.; Krylov, E.V.; Nazarov, E.G., Chemical effects in the separation process of a differential mobility/mass spectrometer system, *Anal. Chem.* 2010, 82(5),1867–1880.

43. Good, A.; Durden, D.A.; Kebarle, P., Ion–molecule reactions in pure nitrogen and nitrogen containing traces of water at total pressures 0.5–4 torr. Kinetics of clustering reactions forming $H^+(H_2O)_n$, *J. Chem. Phys.* 1970, 52, 212–221.

44. Sunner, J.; Nicol, G.; Kebarle, P., Factors determining relative sensitivity of analytes in positive mode atmospheric pressure ionization mass spectrometry, *Anal. Chem.* 1988, 60, 1300–1307.

45. Sunner, J.; Ikonomou, M.G.; Kebarle, P., Sensitivity enhancements obtained at high temperatures in atmospheric pressure ionization mass spectrometry, *Anal. Chem.* 1988, 60, 1308–1313.

46. Wang, Y.F., Effects of moisture and temperature on mobility spectra of organic chemicals, MS thesis, New Mexico State University, Las Cruces, June 1999.

47. Kojiro, D.R.; Cohen, M.J.; Stimac, R.M.; Wernlund, R.F.; Humphry, D.E., Takeuchi, N., Determination of C1–C4 alkanes by ion mobility spectrometry. *Anal Chem.* 1991, 63, 2295–2300.

48. Bell, S.B.; Ewing, R.G.; Eiceman, G.A.; Karpas, Z., Characterization of alkanes by atmospheric pressure chemical ionization mass spectrometry and ion mobility spectrometry, *J. Am. Soc. Mass Spectrom.* 1994, 5, 177–185.
49. Zhou, Q., Fragmentation of gas phase ions at one atmosphere in ion mobility spectrometry, MS thesis, New Mexico State University, Las Cruces, May 2001.
50. Bell, S.E.; Nazarov, E.G.; Wang, Y.F.; Eiceman, G.A., Classification of ion mobility spectra by chemical moiety using neural networks with whole spectra at various concentrations, *Anal. Chim. Acta* 1999, 394, 121–133.
51. Bell, S.E.; Nazarov, E.G.; Wang, Y.F.; Rodriguez, J.E.; Eiceman, G.A., Neural network recognition of chemical class information in mobility spectra obtained at high temperatures, *Anal. Chem.* 2000, 72, 1192–11911.
52. Eiceman, G.A.; Nazarov, E.G.; Rodriguez, J.E., Chemical class information in ion mobility spectra at low and elevated temperatures, *Anal. Chim. Acta* 2001, 433, 53–70.
53. Mäkinen, M.; Sillanpää, M.; Viitanen, A-K.; Knap, A.; Mäkelä, J.M.; Puton J., The effect of humidity on sensitivity of amine detection in ion mobility spectrometry, *Talanta* 2011, 84, 116–21.
54. Ewing, R.E., Stone, J.A.; Eiceman, G.A., Heterogeneous proton bound dimers in IMS, *Int. J. Mass Spectrom.* 1999, 193, 57–68.
55. Ewing, R.E.; Stone, J.A.; Eiceman, G.A., Proton bound cluster ions in ion mobility spectrometry, *Int. J. Mass Spectrom.* 1999, 193, 57–68.
56. Ewing, R.E., Kinetic decomposition of proton bound dimer ions with substituted amines in ion mobility spectrometry, Ph.D. dissertation, New Mexico State University, Las Cruces, December 1996.
57. An, X.; Eiceman, G.A.; Räsänen, R.-M.; Rodriguez, J.E.; Stone, J.A. Dissociation of proton bound dimers of ketones in asymmetric electric fields with differential mobility spectrometry and in uniform electric fields with ion mobility spectrometry, 2012. (submitted)
58. Eiceman, G.A.; Shoff, D.B.; Harden, C.S.; Snyder, A.P., Fragmentation of butyl acetate isomers in the drift region of an ion mobility spectrometer, *Int. J. Mass Spectrom. Ion Proc.* 1988, 85, 265–275.
59. Zhou, Q., Fragmentation of gas phase ions of one atmosphere in ion mobility spectrometry, master's thesis, New Mexico State University, Las Cruces, May 2001.
60. May, J.C.; Russell, D.H., A mass-selective variable-temperature drift tube ion mobility-mass spectrometer for temperature dependent ion mobility studies, *J. Am. Soc. Mass Spectrom.* 2011, 22(7),1134–1145.
61. Kendler, S.; Lambertus, G.R.; Dunietz, B.D.; Coy, S.L.; Nazarov, E.G.; Miller, R.A.; Sacks, R.D., Fragmentation pathways and mechanisms of aromatic compounds in atmospheric pressure studied by GC-DMS and DMS-MS, *Int. J. Mass Spectrom.* 2007, 263, 137–147.
62. An, X.; Stone, J.A.; Eiceman, G.A., Gas phase fragmentation of protonated esters in air at ambient pressure through ion heating by electric field in differential mobility spectrometry and by thermal bath in ion mobility spectrometry, *Int. J. Mass Spectrom.* 2011, 303, 181–190.
63. An, X.; Stone, J.A.; Eiceman, G.A., A determination of the effective temperatures for the dissociation of the proton bound dimer of dimethyl methylphosphonate in a planar differential mobility spectrometer, *Int. J. Ion Mobil. Spectrom.* 2010, 13, 25–36
64. Wilks, A. A consideration of ion chemistry encountered on the microsecond separation timescales of ultra-high field ion mobility spectrometry, 20th annual conference, International Society for Ion Mobility Spectrometry, Edinburgh, Scotland, July 24–28, 2011.

65. Davis, E.J. Dwivedi, P.; Tam, M.; Siems, W.F.; Hill, H.H., Jr., High-pressure ion mobility spectrometry, *Anal. Chem.* 2009, 81, 3270–3275.
66. Tabrizchi, M.; Rouholahnejad, F., Pressure effects on resolution in ion mobility spectrometry, *Talanta* 2006, 69(1), 87–90.
67. Tabrizchi, M.; Rouholahnejad, F., Comparing the effect of pressure and temperature on ion mobilities, *J. Phys. D: Appl. Phys.* 2005, 38, 857.
68. Nazarov, E.G.; Coy, S.L.; Krylov, E.V.; Miller, R.A.; Eiceman, G.A., Pressure effects in differential mobility spectrometry, *Anal. Chem.* 2006, 78(22), 7697–7706.
69. Wu, C.; Siems, W.F.; Asbury, G.R.; Hill, H.H., Jr., Electrospray ionization high-resolution ion mobility spectrometry-mass spectrometry, *Anal. Chem.* 1998, 70, 4929–4938.
70. Leonhardt, J.W.; Rohrbeck, W.; Bensch, H., A high resolution IMS for environmental studies, Fourth International Workshop on Ion Mobility Spectrometry, Cambridge, UK, August 9–16, 1995.
71. Dugourd, P.; Hudgins, R.R.; Clemmer, D.E.; Jarrold, M.F., High-resolution ion mobility measurements, *Rev. Sci. Inst.* 1997, 68, 1122–1129.
72. Viehland, L.A.; Mason, E.A., Transport properties of gaseous ion over a wide energy range, *At. Data Nucl. Data Tables* 60, 37–95, 1995.
73. Krylov, E.; Nazarov, E.G.; Miller, R.A.; Tadjikov, B.; Eiceman, G.A., Field dependence on mobilities for gas phase protonated monomers and proton bound dimers of ketones by planar field asymmetric waveform ion mobility spectrometer (PFAIMS), *J. Phys. Chem. A* 2002, 106, 5437–5444.
74. Eiceman, G.A.; Shoff, D.B.; Harden, C.S.; Snyder, A.P.; Martinez, P.M.; Fleischer M.E.; Watkins, M.L., Ion mobility spectrometry of halothane, enflurane, and isoflurane anesthetics in air and respired gases, *Anal. Chem.* 1989, 61, 1093–1099.
75. Karpas, Z.; Guaman, A.V.; Calvo, D.; Pardo, A.; Marco, S., The potential of ion mobility spectrometry (IMS) for detection of 2,4,6 trichloroanisole (2,4,6-TCA) in wine, *Talanta* 2012, 93, 200–205.

12 Detection of Explosives by IMS

12.1 GENERAL COMMENTS ON DETECTION OF EXPLOSIVES

The past decade may be characterized by an intense research effort to improve the sensitivity and reliability of ion mobility-based instruments for detection of explosives and to expand the inventory of detectable explosive substances. In addition, improvements in sampling techniques have been proposed, standards for calibration of ion mobility spectrometry (IMS) instruments and for method verification have been developed, and novel instrumental techniques based on laser ablation or electrospray ionization have been advanced. This has occurred due to the rising need for rapid, efficient, and reliable detectors of explosives that are used in acts of terrorism worldwide, as stated in many occasions (see Frame 12.1). Some idea about the scale of this trend can be gained from a recent report: "Austin, TX, 15 September, 2011—Sales of Explosives, Weapons, and Contraband (EWC) Detection equipment to the world's airport authorities, amassed a significant $834.9 million in 2010, according to a recent study published by IMS Research, a leading provider of market research in the homeland security industry."

Common commercial explosives consist of nitro compounds, which are highly electronegative and form stable negative ions under conditions of ambient pressure ionization. In fact, the general public is likely to encounter an IMS device, often unaware, as airline passengers being screened for explosives. Several manufacturers are active in this field, particularly after the attack on the Twin Towers at the World Trade Center in New York City on September 11, 2001. After this event, awareness of global terrorism increased, although that particular incident had no direct relationship with explosives, and no explosive detector would have prevented the attack.

In the development of IMS analyzers for explosive detection, stages of technology can be identified. A burst of activity occurred with government contracts after the initial discovery of the capabilities of IMS, and the emphasis was on detecting volatile compounds such as ethylene glycol dinitrate (EGDN), nitroglycerine (NG), and trinitrotoluene (TNT). These are the most volatile explosive compounds or related chemicals in the family of nitrated organic compounds, as listed in Table 12.1. Gradually, the threat to commercial aviation safety from explosives with lower vapor pressures, such as RDX (cyclonite), PETN (pentaerythritol tetranitrate), HMX (octogen), and their composites, such as C4 and Semtex, was understood, and techniques for sampling such compounds with heated inlet systems to transfer sample to the analyzer were developed toward the end of the 1980s. The new

FRAME 12.1   EXCERPTS FROM A REPORT FOR CONGRESS

The following is from Dana A. Shea and Daniel Morga, "Detection of Explosives on Airline Passengers: Recommendation of the 9/11 Commission and Related Issue, CRS Report for Congress, Order Code RS21920, Updated April 26, 2000":

Impact on Screening Time. When multiplied by the large number of airline passengers each day, even small increases in screening times may be logistically prohibitive. The TSA goal for passenger wait time at airports is less than 10 minutes, and screening systems reportedly operate at a rate between 7 to 10 passengers per minute.

A different research challenge is the detection of novel explosives. Detectors are generally designed to look for specific explosives, both to limit the number of false or innocuous positives and to allow a determination of which explosive has been detected. As a result, novel explosives are unlikely to be detected until identifying characteristics and reference standards have been developed and incorporated into equipment designs. Unlike imaging techniques for detecting bulk quantities of explosives, trace analysis provides no opportunity for a human operator to identify a suspicious material based on experience or intuition.

Liquid explosives are a novel threat that has been of particular interest since August 2006, when British police disrupted a plot to bomb aircraft using liquids. The DHS is evaluating technologies to detect liquid explosives. Its efforts are mainly focused on bulk detection, such as scanners to test the contents of bottles. Like solid explosives, however, liquids might be found through trace detection, if the trace detection system is designed to look for them.

Several commercial IMS-based analyzers are competing for a share in this growing market segment. Thanks to the high sensitivity of IMS technology toward several types of explosives, applications have been made in detecting hidden explosives and identifying suspect materials for postdetonation investigations and even for detecting explosives in water and underwater.

types of explosives that have been seized in recent acts of terrorism, or in attempts that have been prevented, are diversifying away from nitrogenous compounds and include peroxide-based explosives such as triacetone triperoxide (TATP) or hexamethylene triperoxide diamine (HMTD) and binary liquids that when mixed form sensitive powerful explosives (see more details in Section 12.7). These create new challenges for explosive detection technology and require modifications of IMS to meet these challenges.

However, the classic analyzers were suited only for screening inanimate items, such as hand luggage and cargo. Portals for rapid screening of humans at controlled entry points such as airport boarding gates, secure buildings, or restricted zones are under review. Naturally, the throughput of screened passengers must be high, and the technology must be seen as noninvasive, with a balance between public safety and individual privacy. There is still an ongoing debate about the ethics of using backscatter x-rays and millimeter wave screening devices for detection of concealed objects in and under the clothing of travelers as these create an image that may invade passenger privacy.

An important consideration with the detection of explosives is elimination of false negatives. Although a false-negative response for detecting drugs may not directly affect the lives of those involved, failure to detect contraband explosives may actually endanger the safety or lives of hundreds of people, such as passengers in an airplane or on a train. Therefore, the performance criteria placed on explosive detection systems are demanding.

Explosive detection methods with high reliability must have a low false-positive rate (below 2% is the recommended goal) and a practically zero false-negative rate. In addition, to be certified the system should be sensitive and fast enough to detect the explosives rapidly so that ideally 7–10 passengers could be screened in a minute (Frame 12.1).

The requirements for an explosive detector were summarized by Lucero[1] three decades ago: Two pounds of explosives should be detected in less than 6 s. The use of volatile compounds as taggants to enable detection of explosives by IMS was investigated.[2,3] Although the advantages of tagged explosives for counterterrorism are self-evident, the proposal to tag all explosives has not been accepted by all manufacturers of explosives (and probably will not be accepted in the future). In addition, improvised explosive devices (IEDs), using homemade explosives, have been found in many recent terrorist attempts. This and the fact that there are stocks of untagged explosives render reliance solely on taggants for detection of explosives impractical.

Although luggage may be examined for explosives by radiation techniques based on neutron activation or bombardment of the object with high-energy x-rays, people must be screened by methods that are harmless and nonintrusive. These are mainly techniques based on sniffing or gentle sampling of particles attached to the subjects' clothing or body (e.g., in walk-through portals and by handheld instruments). The detection of relatively volatile explosives, such as EGDN, NG, or even TNT, the last having a vapor pressure above 10^{-6} torr at 25°C (see Table 12.1), may be done by sampling vapors or particulate matter from the subject or suspect piece of

TABLE 12.1
Vapor Pressure of Common Explosives

Name	Class	Molecular Weight	Vapor Pressure (torr)
EGDN, ethylene glycol dinitrate	Aliphatic	152	4.8×10^{-2}
NG, nitroglycerin	Aliphatic	227	2.3×10^{-4}
DNT, dinitrotoluene	Aromatic	182	1.1×10^{-4}
TNT, trinitrotoluene	Aromatic	227	4.5×10^{-6}
RDX, cyclonite	Cycloaliphatic	222	1.1×10^{-9}
PETN, pentaerythritol tetranitrate	Aliphatic	316	3.8×10^{-10}

Source: Conrad, F.J., Explosives detection—the problem and prospects, *Nucl. Mater. Manage.* 1984, 13, 212.

luggage. However, detection of the explosive constituents of plastic explosives such as RDX or PETN, which have room temperature vapor pressures below 10^{-8} torr, or of well-sealed volatile explosives requires a different sampling strategy. This could be preconcentration of the vapors or collection of microparticles containing the adsorbed explosive vapors before transfer into the detector for heating and analysis. Thus, the first barrier for detecting explosives is that of sampling—bringing enough molecules of the compound into the IMS. One approach for screening people suspected of handling explosives or suspect objects was presented in which a gas jet entrains particles and carries them to a Teflon filter, from which vapors are aspirated into an IMS for analysis.[4]

In addition to detection of explosives, handheld IMS devices may be applied to postdetonation identification of the type of explosive used. This can be done on site, with the IMS serving as a qualitative analytical instrument, or using IMS to screen the debris and select the pieces with traces of the explosive for more detailed and precise analysis at the forensic laboratory.[5–7] The detection of traces of explosives as pollutants in the environment is described elsewhere.

12.2 THE CHEMISTRY UNDERLYING DETECTION OF EXPLOSIVES BY IMS

The ionization chemistry for nitrogenous explosives is governed by reactions in the negative polarity, although in some cases positive ion polarity is favored, particularly for explosives that do not contain a nitro group. In negative polarity, ions of explosive compounds are formed in the reaction region mainly by the reaction mechanisms labeled as charge transfer, proton abstraction, association (or ion attachment), and fragmentation. The reactions between a reactant ion, such as O_2^-, and electronegative molecules may involve charge transfer from the reactant ion (Equation 12.1) to the neutral molecule, which may then dissociate and form a more stable fragment ion, F^-:

Charge transfer and fragmentation

$$M + O_2^- \rightarrow M * O_2^- \rightarrow F^- + \text{neutrals} + O_2 \qquad (12.1)$$

Other reaction channels may involve proton abstraction from the analyte molecule, which is essentially proton transfer to the reactant ion (Equation 12.2), forming an $(M\text{-}H)^-$ ion.

Proton abstraction

$$M + O_2^- \rightarrow (M - H)^- + HO_2 \qquad (12.2)$$

More common are association or ion attachment reactions (Equation 12.3), which can be followed by dimerization or clustering (Equation 12.4) with elevated concentrations of M and collisional stabilization:

Ion attachment or association

$$M + O_2^-(H_2O)_n \rightarrow M * O_2^-(H_2O)_{n-1} + H_2O \qquad (12.3)$$

Dimerization

$$M + M * O_2^- \rightarrow M_2 * O_2^- \qquad (12.4)$$

Charge transfer and proton abstraction reactions have been reported only in a few cases, such as the reaction of O_2^- with TNT in air.[8-11] Often, several association and attachment reaction channels occur simultaneously and compete with one another so that the explosive molecules attach to the different ions that are present in the ionization chamber, forming several different negative product ions. Commonly, the ionization process is controlled by adding a reagent gas, generally chloride or bromide ions, which then form well-defined product ions, as proposed by several investigators,[12-15] although in some cases improved results were obtained without the chloride reagent.[16] In other pathways, fragment ions such as NO_2^- or NO_3^- are formed through dissociative charge transfer and may then associate with another parent molecule. Although any distribution of charge among several ions damages the limits of detection, a variety of product ions can be helpful in confirming the presence of explosives and reducing the rate of false-positive responses. High-resolution negative ion mobility spectra of several nitrogenous explosives are shown in Figure 12.1.

12.3 SAMPLING AND PRECONCENTRATION TECHNIQUES FOR DETECTION OF EXPLOSIVES

The main objective of all sampling techniques is to transport enough molecules of the explosive compound, if present, from the suspect object to the analytical device. The technique deployed to draw a sample that can be used for confirming the presence of explosives depends on the type of analytical device (handheld, desktop, or portal); the nature of the explosive (high or low vapor pressure, liquid or solid); the object being screened (a person, hand luggage, or large items like a shipping container); and the purpose of the sample collection (environmental screening, postdetonation debris, or a searching a site where explosives may have been handled). In the past, explosive detection relied either on "sniffing" of vapors emanating from the explosive into the ambient air or on swiping the suspect object with a proper sampling medium for collection of particles, which are subsequently thermally desorbed in the IMS sample introduction system. However, modern techniques were developed that extend the range of methods used for detection of explosives. For example, with the advent of IMS devices that combine laser desorption and ionization, or laser ablation, solid objects may be sampled directly and screened for explosives, and liquid samples may be tested directly with an IMS equipped with an electrospray ionization inlet. Other sampling devices include

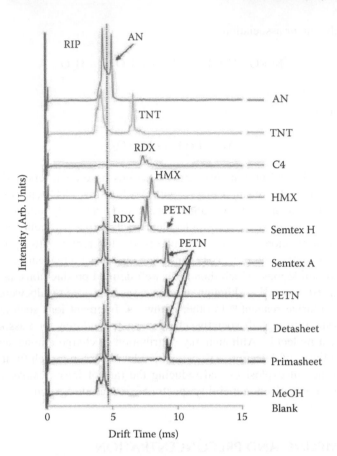

FIGURE 12.1 The mobility spectra of several types of explosives. (From Hilton et al., Improved analysis of explosives samples with electrospray ionization-high resolution ion mobility spectrometry (ESI-HRIMS), *Int. J. Mass Spectrom.* 2010, 298, 64–71. With permission.)

means for preconcentration of the explosive vapors from ambient air or selectively extracting the explosive molecules from liquid samples like water bodies or even urine. A survey of the different approaches to detect explosives is presented in the following sections.

The use of solid phase microextraction (SPME) devices to concentrate vapors of explosives has been investigated intensively by Almirall's group, who developed and tested different SPME devices.[18–20] A dynamic planar SPME was coupled to an IMS for field sampling of vapors of drugs and explosives from a large volume of air.[20] The SPME consisted of a high-surface-area sol-gel polydimethylsiloxane (PDMS) coating for absorption of the vapors followed by direct thermal desorption into an IMS. The reported sensitivity enabled the detection of concentrations of parts per trillion of the analytes when 3.5 L of air were sampled over the course of 10 s (corresponding to absolute detection of less than a nanogram). This dynamic planar SPME (PSPME) was used to sample the headspace of Pentolite, detecting 0.6 ng of TNT, and of several smokeless powders, with detection limits of about

30 ng for 2,4-DNT and 11–74 ng of diphenylamine (DPA). Methods for sampling the presence of taggants in the vapor phase and preconcentrating them with a SPME device were also studied by the same group.[20]

12.4 MEASUREMENT WITH HANDHELD DEVICES, PORTABLE INSTRUMENTS, AND PORTALS

Although not widely deployed, handheld IMS devices for detecting explosives, mainly by "sniffing," can be used much like handheld metal detectors (or wands) to screen people entering a controlled area by passing the device close to the subject's body, clothes, and belongings without actually touching the subject. These devices may also be used by security patrols to examine suspect objects and verify "sterile zones" in public places. The efficiency of this approach is limited to detection of unwrapped volatile explosives or traces of these on the exterior of the suspect object but has the advantage of deterring terrorists and instilling a feeling of security among the public. In addition, such devices may be readily rushed to the scene to enhance security measures. Some of the older commercial devices of this kind include the GE Interlogix VaporTracer (shown in Figure 12.2) and the Smiths Detection Sabre 2000. Some examples of the current generation of explosive detectors based on IMS technology are shown in Figure 12.3. The ItemiserDX and the MobileTrace are made by Morpho Detection (Safran), which acquired the GE line of IMS-based products, and the Sabre 5000, MMTD, and IONSCAN 500DT (based on the IONSCAN) are made by Smiths Detection.

Desktop instruments are usually stationed at control points such as airport boarding gates and rely on human operators, who use swipes or air suction (like vacuum cleaners) techniques to trap particles on a filter, which is then inserted into the inlet of the analyzer and placed on the anvil heater unit or the sample is directly heated. Sampling of this type is only marginally intrusive as it requires gentle physical contact with the suspect subject or object. These instruments are often combined with, or positioned adjacent to, an x-ray machine so that the operator may be able to immediately test any object that appears as suspicious in the imaging device (see Frame 12.2). Although larger than the handheld analyzer, these explosive detectors are portable and may be moved about or relocated conveniently to a controlled area for additional capability in screening.

Portals for explosive detection should be like those used for metal detection: screen people rapidly and with minimal intrusiveness as they pass through a portal.[21–24] Portals have the capability to draw an air sample from a person's whole body, preconcentrate the vapors, and perform an analysis by IMS within a few seconds. Portals may also employ jets of air and gentle physical contact to detach particles, trap them on a filter, and thus enhance the sensitivity of the detector.[21] In principle, a security checkpoint with such a portal could be designed to lock and trap a would-be bomber and limit the damage that may be inflicted by a suicide bomber.

In one design, marketed as the Sentinel II by Smiths Detection,[22] streams of air are drawn downward to the bottom of the portal into a preconcentrator and IMS detector

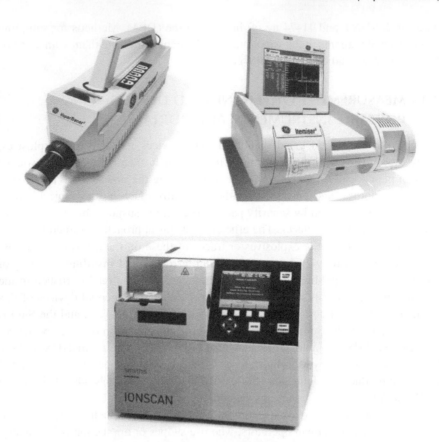

FIGURE 12.2 Three of the older explosive detection devices: the VaporTracer, a handheld explosive analyzer (made by GE Interlogix) (top); portable explosive analyzers: the Itemiser made by GE Interlogix (middle) and the IONSCAN made by Smiths Detection (bottom from http://www.smithsdetection.com).

(Figure 12.4). An automatic portal for luggage, which can possibly be adapted for inspection of people, is described in the next section.

12.5 RESEARCH AND OPERATIONAL EXPERIENCE

In this section, some of the operational experiences and recent developments for explosives detection are discussed. Most of the data in this section were published by law enforcement agencies or supplied by manufacturers of commercial IMS devices. Thus, there is an emphasis on practical aspects and field tests. The descriptions are based mainly on the manufacturers' claims and brochures, except if objective reports on comparative evaluations and field tests were available. The common or conventional explosives that contain nitro functional groups can be divided into two categories: relatively volatile substances such as NG and EGDN and low-vapor-pressure materials such as RDX, PETN, and HMX. The vapor pressure of TNT is between these two groups. There are several commercial and military composites that are mixtures

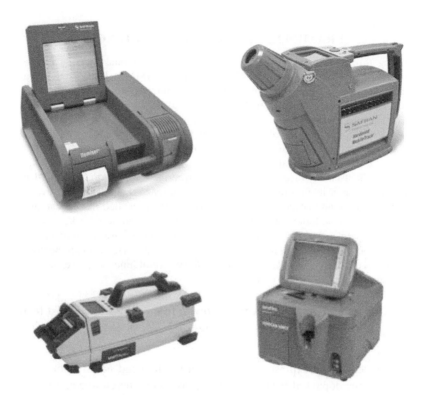

FIGURE 12.3 Some modern IMS-based commercial explosive detectors. Top: The ItemiserDX and Hardened MobileTrace (Morpho Detection, Safran) (from http://www.morpho.com/detection/see-all-products/trace-detection/). Bottom: The IONSCAN 500DT and the MMTD (Multi-Mode Threat Detector) (Smiths Detection) (from http://www.smiths-detection.com).

of two or more of these, such as Compositions B and C and Semtex (three types are available: H, A, 10). New types of alternative explosives, including homemade compounds that can be easily prepared from available materials, have been involved in acts of terrorism. Among those are ammonium nitrate mixed with fuel oil (ANFO) and liquid nitromethane that was used in the Oklahoma City bombing; and TATP and urea nitrate (UN), which have been used principally by Palestinians and others. A relatively new type of homemade explosive is 1,3,5-trinitroso-1,3,5-triazacyclohexane (also called R-salt), which can also be detected by IMS.

Some of the common explosives mentioned can be readily detected by IMS operated in the negative mode as shown in Figure 12.1, with subnanogram detection limits in laboratory settings reported by several investigators and manufacturers. Most commercial instruments for detection of these substances are operated at elevated temperatures (from about 100°C to 200°C), make use of a doping agent to control the reactant ion chemistry, and can be used for detection of vapors ("sniffing") or trapped particles with a filter insertion inlet from which the vapors are released through thermal desorption. Under these conditions, the major mechanisms for product ion formation are clusterization and dimerization (Equations 12.3 and 12.4).

FRAME 12.2   PERSONAL EXPERIENCE

Going through Newark airport, on the way to the annual meeting of the International Society for Ion Mobility Spectrometry (ISIMS), I saw a red screen flashing and heard an audible alarm coming from one of the IMS instruments that had supposedly detected traces of explosives on the laptop of a passenger. The passenger was then asked to open his hand luggage and take out all items. Every single item he handled that was swiped and tested also gave an explosives alarm. After calling the supervisor, several repeated tests and a brief interrogation; the passenger was allowed to continue his travel. I approached him and asked if he used any medication (like nitroglycerine for a heart condition) that could be responsible for the false alarm. He mentioned that he traveled several times a month and never had any problems, but then recalled that he had just changed the hand lotion he used for a persistent skin condition. I asked him to show me the new lotion and saw that one of the ingredients was a chloroanisole compound that could well be responsible for the false alarm. Though inconvenient and even embarrassing, it is better to delay several innocent passengers for a short while than to miss one bomb.

Zeev Karpas, Ion mobility spectrometry: a tool in the war against terror,
Bull. Isr. Chem. Soc. 2009, 24, 26–30

As identification of the explosives is based on the reduced mobility of the ions in the spectrum, the dopant affects the product ions and is therefore a consideration in detection algorithms. Several of the older studies on the suitability of IMS for the detection and identification of explosives arising early in the development of modern analytical IMS have been summarized in a review article,[25] and later advances have been described.[26] Some of these studies were concerned with the ion chemistry of explosive vapors, although attempts to determine minimum detectable limits were made in others.

A word of caution is needed when discussing or conducting quantitative measurements with explosives, particularly in the vapor phase. Most explosive substances are highly polar molecules that tend to absorb on surfaces or undergo thermal decomposition.[27] Thus, there may be a significant difference between the amount of vapor released from the explosive source,[27–29] say a heated permeation tube, and the amount actually reaching the analyzer. Consequently, the actual concentrations of explosives detected may be lower than calculated on the basis of weight loss from a vapor generator. Later, the missing material or decomposition products may reappear when displaced or desorbed from the surfaces of the generator or the analyzer. Quantitative calibration of the IMS response may be determined best by depositing a known amount of the substance (usually from a calibrated solution of the explosive in a volatile organic solvent) on a suitable filter and desorbing the sample directly into the IMS drift tube from the heated inlet. This is sometimes referred to as "dry transfer."

Several studies attempted to address the problem of detection of explosives and their degradation or dissociation products in water. These are only briefly discussed

FIGURE 12.4 Prototype portals for screening humans for explosives. Top: EntryScan (Morpho Detection; http://www.morpho.com) high-throughput, nonintrusive, walk-through portal for rapid detection of explosives and narcotics. Microscopic traces of explosives can easily be detected and identified. Bottom: Sentinel II for screening for explosive residues: Streams of air are used to sample the subject's body, and the air is drawn through ports at the bottom of the portal into a preconcentrator and IMS detector (Smiths Detection; http://www.smithsdetection.com).

here as they are described elsewhere in this monograph. In one study, stir bar sorptive extraction (SBSE) was used to concentrate traces of TNT and RDX from water, which was followed by thermal desorption into an IMS. The reported limits of detection were 0.1 ng/mL for TNT and 1.5 ng/mL for RDX, and the process could be completed in less than 1 min.[30] An interesting approach for detection of common military explosives and dissociation products in liquid and vapor phase based on secondary electrospray ionization-ion mobility spectrometry (SESI-IMS) was described by Tam and Hill.[31] The effects of temperature and use of dopants were investigated, and detection limits in the range of parts per billion were reported.

Combining high-speed gas chromatography (GC) with differential mobility spectrometry (DMS) was used to determine mixtures of several explosives, including PETN, RDX, Tetryl, and peroxides, in solution. The use of GC for preseparation of the sample improved the DMS response, and distinct patterns were found for each of the compounds. The authors concluded that additional selectivity may be achieved by use of several differential mobility detectors operated in parallel or series with characteristic separation voltages.[32]

In another attempt to develop devices for selective, passive preconcentration of explosives, vapor-deactivated quartz fiber filters impregnated with metal β-diketonate polymers were used.[33] The Lewis acidic polymers selectively interact with Lewis base analytes, such as explosives. The uptake kinetics of TNT and RDX vapor from a saturated atmosphere were characterized through passive equilibrium sampling. Approximately 5 ng of RDX were collected on the filter at the end of 1 month, and this amount was readily detected by the IMS.

12.6 WALK-THROUGH PORTALS AND SYSTEMS FOR LUGGAGE SCREENING

The widespread deployment of walk-through portals for humans or for automatic screening of luggage is pending. Two of the major manufacturers of IMS equipment, Smiths Detection (formerly Barringer Research) and Morpho Detection (Safran) (which acquired GE Interlogix and was formerly called Ion Track Instruments) have put forth experimental models for testing in realistic scenarios.

Barringer Research had produced a series of portable IMS-based instruments for detection of drugs and explosives, known as the IONSCAN, and a handheld analyzer. Similarly, Ion Track Instruments had built an instrument family based on IMS to meet the same requirements for aviation safety. The core technology for these instruments is integrated into the prototype luggage-screening systems[34] and portals for human screening.[35]

A few articles and technical reports on screening uses for IMS analyzers have been made available and disclose some of the features of analysis. Only some of the more significant facets are discussed here. In routine operation, a powerful vacuum cleaner collects microscopic particles, which are then trapped on a filter. The filter is inserted into a heater connected directly to the IMS, and desorbed vapors are transported by a carrier gas into the drift tube (Figure 6.4). Detection limits of 50 to 350 pg, corresponding to the amount collected on the filter, were claimed for most common explosives, including plastic explosives. With the support of the Federal Aviation Administration, an automatic trace detection system for the detection of explosive vapors and particles in luggage had been tested.[34] The system consisted of a conveyor belt that carried the luggage into a tunnel where particles were picked up by an array of 225 flexible sampling tubes and carried to a preconcentrator unit.

A coated mesh vapor concentrator further increased the concentration of the vapors, which were then analyzed by an IONSCAN. Meanwhile, the vapors given off by the luggage were collected by another preconcentrator unit and released into a second set of coated mesh concentrators and an IMS instrument. The drift tube and inlet temperatures for vapor detection were set at 60°C and 110°C, respectively, and for particle detection at 115°C and 245°C, respectively. The overall time for each piece of luggage was 20 s: 7 s for sampling, 3 s for desorption and rotation of the table, and 10 s for particle desorption and analysis. However, the actual throughput was higher as the next bag entered the tunnel while the first one was being sampled, and then two more bags followed while the analysis of the first two was being carried out. Thus, the throughput of the system was reported as 720 bags an hour, with the IMS performing six analyses per minute.[34] A large-volume (LV) IMS analyzer was produced specifically for detecting explosives and tested at Sandia Laboratories. The results were summarized in a technical report.[22] The system tested was comprised of a walk-through portal equipped with a purge gas flowing at rates of up to 2,300 L/min and a large-diameter (with a cross section of 100 cm^2) IMS drift tube capable of taking a carrier gas flow of 12.6 L/min (two orders of magnitude higher than regular IMS drift tubes). The main findings in this study were that 0.5 g of TNT placed in a plastic bag inside the portal and mixtures of TNT–RDX (composition B) could be detected. However, pure RDX or PETN was not detected. Limits of detection were found to be close to those of a conventional IMS with preconcentration.

In another study, 17 of the most common suspected interferents for detection of TNT by IMS technology in airport scenarios were investigated.[36] The reactant ion was chloride, the most common reactant ion in explosive detectors. Ten of the interferents showed no IMS response, and of the other seven, only two presented a problem: 4,6-Dinitro-o-cresol had a close reduced mobility value and 2,4-dinitrophenol competed with ionization of TNT. The personal account given in Frame 12.2 shows that some chemicals that give a false-positive response may come from unexpected origins, such as hand lotion or other common cosmetics and medications.

12.7 HOMEMADE AND ALTERNATE EXPLOSIVES

As mentioned, new types of homemade explosives or alternates to common nitrated explosives have appeared in several terror scenes and pose a threat to public safety (see Frame 12.3). To counteract these threats, instrumentation has to be calibrated, data systems have to be revised, and some modifications in operational parameters may be necessary. Some of the major manufacturers of explosive detectors have characterized operational conditions for the detection of several alternative explosives without severe deterioration of performance for the standard explosives. In one example, the temperature of the drift tube was lowered to 169°C, and the desorber was set to 220°C. In addition, the instrument was operated in dual mode (i.e., measurement of the negative and positive mobility spectra almost simultaneously). With these modifications, operational detection limits (ODLs), defined by the authors as the mass of target substance that was required to produce an alarm at a given detection threshold, of 50 to 100 ng were established for TATP, ammonium nitrate, and gunpowder.[37] These detection limits have been improved with the current generation of explosive detectors, especially for TATP. Under these conditions, the response and detection limit for TNT is degraded by a factor of two, but those for RDX and NG are barely affected; the detection limit for PETN was improved.[37]

FRAME 12.3 *NEW SCIENTIST, ANALYSIS*

EXPLOSIVE DETECTION TECHNOLOGIES

17:24 10 AUGUST 2006 BY WILL KNIGHT

LIQUID EXPLOSIVES

But the alleged terror plot may highlight the need for even more sophisticated scanning technologies, as according to US security officials it involved liquid explosives, perhaps designed to evade current security checks. Paul Wilkinson, director of the Centre for the Study of Terrorism and Political Violence at the University of St Andrews in Fife, UK, told *New Scientist* that terrorists are *increasingly diversifying away from nitrogenous explosives.*

In a study of TATP by IMS and tandem mass spectrometry, the peaks in the mobility spectrum were identified. A peak at mass 223 Da was identified as arising from TATP.[38] In the mobility spectrum, a cluster of three peaks was observed, and a large increase in peak intensity was observed after dissolution of TATP in toluene. The reduced mobility of the main peak was reported as 2.71 cm^2/V s. This high-mobility value must correspond to an ion with low molecular weight, probably a nitrate ion, and not to the molecular ion. A recent study has indeed found a quasi-molecular peak with a mass of 240 Da in the positive ion mobility spectrum of TATP corresponding to an adduct with ammonia, and the intensity of the peak decreased when the ammonia concentration was increased from 4.7 to 8.1 ppmv (Figure 12.5). The reduced mobility of this peak was low, in agreement with the prediction based on the mass of the adduct ion.[39]

Two other studies concerned with the detection of improvised explosives in positive and negative polarity were presented at ISIMS meetings in 2001 and 2003.[40,41] In the former, response of the IMS to TATP was found in both polarities, whereas in the latter study ions arising from TATP were found only in positive-mode operation. The humidity level, which had to be kept below 100 ppmv, affected the detection of TATP and HMTD, the peroxide compounds.[41]

12.8   STANDARDS FOR CALIBRATION OF EXPLOSIVE DETECTORS

A comprehensive discussion of the problems associated with calibration of explosive detectors was presented at the 2004 conference of the ISIMS by a group from the National Institute for Standards and Technology (NIST).[42] A solution based on the deposition of droplets from a simple ink-jet printer was demonstrated. By replacing the standard ink with a solution of the explosive compound in a volatile solvent and by slightly modifying the printer spots, or dots, the desired amounts of the analyte could be placed on a piece of paper or on different test surfaces, like luggage handles or floppy disks. By this method, highly reproducible standards were produced. Homemade standards can be prepared by placing a known volume of a solution with known concentrations of the analyte by pipette on a surface and allowing the

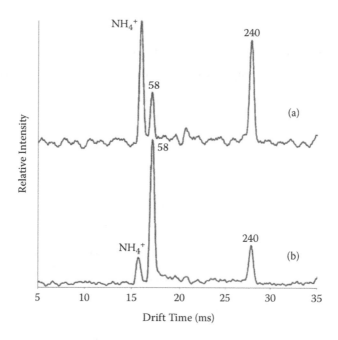

FIGURE 12.5 IMS of a 0.2-μL stock TATP at 80°C (A) with 4.7-ppmv ammonia and (B) with 8.1-ppmv ammonia. (From Ewing; Waltman; and Atkinson, Characterization of triacetone triperoxide by ion mobility spectrometry and mass spectrometry following atmospheric pressure chemical ionization, *Anal. Chem.* 2011, 83, 4838–4844. With permission.)

solvent to evaporate (dry transfer method). This would be practical for soluble types of explosives and by use of common solvents like acetone, alcohol, or acetonitrile.

As mentioned, the generation of controlled concentrations of vapors of explosives is difficult due to the polarity of the compounds, their tendency to absorb on surfaces, and for sensitive compounds to undergo degradation. A portable explosive vapor generator was described, and the problems connected with this were discussed.[43] The basic idea underlying this device is to allow a droplet generated by an ink-jet printer like the type described to evaporate in an enclosed volume and to sample the vapors.

12.9 DATABASE FOR EXPLOSIVES

In Table 12.2, the reduced mobility values of most of the common explosives are shown. There are several different reduced mobility values for some of the compounds associated with unique product ions, which were not always identified by mass spectrometry. Clustering of target molecules with ions, as shown in the reaction in Equation 12.3, plays an important role in explosive detection. A comprehensive study showed that the prominent ion was an adduct or cluster between the molecule and reactant ions. These association processes were strongly temperature dependent and were abundant at a relatively low temperature (50°C).

TABLE 12.2
Reduced Mobility Values of Common Explosives

Name	Ion Identity (assumed)	MDL	K_o (cm²/Vs)	Parameters
Nitroaromatics				
MNT			1.74	Air, 166°C
			1.81	
			2.40	
2,4 DNT			1.68, 2.10	Air, 200°C
3,4 DNT			1.54	Air, 50°C
2,6 DNT			1.67	Air, 250°C
TNT		200 pg	1.45	Air, 166°C
			1.49	
			1.54	
			1.59	
Nitroaliphatics				
Dynamite			2.10, 2.48	Air, 200°C
NG	NG·Cl⁻, NG·NO₃⁻	50 pg	1.32, 1.34	Air, 150°C
		200 pg	1.28	
EGMN	NO₃⁻		2.46	Air, 150°C
EGDN	NO₃⁻		2.46	Air, 150°C
Low vapor pressure				
HMX			1.30, 1.25	Air, 250°C
RDX	RDX	200 pg	1.48, 1.39	Air, 250°C
	RDX·Cl⁻, RDX·NO₃⁻	800 pg	1.31	
	(RDX)₂Cl⁻	1,000 pg	0.95	
PETN	PETN		1.48 1.21	Air, 166°C
	PETN(–H)	80 ng	1.145	
	PETN Cl⁻	200 pg	1.10	
	PETN·NO₃⁻	1,000 pg		
Tetryl			1.45 1.62	Air, 250°C
Comp B			1.57, 1.70, 1.81	Air, 200°C
Improvised				
TATP	TATP·NH₄⁺	~µg	1.36	100°C (Marr)
HMTD	HMTD·H⁺		1.50	100–130°C
Black gunpowder	N(CH₂O)₃H·Cl⁻		1.88	150°C

Source: Fetterolf, D.D.; Clark, T.D., Detection of trace explosive evidence by ion mobility spectrometry, *J. Forens. Sci.* 1993, 38, 28–39 (Barringer IONSCAN 200); Marr, A.J.; Groves, D.M., Ion mobility spectrometry of peroxide explosives TATP and HMTD, *Int. J. Ion Mobil. Spectrom.* 2003, 6, 59–61.

Note: Air, the drift gas, ionization source flow, and the sample vapors were carried by air. Cl⁻, chloride reactant ion present.

MDL, minimum detectable level; EGMN, ethylene glycol mononitrate; MNT, mononitrotoluene.

REFERENCES

1. Lucero, D.P., User requirements and performance specifications for explosive vapor detection systems, *J. Test. Eval.* 1985, 13, 222–233.
2. Wernlund, R.F.; Cohen, M.J.; Kindell, R.C., The ion mobility spectrometer as an explosive taggant detector, in *Proceedings of the New Concepts Symposium and Workshop on Detection Identification of Explosives*, Reston, VA, 1978, p. 185; Perr, J.M.; Furton, K.G.; Almirall, J.R., Solid phase microextraction ion mobility spectrometer interface for explosive and taggant detection, *J. Sep. Sci.* 2005, 28, 177–183.
3. Ewing, R.G.; Miller, C.J., Detection of volatile vapors emitted from explosives with a handheld ion mobility spectrometer, *Field Anal. Chem. Technol.* 2001, 5, 215–221
4. Phares, D.J.; Holt, J.K.; Smedley, G.T.; Flagan, R.C., Method for characterization of adhesion properties of trace explosives in fingerprints and fingerprint simulations, *J. Forensic Sci.* 2000, 45, 774–784.
5. Spangler, G.E.; Carrico, J.P.; Kim, S.H., Analysis of explosives and explosive residues with ion mobility spectrometry (IMS), in *Proceedings of the International Symposium on Detection of Explosives*, Quantico, VA, 1983, p. 267.
6. Fetterolf, D.D.; Clark, T.D., Detection of trace explosive evidence by ion mobility spectrometry, *J. Forensic Sci.* 1993, 38, 28–39.
7. Garofolo, F.; Migliozzi, V.; Roio, B., Application of ion mobility spectrometry to the identification of trace levels of explosives in the presence of complex matrixes, *Rapid Commun. Mass Spectrom.* 1994, 8, 527–532.
8. Karasek, F.W.; Tatone, O.S.; Kane, D.M., Study of electron capture behavior of substituted aromatics by plasma chromatography, *Anal. Chem.* 1973, 45, 1210–1214.
9. Karasek, F.W.; Denney, D.W., Detection of 2,4,6-trinitrotoluene vapors in air by plasma chromatography, *J. Chromatogr.* 1974, 93, 141–147.
10. Karasek, F.W., Detection of TNT in air, *Research/Development* 1974, 25, 32–34.
11. Spangler, G.E.; Lawless, P.A., Ionization of nitrotoluene compounds in negative ion plasma chromatography, *Anal. Chem.* 1978, 50, 884–892.
12. Danylewych-May, L.L., Modifications to the ionization process to enhance the detection of explosives by IMS, paper presented at the Proceedings of the First International Symposium on Explosion Detection Technology, Atlantic City, NJ, November 1991, Paper C-10.
13. Proctor, C.J.; Todd, J.F.J., Alternative reagent ions for plasma chromatography, *Anal. Chem.* 1984, 56, 1794–1797.
14. Spangler, G.E.; Carrico, J.P.; Campbell, D.N., Recent advances in ion mobility spectrometry for explosives vapor detection, *J. Test. Eval.* 1985, 13, 234–240.
15. Lawrence, A.H.; Neudorfl, P., Detection of ethylene glycol dinitrate vapors by ion mobility spectrometry using chloride reagent ions, *Anal. Chem.* 1988, 60, 104–109.
16. Daum, K.A.; Atkinson, D.A.; Ewing, R.G.; Knighton, W.B.; Grimsrud, E.P., Resolving interferences in negative mode ion mobility spectrometry using selective reactant ion chemistry, *Talanta* 2001, 54, 299–306.
17. Hilton, C.K.; Krueger, C.A.; Midey, A.J.; Osgood, M.; Wu, J.; Wu, C., Improved analysis of explosives samples with electrospray ionization-high resolution ion mobility spectrometry (ESI-HRIMS), *Int. J. Mass Spectrom.* 2010, 298, 64–71.
18. Lai, H.; Guerra, P.; Joshi, M.; Almirall, J.R., Analysis of volatile components of drugs and explosives by solid phase microextraction-ion mobility spectrometry, *J. Sep. Sci.* 2008, 31, 402–412.
19. Guerra, P.; Lai, H.; Almirall, J. R., Analysis of the volatile chemical markers of explosives using novel solid phase microextraction coupled to ion mobility spectrometry, *Sep. Sci.,* 2008, 31, 2891–2898.

20. Guerra-Diaz, P.; Gura, S.; Almirall, J.R., Dynamic planar solid phase microextraction-ion mobility spectrometry for rapid field air sampling and analysis of illicit drugs and explosives, *Anal. Chem.,* 2010, 82, 2826–2835.

21. Gowadia, H.A.; Settles, G.S., The natural sampling of airborne trace signals from explosives concealed upon the human body, *J. Forensic Sci.* 2001, 46, 1324–1331.

22. Schellenbaum, R.L.; Hannum, D.W., Laboratory evaluation of the PCP large reaction volume ion mobility spectrometer (LRVIMS), Report SAND-89-0461, 1990, 37.

23. Elias, L.; Neudorfl, P., Laboratory evaluation of portable and walk-through explosives vapor detectors, Report NAE-LTR-UA-104, CTN-91-60015, 1990.

24. Wu, C.; Steiner, W.E.; Tornatore, P.S.; Matz, L.M.; Siems, W.F.; Atkinson, D.A.; Hill, H.H., Construction and characterization of a high-flow, high-resolution ion mobility spectrometer for detection of explosives after personnel portal sampling, *Talanta* 2002, 57, 123–134.

25. Karpas, Z., Forensic science applications of ion mobility spectrometry (IMS), a review, *Forensic Sci. Rev.* 1989, 1, 103–119.

26. Ewing, R.G.; Atkinson, D.A.; Eiceman, G.A.; Ewing, G.J., A critical review of ion mobility spectrometry for the detection of explosives and explosive related compounds, *Talanta,* 2001, 54, 515–529.

27. Eiceman, G.A.; Preston, D.; Tiano, G.; Rodriguez, J.; Parmeter, J.E., Quantitative calibration of vapor levels of TNT, RDX, and PETN using a diffusion generator with gravimetry and ion mobility spectrometry, *Talanta* 1997, 45, 57–74.

28. Cohen, M.J.; Wernlund, R.F.; Kindel, R.C., An adjustable vapor generator for known standard concentrations in the fractional parts per billion range, in *Proceedings of the New Concepts Symposium Workshop on Detection and Identification of Explosives,* Reston, VA, 1978, p. 41.

29. Davies, J.P.; Blackwood, L.G.; Davis, S.G.; Goodrich, L.D.; Larson, R.A., Design and calibration of pulsed vapor generators for 2,4,6-trinitrotoluene, cyclo-1,3, 5-trimethylene-2,4,6-trinitramine, and pentaerythritol tetranitrate, Idaho National Engineering Laboratory, *Anal. Chem.* 1993, 65, 3004–3009.

30. Lokhnauth, J.K.; Snow, N.H., Stir-bar sorptive extraction and thermal desorption-ion mobility spectrometry for the determination of trinitrotoluene and 1,3,5-trinitro-1,3,5-triazine in water samples, *J. Chromatog. A* 2006, 1105, 33–38.

31. Tam, M.; Hill, H.H., Secondary electrospray ionization-ion mobility spectrometry for explosive vapor detection, *Anal. Chem.* 2004, 76, 2741–2747.

32. Cagan, A.; Schmidt, H.; Rodriguez, J.R.; Eiceman, G.A., Fast gas chromatography-differential mobility spectrometry of explosives from TATP to Tetryl without gas atmosphere modifiers, *Int. J. Ion Mobil. Spectrom.* 2010, 13, 157–165.

33. Harvey, S.D.; Ewing, R.G.; Waltman, M.J., Selective sampling with direct ion mobility spectrometric detection for explosives analysis, *Int. J. Ion Mobil. Spectrom.* 2009, 12, 115–121.

34. Fricano, L.; Goledzinowski, M.; Jackson, R.; Kuja, F.; May, L.; Nacson, S., An automatic trace detection system for the detection of explosives' vapors and particles in luggage, *Int. J. Ion Mobil. Spectrom.* 2001, 4, 22–26.

35. Smiths Detection, IONSCAN Sentinel II, http://www.smithsdetection.com/Sentinel.php; Safran, EntryScan®, http://www.morpho.com/detection/see-all-products/trace-detection/entryscan-r/.

36. Matz, L.M.; Tornatore, P.S.; Hill, H.H., Evaluation of suspected interferents for TNT detection by ion mobility spectrometry, *Talanta* 2001, 54, 171–179.

37. McGann, W.J.; Haigh, P.; Neves, J.L., Expanding the capability of IMS explosive trace detection, *Int. J. Ion Mobil. Spectrom.* 2002, 5, 119–122.

38. Buttigieg, G.A.; Knight, A.K.; Denson, S.; Pommier, C.; Denton, M.B., Characterization of the explosive triacetone triperoxide and detection by ion mobility spectrometry, *Forensic Sci. Int.* 2003, 135, 53–59.

39. Ewing, R.G.; Waltman, M.J.; Atkinson, D.A. Characterization of triacetone triperoxide by ion mobility spectrometry and mass spectrometry following atmospheric pressure chemical ionization, *Anal. Chem.* 2011, 83, 4838–4844.

40. McGann, W.J.; Goedecke, K.; Becotte-Haigh, P.; Neves, J.; Jenkins, A., Simultaneous dual-mode IMS detection system for contraband detection and identification, *Int. J. Ion Mobil. Spectrom.* 2001, 4, 144–147.

41. Marr, A.J.; Groves, D.M., Ion mobility spectrometry of peroxide explosives TATP and HMTD, *Int. J. Ion Mobil. Spectrom.* 2003, 6, 59–61.

42. Gillen, G.; Fletcher, R.; Verkouteren, J.; Klouda, G.; Zeissler, C.; Evans, A.; Davis, D.; Santiago, M.; Verkouteren, M., Advanced metrology to support IMS trace explosive detection: NIST capabilities and progress, ISIMS Conference, Gatlinburg, TN, July 26–29, 2004, http://www.cstl.nist.gov/div837/Division/outputs/Explosives/ISIMS_Gatlinburg.pdf.

43. Antohe, B.V.; Hayes, D.J.; Taylor, D.W.; Wallace, D.B.; Grove, M.E.; Christison, M., Vapor generator for the calibration and test of explosive detectors, in *2008 IEEE Conference on Technology Homeland Security*, Waltham, MA, May 12–13, 2008, pp. 384–389.

44. Conrad, F.J., Explosives detection—the problem and prospects, *Nucl. Mater. Manage.* 1984, 13, 212.

13 Chemical Weapons

13.1 INTRODUCTION AND GENERAL COMMENTS ON DETECTION OF CHEMICAL WARFARE AGENTS

Several review papers on the application of ion mobility spectrometry (IMS) and differential mobility spectrometry (DMS) for detection of chemical warfare agents (CWAs) and their simulants have been published during the last decade.[1–6] The comprehensive review by Sferopoulos on detector technologies and commercial devices for detection of CWAs is especially noteworthy as it surveyed all the different techniques and included a chapter on IMS-based instrumentation.[6]

Chemical weapons are defined by the Organisation for the Prohibition of Chemical Weapons (OPCW) as "anything specifically designed or intended for use in direct connection with the release of a chemical agent (CA) to cause death or harm."[7]

CWAs are chemical compounds that may severely incapacitate personnel at low concentrations and at slightly higher concentrations can be lethal, even after a short exposure period through inhalation or skin exposure. The CWAs can be dispersed in different forms, including solid, liquid, gas, vapor, and aerosol, so detection systems should be able to function effectively in a variety of operational scenarios.

CWAs can be divided into two main groups: nerve agents (sometimes classified as G agents or the more persistent V agents), which disrupt the normal functioning of the central nervous system (CNS) or peripheral nervous system, and blister agents (classified as vesicants or H agents), which affect tissues and may cause severe injuries, especially to lung tissues, and even lead to death. The nerve agents are generally organophosphorus compounds, while blister agents are usually chloride-containing organosulfur or arsenic compounds. There are also other types of chemical agents, such as choking agents, which may cause asphyxia, and blood agents. A list of the common CWAs is given in Table 13.1.

The favorable response of IMSs in the positive ion mode to organophosphorus compounds was reported in the 1970s.[8,9] Since then, several groups have studied the response of specific IMS-based instrument configurations to CWAs or their degradation products in the environment and to simulants of CWAs.[10–30] In some of the studies, the IMS was combined with a chromatographic column to separate the complex mixture of compounds before they entered the drift tube.[5,12,22] The studies of CWA degradation products in the environment[14–20] generally involved preconcentration from liquid samples with a solid phase microextraction (SPME) fiber or soil samples,[20,28] electrospray ionization (ESI) with a high-resolution IMS drift tube,[18,19] or even direct aspiration of the liquid sample.[15–19]

TABLE 13.1

List of Classical CWAs: Agent Class, Name, and Abbreviation

Nerve　　Tabun, GA; sarin, GB; soman, GD; ethyl sarin, GE; cyclosarin, GF; O-ethyl-S-diisopropyl amino methyl; methylphosphonothiolate, VX; S-(diethyl amino)ethyl O-ethyl ethylphosphonothioate, VE; Amiton or Tetram, VG; phosphonothioic acid, methyl-, S-(2-(diethyl amino)ethyl) O-ethyl ester, VM

Vesicants　　Sulfur mustard, H, HD; nitrogen mustard, HN-1, HN-2, HN-3; lewisite, L; mustard-lewisite, HL; phenyldichloroarsine, PD; phosgene oxime, CX; hydrogen cyanide, AC

Blood　　Cyanogen chloride, CK; arsine, SA

Choking　　Chlorine, Cl; phosgene, CG; diphosgene, DP; chloropicrin, PS

Source:　Sferopoulos, R., A review of chemical warfare agent (CWA) detector technologies and commercial-off-the-shelf items, Defence Science and Technology Organisation, DSTO-GD-0570, Australia, 2008.

13.2　THE ION CHEMISTRY UNDERLYING DETECTION OF CHEMICAL WEAPONS

As mentioned, nerve agents designed for use as CWAs are mostly derivatives of organophosphorus compounds that have high proton affinities.[31] Thus, the sensitivity and specificity of response and the high speed of measurement make mobility spectrometers superb instruments for in-field or on-site determinations. In addition, blister agents and related breakdown products can be detected in the negative polarity.

Positive gas phase ion chemistry governs the reactions of nerve agents or simulants such as DMMP (dimethyl methyl phosphonate) in the IMS ionization source. The main reaction pathway between the reactant ion that is a protonated dimer of the dopant $R_2H^+(H_2O)_n$ [where R could be acetone $(CH_3)_2CO$] by an organophosphorus molecule G involves displacement of an R molecule by G to form a heterogeneous proton-bound dimer $GRH^+(H_2O)_{n-1}$ (Equation 13.1).

Displacement of dopant molecule:

$$G + R_2H^+(H_2O)_n \rightarrow GRH^+(H_2O)_{n-1} + R + H_2O \qquad (13.1)$$

If the concentration of G is increased, this could be followed by the formation of a homogeneous proton-bound dimer $G_2H^+(H_2O)_{n-1}$ (Equation 13.2).

Second displacement of dopant molecule:

$$G + GRH^+(H_2O)_n \rightarrow G_2H^+(H_2O)_{n-1} + R + H_2O \qquad (13.2)$$

To control the ionization processes and reduce interfering reactions, a reagent gas that forms well-defined reactant ions may be added. Thus, R might be one of several chemicals, including ammonia or acetone.[32–34] In principle, the ion $G_2H^+(H_2O)_{n-1}$ will be formed if the association between G and H^+ is strong.[33]

Furthermore, proton-bound trimer ions $G_3H^+(H_2O)_n$ and higher clustered ions can form in the reaction region where ions and sample neutrals mix. When such ions are extracted from the reaction region into the drift region of the IMS, largely free of sample neutrals, decomposition of the trimer ions is rapid. Such ions exhibit lifetimes too short to be measured in the drift tube at ambient temperature.

Blister agents that generally contain halogen atoms (usually chlorine atoms), such as mustard gas (HD), have a low proton affinity and therefore do not form stable positive ions at ambient pressure. However, these chemicals (termed H agents in military codes but called B in this section) can form adduct ions, that is, $B*O_2^-(H_2O)_{n-1}$ in the negative polarity as shown in Equation 13.3:

Adduct ion formation in negative mode:

$$B + O_2^-(H_2O)_n \rightarrow B*O_2^-(H_2O)_{n-1} + H_2O \tag{13.3}$$

Elevated moisture levels may interfere with this reaction and enhance the reverse reaction of Equation 13.3, namely, the dissociation of the adduct ion. In such cases, the final ion observed in the mobility spectrum may be the reactant ion, leading to a false-negative response. Therefore, as mentioned, a confidence testing with a simulant should be performed before attempting to detect these blister agents.

13.3 SAMPLING AND PRECONCENTRATION TECHNIQUES

Different analytical strategies have been employed to increase the sensitivity and specificity of IMS-based instruments for detection of CWAs and toxic industrial chemicals (TICs). Dopants, such as acetone, are utilized to enhance the performance of IMS instruments[32–36] in several commercial instruments.[35] A different approach to increase performance was to operate at pressures below[37] or above[38] the ambient pressure. In the former case, it was reported that the reduced pressure doubled the resolution of the DMS, thus improving the specificity of the device.[37] In the latter case, the IMS was operated at pressures up to 4,560 torr (6 bar) for detection of DMMP (commonly used as a simulant for nerve agents) and other compounds. Resolution increased, to a certain extent, with pressure (and the electric field strength), but once clustering effects became dominant, the resolution did not increase as theory predicted.[38]

The combination of a SPME device with a pyrolysis gas chromatographic (GC) IMS system improved the limit of detection of tributylphosphate (TBP, which served as a simulant) in water by a factor of 20 compared to the same system without the SPME device.[39] SPME fibers were also used to sample headspace vapors of several types of nerve agents, and the fibers were introduced directly into a modified ESI source for subsequent detection by IMS and mass spectrometry (MS).[26] A SPME–IMS system, with thermal desorption, was also used to screen soil samples for precursor and degradation products of CWAs, and it was found that fibers of polydimethylsiloxane (PDMS) were superior to PDMS-divinylbenzene fibers.[28]

Direct sampling of water sources or other liquid samples with an ESI inlet coupled to an IMS for detection of CWAs, their precursors, or degradation products has been investigated.[18,26,27,40] In one study, six G-series nerve agents or their precursors were measured with an IMS operated at ambient pressure in negative mode that was coupled to a time-of-flight (TOF) MS and ESI inlet that was used for sample introduction and ionization (ESI ion mobility [IM] [TOF] MS).[40]

13.4   RESEARCH, OPERATIONAL EXPERIENCE, AND HISTORICAL PERSPECTIVE OF INSTRUMENTATION

Several examples of the research involving detection of CWAs, their precursors, and degradation products were given in the previous discussion. For obvious reasons, the open literature does not contain reports about operational experience with CWAs and the performance of detection systems even in exercises. With regard to the actual detection limits for CWAs of IMS-based detectors, one can refer to the specifications presented by the manufacturers, which usually state that performance conforms to the operational requirements without giving actual substantiated quantitative results. A historical perspective of the development of CWA detectors based on IMS technology is given here and is followed by a state of the art overview.

Programs to develop IMS-based detectors and monitors for CWAs were undertaken with the support of defense establishments in several governments, notably the United Kingdom and United States. In the United States, research toward the development of an automatic chemical agent detector and alarm (ACADA) was begun[33] but later discontinued. Efforts were resumed by Smiths Detection, and the result has been placed into service as the GID-3™ and its later models, like the GID-M.[41] The intention for this analyzer is to continuously monitor the presence of CWAs in ambient air and sound the alarm when threshold values are exceeded. The GID-3 and all other IMS-based analyzers are regarded as point sensors and cannot determine chemical compositions at distances greater than a few meters in the absence of breezes.

A handheld chemical agent monitor (CAM) was successfully produced by Graseby Dynamics, Limited.[35] in the United Kingdom for use in postattack scenarios and was intended for an individual to screen surfaces, equipment, and personnel to verify decontamination before further contact. The response of the CAM would then guide decisions regarding handling and decontamination of the object or person. The CAM analyzer has been regularly upgraded, and an improved model is now deployed as the i-CAM. Handheld IMS analyzers for use in battlefield environments were developed also in Germany by Bruker–Saxonia as the RAID™ (Rapid Alarm and Identification Device) series and the IMS2000.[42] A general resemblance exists between these handheld analyzers, although technical specifications differ slightly. In Finland, Environics Oy produced a series of mobility-based analyzers with the aspirator designs for CWA monitoring, including the IMCELL™ MGD-1 and its later models.[15] Finally, a pocket-size CWA monitor for individual soldiers has been developed by Smiths Detection and deployed with the British Armed Forces. All of these instruments were designed for users who have limited training and experience with analytical measurement. Moreover, the environmental conditions where these detectors are deployed could be hostile and demanding. Thus, the analyzers had to be rugged, simple to

operate, and practically maintenance free for long periods. In practice, a soldier challenges the analyzer with a simulant (a confidence tester) before screening for CWAs
to validate the proper operation and performance of the instrument.

Over 20 years ago, it was found that the sensitivity and specificity of the earlier generation of instruments toward nerve agents could be enhanced by the use of acetone-based
reagent ion chemistry.[32] As explained previously, in the positive ion mode the acetone
molecules readily protonate and associate to produce protonated acetone monomers
and dimers, thus effectively eliminating many interferents while not losing response
to organophosphorus nerve agents that have proton affinities above that of acetone[31]
(196 kcal/mol). Acetone does not significantly affect negative ion chemistry and does
not prevent blister agent detection with a drift tube containing acetone as a reagent gas.
The exact data on the sensitivity of the military units for nerve and blister agents are
not available from research articles or objective studies, apart from statements of com-
pliance with established NATO (North Atlantic Treaty Organization) requirements.
However, in some brochures of the manufacturers of IMS equipment, the performance
is listed (Table 13.2). For example, in the brochure of Bruker's handheld IMS2000,[42]
the range for the nerve agents GB (sarin) and GA (tabun) is given as 20 to 600 $\mu g/m^3$.
In the listing of the minimum detectable limits (MDLs) of the civilian GP-IMS and
FP-IMS instruments made formerly by ETG, Incorporated, 5 and 10 ppb are given as
the MDLs for nerve and blister agents, respectively. Information about field tests of
these instruments is also not available to the general public.

The high sensitivity of IMS devices toward CWAs is true also for differential
mobility. For example, MDLs for GB and phosgene were reported as 8 and 4 ppt

TABLE 13.2
Limits of Detection for Chemical Warfare Agents

Chemical Warfare Agent	Low Alarm Limit (ppm)	Low Alarm Limit (mg/m³)
GA (tabun), GB (sarin)	0.2	0.02
GD (soman), GF (cyclo-sarin)/VX	0.1	0.01
HD (sulfur mustard)	0.312	2.00
L (lewisite)	0.242	2.00
Blood agents	—	30.0
Ethylene oxide	100	180
Acrylonitrile	100	213
Hydrogen sulfide	10	14
Arsine	5	16
Ammonia	400	278
Phosphor trichloride	25	140
Carbon disulfide	500	1,557
Allyl alcohol	40	95

Source: From AFC International, ChemRae chemical warfare agent detector from Rae
       System, http://www.afcintl.com/product/tabid/93/productid/123/sename/chemrae-
       chemical-warfare-agent-detector-from-rae-system/default.aspx.

(by volume), respectively, by the now-defunct Field Ion Spectrometer formerly marketed by Mine Safety Appliance, Incorporated.[43]

Deployment of IMS analyzers on unmanned airborne vehicles (UAVs) for detection of clouds of CWAs has also been demonstrated with simulants.[44] When a point sensor is made mobile, the capabilities of the analyzer are augmented by providing chemical determinations at points within a large area.

13.5   STATE-OF-THE-ART COMMERCIAL INSTRUMENTS, STANDARDS, AND CALIBRATION

In this section, we present some of the commercially available detectors for CWAs and toxic chemicals that are based on IMS or DMS. These include a large variety of devices ranging from handheld, or even pocket-size, instruments to larger fixed-point monitors that can even be placed in the ventilation system of buildings. The devices mentioned here are for demonstration purposes and are not to be understood as recommended by us.

Smiths Detection has several devices that are based on IMS technology that are suitable for detection of explosives, narcotics, CWAs, and TICs. These detectors come in different sizes—from pocket size to handheld instruments, portable instruments to larger continuous point monitors (Figures 13.1 and 13.2). Some are specifically designed to detect and identify toxic chemical vapors in ambient air, while others use different sampling methods to detect particulate matter and aerosols. In general, several of the devices can be linked to provide an area network for detection of chemical threats. The CAM is a handheld monitor for the detection of nerve or blister agents or liquid agent contamination. The CAM is probably the most popular IMS-based product, with over 70,000 devices produced. The CAM is deployed and in service with armed forces, civil defense personnel, and first-responder organizations worldwide. It can continuously and automatically scan between positive mode for nerve agent detection and negative mode for detection of blister, blood, and choking agents. It was originally designed for verification of decontamination of personnel, vehicles, and objects in areas suspected of being contaminated by CWAs. For example, according to the manufacturer, the continuous Centurion II monitor is suitable for detection and alarming the nerve and blister agents such as tabun, sarin, soman (GD), cyclosarin (GF), agents VX (methylphosphonothiolate) and VXR, cyanogen chloride (CK), nitrogen mustard, and mustard. A portable version that can be operated on a vehicle or by dismounted troops is the GID-3, and the LCD-Nexus model is suitable for detection of CWAs and TICs in harsh environments. The sensitivity of the handheld Multi-Mode Threat Detector (MMTD) is reportedly in the low-nanogram range for particles and in the low parts-per-million range for vapors. The detailed limits of detection are not given on the Web site (http://www.smithsdetection.com/mmtd.php), but they are usually within the specifications of the military definitions.

Bruker Daltonics has also developed chemical monitors based on IMS technology that cover CWA and TIC detection tasks (Figure 13.3). The RAID series includes the portable RAID-M 100 detector, which provides very low detection limits, and the RAID-XP, which has chemical and radiological detection in one system, as well as the

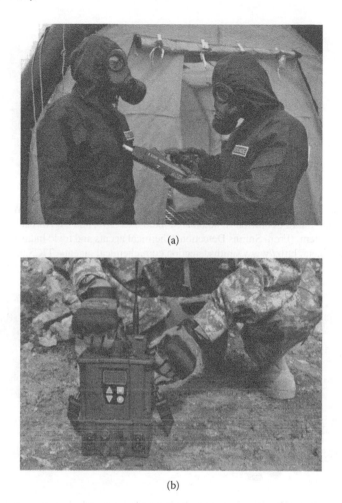

(a)

(b)

FIGURE 13.1 (a) Checking for chemical agents with the CAM. (b) Vapor sampling for detection of CWAs. (From Smiths Detection, Chemical agents and toxic industrial chemicals detection equipment, http://www.smithsdetection.com/chemical_agents_TICS.php.)

RAID-AFM (Automated Facility Monitor) especially designed for critical infrastructure and facilities monitoring and the RAID-S2 designed for all classes of naval vessels.

The ChemPro 100 is Environics' unique open-loop IMS sensor that uses a Nuclear Regulatory Commission (NRC) exempt [241]Am ionization source (Figure 13.4). The detection limits for the nerve gases GA, GB, GD, GF, and VX are 0.1 mg/m³, 2 mg/m³ for the vesicant precursors sulfur mustard (HD) and lewisite (L), and 20 mg/m³ hydrogen cyanide (AC) and CK. The improved version, ChemPro 100i, combines the IMS sensor with six new chemical sensors and two semiconductor sensors, improving the sensitivity for nerve agents by a factor of 2.5 (to 0.04 mg/m³). The rate of false alarms is also reduced. Environics also has a fixed-site monitor called ChemProFX.

Based on the ChemPro 100 model of Environics, ChemRae (San Jose, CA) developed a handheld detector that reportedly provides enhanced selectivity and

FIGURE 13.2 Left: Small, lightweight, continuous, real-time detector of CWAs and toxic chemicals. Right: Sabre Centurion II, automated fixed-site CWA and TIC threat air monitoring and detection system. (From Smiths Detection, Chemical agents and toxic industrial chemicals detection equipment, http://www.smithsdetection.com/chemical_agents_TICS.php.)

FIGURE 13.3 Right: RAID-S2, a dependable rapid alarm and identification device. Left: RAID M-100 extensive portable capability. (From Bruker Daltonics, Chemical detection, http://www.bdal.com/products/mobile-detection/chemical.html.)

sensitivity and can be used as a stand-alone portable monitor or integrated into an AreaRae network (Figure 13.4).

General Dynamics (now ChemRing) has produced handheld detectors for CWAs based on DMS called JUNO. The technology is claimed to be better than the traditional (linear field drift tube) IMS devices with regard to selectivity and sensitivity. JUNO can be used with a preconcentrator and detect CWAs at miosis levels and enables users to monitor personal chemical agent exposure levels and confirm decontamination effectiveness.

IUT Berlin produces the IMS Mini-200, which is a portable multigas detector for TICs and CWAs developed by IUT (Figure 13.5). Toxic gases can be detected and identified without any enrichment directly in situ at a very low concentration level. It can be operated either in manual or in automatic positive and negative detection modes.

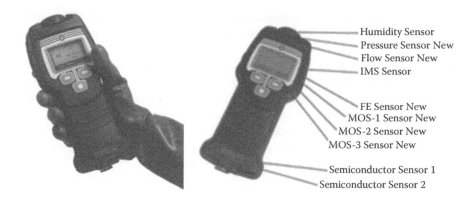

FIGURE 13.4 Left: ChemRae Chemical Warfare Agent Detector from Rae System (from AFC International, ChemRae Chemical Warfare Agent Detector from Rae System, http://www.afcintl.com/product/tabid/93/productid/124/sename/chempro-100-ims-chemical-warfare-agent-detector-from-environics/default.aspx). Right: Chempro 100i IMS improved chemical detector from Environics (from http://www.afcintl.com/product/tabid/93/productid/413/sename/chempro-100i-ims-improved-chemical-detector-from-environics/default.aspx).

FIGURE 13.5 Left: IMS Mini-200, a portable IMS for the detection of toxic gases and CWAs (from IUT Berlin, NEW: IMS Mini-200, http://www.iut-berlin.info/7.0.html?&L=1). Right: The JUNO® hand-held chemical detector developed and produced by Chemring Detection Systems (CDS) for detecting, identifying, quantifying, and alerting to the presence of chemical vapors. (http://www.chemringds.com/Products/ChemicalDetection/JUNO1/)

13.6 SUMMARY

The performance criteria for CWA and TIC detectors include specificity (selective response to toxic compounds), sensitivity (the ability to detect concentrations that are below the threshold of damage or injury), response time (give a warning before the onset of damage), and low false alarm rates (false negative could endanger people, while false positive could lead to panic and loss of reliability). Considering the various threats from CWAs and toxic chemicals in general and the low concentrations at which these may cause injuries to unprotected troops or civilian populations, the requirements from a detection system are very severe. In her summary, Sferopoulos cautioned that "there is no single detector that has all the desired capabilities and performance functions" (page 12).[6]

IMS technology appears to provide one of the best responses to these requirements. In the positive ion mode, particularly as far as nerve agents based on organophosphorus compounds are concerned due to its selectivity (with the aid of suitable dopants), sensitivity (sub-part-per-billion range for vapors, gas and aerosols), and rapid response time. As far as blister agents, blood agents, and choking agents are concerned, IMS technology can also provide a reasonable response in the negative ion mode or combining positive and negative ion modes.

REFERENCES

1. Kolakowski, B.M.; Mester, Z., Review of applications of high-field asymmetric waveform ion mobility spectrometry (FAIMS) and differential mobility spectrometry (DMS), *Analyst* 2007, 132, 842–864.
2. Seto, Y.; Kanamori-Kataoka, M.; Tsuge, K.; Ohsawa, I.; Maruko, H.; Sekiguchi, H.; Sano, Y.; Yamashiro, S.; Matsushita, K.; Sekiguchi, H.; et al., Development of an on-site detection method for chemical and biological warfare agents, *Toxin Rev.* 2007, 26, 299–312.
3. Makinen, M.A.; Anttalainen, O.A.; Sillanpaa, M.E.T., Ion mobility spectrometry and its applications in detection of chemical warfare agents, *Anal. Chem.* 2010, 82, 9594–9600.
4. Hill, H.H.; Steiner, W.E., Ion mobility spectrometry for monitoring the destruction of chemical warfare agents, Edited by Kolodkin, V.M.; Ruck, W., Ecological Risks Associated with the Destruction of Chemical Weapons, *Proceedings of the NATO Advanced Research Workshop on Ecological Risks Associated with the Destruction of Chemical Weapons*, Lueneburg, Germany, October 22–26, 2003, 2006, pp. 157–166.
5. Buryakov, I.A., Express analysis of explosives, chemical warfare agents and drugs with multicapillary column gas chromatography and ion mobility increment spectrometry, *J. Chromatog. B Anal. Technol. Biomed. Life Sci.* 2004, 800, 75–82.
6. Sferopoulos, R., A review of chemical warfare agent (CWA) detector technologies and commercial-off-the-shelf items, Defence Science and Technology Organisation, DSTO-GD-0570, Australia, 2008.
7. Organisation for the Prohibition of Chemical Weapons (OPCW), Fact Sheet 4: What Is a Chemical Weapon? The Hague, The Netherlands. 2000. http://www.opcw.org/about-chemical-weapons/what-is-a-chemical-weapon/
8. Moye, H.A., Plasma chromatography of pesticides, *J. Chromatogr. Sci.* 1975, 13, 285–290.
9. Preston, J.M.; Karasek, F.W.; Kim, S.H., Plasma chromatography of phosphorus esters, *Anal. Chem.* 1977, 49, 1346–1350.
10. Kim, S.H.; Spangler, G.E., Ion-mobility spectrometry-mass spectrometry of two structurally different ions having identical ion mass, *Anal. Chem.* 1985, 57, 567–569.
11. Karpas Z.; Pollevoy, Y., Ion mobility spectrometric studies of organophosphorus compounds, *Anal. Chim. Acta* 1992, 259, 333–338.
12. Dworzanski, J.P.; Kim, M.-G.; Snyder, A.P.; Arnold, N.S.; Meuzelaar, H.L.C., Performance advances in ion mobility spectrometry through combination with high-speed vapor sampling, preconcentration and separation techniques, *Anal. Chim. Acta* 1994, 293, 219–235.
13. Turner, R.B.; Brokenshire, J.L., Hand-held ion mobility spectrometers, *Trends Anal. Chem.* 1994, 13, 275–280.
14. Leonhardt, J.W., New detectors in environmental monitoring using tritium sources, *J. Radioanal. Nucl. Chem.* 1996, 206, 333–339.

15. Tuovinen, K., Paakkanen, H.; Hänninen, O., Determination of soman and Vx degradation products by an aspiration ion mobility spectrometry, *Anal. Chim. Acta* 2001, 440, 151–159.
16. Paakanen, H., About the applications of IMCELLTM MGD-1 detector, *Int. J. Ion Mobil. Spectrom.* 2001, 4, 136–139.
17. Kättö, T.; Paakkanen, H.; Karhapää, T., Detection of CWA by means of aspiration condenser type IMS, in *Proceedings of the Fourth International Symposium on Protection Against Chemical Warfare Agents,* Stockholm, Sweden, National Defense Research Establishment, Department of NBC-defense, 1992, pp. 103–108.
18. Asbury, G.R.; Wu, C.; Siems, W.F.; Hill, H.H., Separation and identification of some chemical warfare degradation products using electrospray high-resolution ion mobility spectrometry with mass-selected detection, *Anal. Chim. Acta* 2000, 404, 273–283.
19. Steiner, W.E.; Clowers, B.H.; Matz, L.M.; Siems, W.F.; Hill, H.H., Jr., Rapid screening of aqueous chemical warfare agent degradation products: ambient pressure ion mobility mass spectrometry, *Anal. Chem.* 2002, 74, 4343–4352.
20. Fällman, Å.; Rittfeldt, L., Detection of chemical warfare agents in water by high temperature solid phase microextraction–ion mobility spectrometry (HTSPME-IMS), *Int. J. Ion Mobil. Spectrom.* 2001, 4, 85–87.
21. Harden, C.S.; Blethen, G.E.; Davis, D.M.; Harper, S.; McHugh, V.M.; Shoff, D.B., Detection and analysis of explosively disseminated CW agents in 400 m^3 chamber, *Int. J. Ion Mobil. Spectrom.* 2001, 4, 13–21.
22. Sielemann, S.; Li, F.; Schmidt, H.; Baumbach, J.I., Ion mobility spectrometer with UV-ionization source for determination of chemical warfare agents, *Int. J. Ion Mobil. Spectrom.* 2001, 4, 81–84.
23. Sielemann, S.; Baumbach, J.I.; Schmidt, H., IMS with non radioactive ionization sources suitable to detect chemical warfare agent simulation substances, *Int. J. Ion Mobil. Spectrom.* 2002, 5, 143–148.
24. Ringer, J.; Ross, S.K.; West, D.J., An IMS/MS investigation of lewisite and lewisite/mustard mixtures, *Int. J. Ion Mobil. Spectrom.* 2002, 5, 107–111.
25. Kolakowski, B.M.; D'Agostino, P.A.; Chenier, C.; Mester, Z., Analysis of chemical warfare agents in food products by atmospheric pressure ionization high field asymmetric waveform ion mobility spectrometry-mass spectrometry, *Anal. Chem.* 2007, 79, 8257–8265.
26. D'Agostino, P.A.; Chenier, C.L., Desorption electrospray ionization mass spectrometric analysis of organo-phosphorus chemical warfare agents using ion mobility and tandem mass spectrometry, *Rapid Commun. Mass Spectrom.* 2010, 24, 1613–1624.
27. Gunzer, F.; Zimmermann, S.; Baether, W., Application of a nonradioactive pulsed electron source for ion mobility spectrometry, *Anal. Chem.* 2010, 82, 3756–3763.
28. Rearden, P.; Harrington, P.B., Rapid screening of precursor and degradation products of chemical warfare agents in soil by solid-phase microextraction ion mobility spectrometry (SPME-IMS), *Anal. Chim. Acta* 2005, 545, 13–20.
29. Steiner, W.E.; Klopsch, S.J.; English, W.A.; Clowers, B.H.; Hill, H.H., Detection of a chemical warfare agent simulant in various aerosol matrixes by ion mobility time-of-flight mass spectrometry, *Anal. Chem.* 2005, 77, 4792–4799.
30. Steiner, W.E.; English, W.A.; Hill, H.H., Separation efficiency of a chemical warfare agent simulant in an atmospheric pressure ion mobility time-of-flight mass spectrometer (IM(tof)MS), *Anal. Chim. Acta* 2005, 532, 37–45.
31. Lias, S.G.; Liebman, J.F.; Levin, R.D., Evaluated gas phase basicities and proton affinities of molecules: heats of formation of protonated molecules, *J. Phys. Chem. Ref. Data* 1984, 13, 695–808.
32. Spangler, G.E.; Campbell, D.N.; Carrico, J.P., Acetone reactant ions for ion mobility spectrometry, paper presented at the Pittsburgh Conference on Analytical Chemistry and Applied Spectroscopy (PittCon). Atlantic City, NJ, 1983, Paper No. 641.

33. Carrico, J.P.; Davis, A.W.; Campbell, D.N.; Roehl, J.E.; Sima, G.R.; Spangler, G.E.; Vora, K.N.; White, R.J., Chemical detection and alarm for hazardous chemicals, *Am. Lab.* 1986, 18, 152–163.

34. Preston, J.M.; Rajadhyax, L., Effect of ion-molecule reactions on ion mobilities, *Anal. Chem.* 1988, 60, 31–34.

35. Brochure of the chemical agent monitor (CAM), Graseby, Watford, UK, http://www.smithsdetection.com/.

36. Ross, S.K.; McDonald, G.; Marchant, S., The use of dopants in high field asymmetric waveform spectrometry, *Analyst* 2008, 133, 602–607.

37. Griffin, M.T., Differential mobility spectroscopy for chemical agent detection, *Proc. SPIE Intern. Soc. Opt. Eng.* 2006, 6218 (Chemical and Biological Sensing VII), 621806/1–621806/10.

38. Davis, E.J.; Dwivedi, P.; Tam, M.; Siems, W.F.; Hill, H.H., High-pressure ion mobility spectrometry, *Anal. Chem.* 2009, 81, 3270–3275.

39. Erickson, R.P.; Tripathi, A.; Maswadeh, W.M.; Snyder, A.P.; Smith, P.A., Closed tube sample introduction for gas chromatography-ion mobility spectrometry analysis of water contaminated with a chemical warfare agent surrogate compound, *Anal. Chim. Acta* 2006, 556, 455–461.

40. Steiner, W.E.; Harden, C.S.; Hong, F.; Klopsch, S.J.; Hill, H.H.; McHugh, V.M., Detection of aqueous phase chemical warfare agent degradation products by negative mode ion mobility time-of-flight mass spectrometry [IM(tof)MS], *J. Am. Soc. Mass Spectrom.* 2006, 13, 241–245.

41. Thathapudi, N.G., The development of a high sensitivity, man-portable chemical detector-GID-M, *Int. J. Ion Mobil. Spectrom.* 2005, 8, 72–76.

42. Brochure of Bruker RAID series and IMS2000 hand-held instruments, Bruker-Saxonia, Germany, http://www.army-technology.com/contractors/nbc/bruker/.

43. Carnahan, B.; Day, S.; Kouznetsov, V.; Tarrasov, A., Development and applications of a traverse field compensation ion mobility spectrometer, in *Fourth International Workshop on Ion Mobility Spectrometry,* Editor Brittain, A., Cambridge, U.K., 1995.

44. Cao, L.; de B. Harrington, P.; Harden, C.S.; McHugh, V.M.; Thomas, M.A., Nonlinear wavelet compression of ion mobility spectra from ion mobility spectrometers mounted in an unmanned aerial vehicle, *Anal. Chem.* 2004, 76, 1069–1077.

14 Drugs of Abuse

14.1 INTRODUCTION AND GENERAL COMMENTS ON DETECTION OF DRUGS

The potential for detecting contraband drugs, especially heroin and cocaine, by ion mobility spectrometry (IMS) was recognized in the 1970s.[1] However, in the last decade or two, more attention has been placed on developing techniques and instrumentation for the detection of so-called recreational drugs like the amphetamines MDMA (3,4-methylenedioxy-methamphetamine) and other derivatives. It should be noted that to increase the capability and performance of the detection systems, sampling and preconcentration techniques received most of the attention, while IMSs themselves did not undergo major changes. Another field of interest for researchers arose from the practical need to distinguish the presence of illicit drugs in environments that contained several potential interferences from other compounds that were there either intentionally or not. Unlike the risks to lives and property involved in obtaining a "false-negative" response in an explosives detector, a false-negative response in detecting illicit drugs does not pose an immediate risk. Thus, there is more of a tolerance in specifications for drug detection systems than for explosive or toxin detectors. The actual application of IMS technology for detection and identification of illicit drugs can be divided into two parts: One is the detection of concealed contraband material, such as MDMA and cocaine, usually hidden in luggage, containers, and illicit laboratories; and the other is screening of suspect substances and identification of illicit materials as opposed to legitimate substances. The following sections focus on the properties and ionization chemistry of such substances in air at ambient pressure, which is important for correct identification; on the sampling and preconcentration methods used to transport enough molecules from the suspected object into the IMS; and on the procedures that were developed for drug bioassay (mainly skin, hair, and urine).

The common illicit drugs can be divided into three main groups: opiates or narcotic substances such as heroin and cocaine, amphetamines and benzodiazepines and their derivatives, and all other unlawful addictive substances. Applications of IMS in the expanding field of pharmaceutical compounds are discussed in detail in Chapter 15.

14.2 THE ION CHEMISTRY UNDERLYING DETECTION OF DRUGS

Several of the common substances used as illicit drugs consist of nitrogen compounds, mainly amides that have a high proton affinity (Figure 14.1). Such compounds form stable positive ions by proton transfer reactions, where M is the illicit

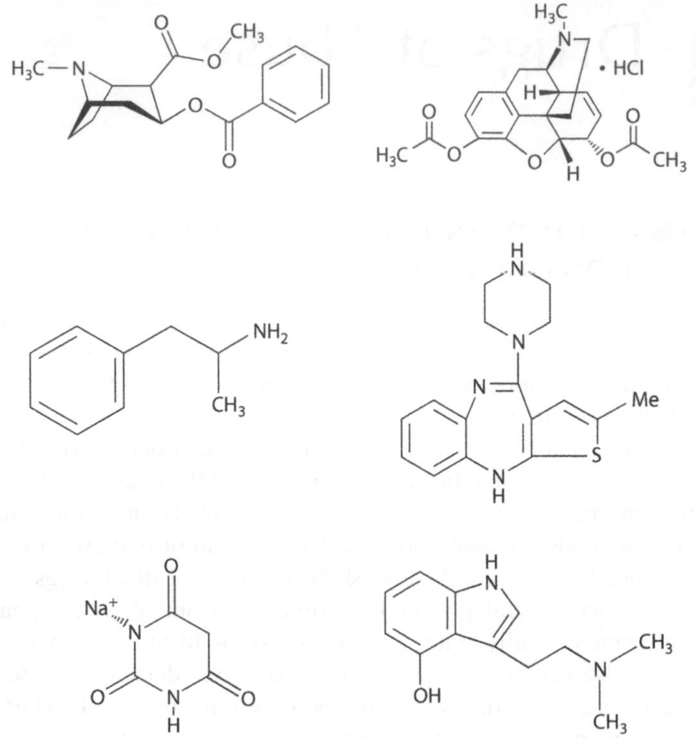

FIGURE 14.1 Molecular structure of some of the common illicit drugs (top left to right bottom): cocaine, heroin hydrochloride, amphetamine, benzodiazepine derivative, sodium barbiturate, and psilocin. Note that all these are nitrogen bases.

drug. Detection can be based on water chemistry (Equation 14.1) or on a dopant with suitable proton affinity, below that of the illicit substances but above that of several potential interferents (R in Equation 14.2):

$$M + H^+ (H_2O)_n \rightarrow MH^+ (H_2O)_{n-1} + H_2O \tag{14.1}$$

$$M + RH^+ (H_2O)_n \rightarrow MH^+ (H_2O)_{n-1} + R + H_2O \tag{14.2}$$

IMS/MS (mass spectrometric) measurements of heroin and cocaine were reported by Karasek et al.,[1] and the major ions formed from heroin were identified as the molecular ion $(M)^+$, a quasimolecular ion $(M-H_2)^+$, and a fragment ion $(M-CH_3CO_2)^+$. In cocaine, the molecular ion $(M)^+$ and two fragment ions $(M-C_6H_5CO_2)^+$ and $(M-C_6H_5CO_2-CH_3CO_2)^+$ were the dominant ions observed in the positive ion mobility spectra.[1] The response in positive mode and negative mode to several types of barbiturates has been reported,[2] and later a Fourier transform IMS with capillary column gas chromatograph (GC) was employed for the determination of barbiturates.[3] Typical positive ion mode mobility spectra of several common illicit drugs are shown

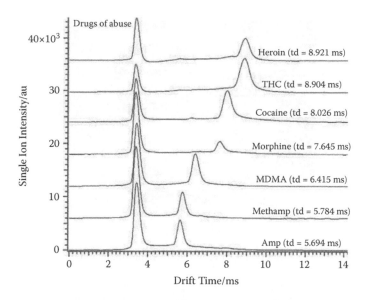

FIGURE 14.2 The mobility spectra of several illicit drugs. (From Kanu, A.B.; Wu, C.; Hill, H.H., Rapid preseparation of interferences for ion mobility spectrometry, *Anal. Chim. Acta* 2008, 610, 125–134. With permission.)

in Figure 14.2.[4] Note that in all these spectra the position of the reactant ion peak remains fixed at 3.5 ms, while the drift time of each of the product ions, arising from the sample of the illicit drug, is well separated.

14.3 SAMPLING AND PRECONCENTRATION TECHNIQUES

The traditional methods for detection of illicit drugs are based either on "sniffing" of vapors emitted from the contraband materials or on swipe samples by which the suspect object (or person) is swabbed with a suitable material (paper, cloth, or a polymer) that is inserted into a heated inlet of the IMS (Figure 14.3, IONSCAN inlet schematic). These sampling methods have been used in several laboratory and operational scenarios, as shown in Section 14.4.

One of the areas where intensive research is taking place is the use of solid phase microextraction (SPME) techniques to pre-concentrate the vapors of the analytes (illicit drugs or explosives)[5–11] or dissolved drugs from biological fluids.[12–15] A schematic representation of a desorption chamber used for a SPME fiber is shown in Figure 14.4,[12] and another variation of a thermal desorption device is shown in Figure 14.5[6] The former device can be coupled not only to a differential mobility spectrometer (DMS)/MS (as shown here) but also to a linear IMS, and the second device was developed for an IMS drift tube. In one study, a device consisting of a transfer line with silicosteel coating on a stainless steel tube with a low thermal mass was used to concentrate camphor vapors or other compounds emitted from eucalyptus leaves and to absorb diazepam and cocaine from an aqueous solution with detection levels of 10 and 50 ng/mL, respectively.[6] The Almirall's group has been particularly

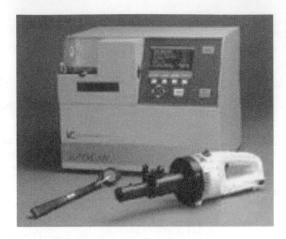

FIGURE 14.3 Smith's Detection IONSCAN 400b sampling media.

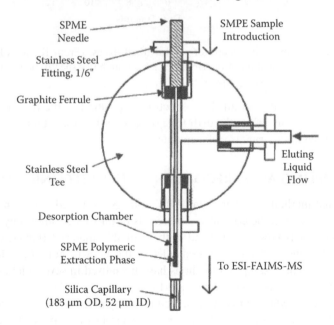

FIGURE 14.4 Schematic representation of solid phase microextraction (SPME) fiber desorption chamber (From McCooeye et al., Quantitation of amphetamine, methamphetamine, and their methylenedioxy derivatives in urine by solid-phase microextraction coupled with electrospray ionization high-field asymmetric ion mobility spectrometry-mass spectrometry, *Anal. Chem.* 2002, 74, 3071–3075. With permission.)

active in the development of devices and techniques for concentrating the vapors from a large volume, or headspace vapors, on a solid absorbent and then thermally emitting them directly into the IMS.[7–11] A planar geometry device (PSPME) coated with PDMS (polydimethylsiloxane) and sol-gel PDMS was used to sample vapors of explosives but is also suitable for illicit drug detection.[8] It was claimed that, due to

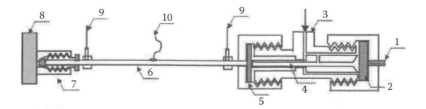

FIGURE 14.5 Schematic diagram of SPME–IMS system: (1) needle guide; (2) septum; (3) T-connection connector; (4) modified SRI GC liner; (5) nut with punched septum; (6) transfer line/desorber; (7) PEEK union; (8) IMS; (9) power lead and clamps; (10) thermocouple. FAIMS, field asymmetric IMS. PEEK, polyether ether ketone. (From Liu et al., A new thermal desorption solid-phase microextraction system for hand-held ion mobility spectrometry, *Anal. Chim. Acta* 2006, 559, 159–165. With permission.)

its surface chemistry, high surface area, and capacity, PSPME provides significant increases in sensitivity over conventional fiber SPME.[8] In a later study from that group, it was demonstrated that an IMS system with this PSPME device was capable of detecting part-per-trillion levels of MDMA in headspace vapors.[11]

Applications of SPME for preconcentration of drugs like amphetamines from urine coupled with electrospray ionization (ESI) IMS[12] or of metabolites[13] and ephedrine[14] from urine and of methamphetamines from human serum[15] were also demonstrated. These are discussed in Section 14.4.

The detection of the illicit drugs in the presence of interfering compounds is especially challenging, and different approaches have been adopted to resolve this problem. One approach relied on preseparation with a short column packed with adsorption packing; the retention time served to reduce false-positive responses.[4] The use of in situ derivatization to resolve the interference from nicotine, which is often abundant in clandestine laboratories that produce metamphetamines, was also proposed.[16] Changing the drift gas, or using a secondary drift gas, has also been applied to resolve interferences from overlapping peaks in the mobility spectrum, thus gaining a factor of two in detection of drugs like THC (tetrahydrocannabinol) and heroin (Figure 14.6), which are unresolved in nitrogen but nicely resolved in CO_2 or nitrous oxide.[17]

14.4 RESEARCH, OPERATIONAL EXPERIENCE, AND INSTRUMENTATION

Lawrence and coworkers at the Canadian National Research Council (NRC) made significant advances in drug detection by IMS in the 1980s by developing new sampling methods and applications.[18–24] Simulated field tests using a detachable sampling cartridge involved searching mail, luggage, and personnel for drugs.[18] It was reported that some potential interferents, including headspace volatiles of coffee and tea, did not affect detection of target drugs. In another test, letters spiked with narcotics were distinguished from negative controls. Sampling close to the zipper of a spiked suitcase and the hands and pockets of an individual who had handled drugs gave distinct signals. However, in these tests, sampling the suitcase externally and away from the

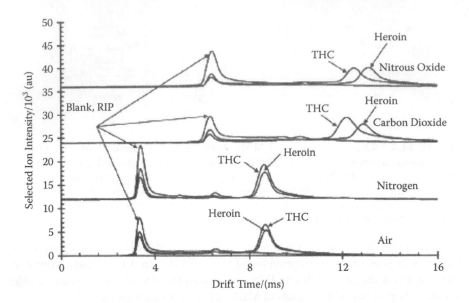

FIGURE14.6 Ion mobility spectra of heroin and tetrahydrocannabinol (THC) in different drift gases: air, nitrogen, carbon dioxide, and nitrous oxide. The spectra illustrate the two compounds can be resolved with high-polarizability drift gases carbon dioxide and nitrous oxide.

zipper did not give a positive response. The most significant results of this test were that false alarms were not received from innocent items, and that near-real-time performance for simultaneous detection of cocaine and heroin was achieved. A similar test to search for heroin and cocaine in various customs scenarios, such as letter mail and cargo containers, was described by Chauhan et al. of the Customs and Excise Authority of Canada.[25] In that study, 4 of the 339 letters examined contained drugs, and all were correctly identified by IMS. Two false-positive results were reported. Both false-positives were for heroin, one of which was found, after GC–MS study, to contain morphine based tar. Of 18 containers tested for cocaine, the IMS gave correct responses in 13 instances and showed 5 false negatives.

In addition to customs and mail scenarios, a method for detecting and identifying drug residues on the skin of subjects was developed. Hand swabs of subjects were obtained and tested by directly inserting the sampling needle into the IMS inlet.[19,20] Several prescription and illicit drugs were successfully detected and identified by this method. The use of hand (fingers and palm) and nostril swabbing for noninvasive preliminary screening of patients arriving with drug overdoses at hospital emergency rooms was described.[20] The reported rate of correctly identifying the illicit or prescription drug (as confirmed later by laboratory tests) was as high as 53% for patients being tested by IMS within less than 30 min of arrival at the emergency room. This is quite an accomplishment for such a simple testing procedure, considering the variety of substances measured and that several of the prescription drugs were actually coated tablets or capsules, thus leaving little or no traces of the active substance on the subject's hands.

Eiceman et al. demonstrated that IMS can be used to verify the content of some common over-the-counter pharmaceuticals.[26] The tablet or capsule was slightly warmed, and headspace vapors were sampled by a handheld IMS; mobility spectra were compared with known spectra to confirm the prescription. As mentioned, the application of IMS for detection and identification of pharmaceuticals, as opposed to illicit drugs, is discussed in Chapter 15.

The use of cartridges or filters with high volumetric sampling rates appears to be the most effective method for drug detection. Due to the low vapor pressure of most contraband drugs, detection should preferably be based on the entrapment of drug microparticles and their evaporation at an elevated temperature in the analyzer inlet. This approach has been adopted by Barringer Research (now Smiths Detection) in the IONSCAN analyzer as shown in Figure 14.2. Preliminary results on drug detection in field tests were described previously and presented by Fetterolf et al. of the Federal Bureau of Investigation (FBI).[27] Detection limits of nanogram quantities were reported for several drugs, and reduced mobility values were measured. Cocaine residues on hands reportedly could be detected 1.5 h after brief contact with the substance, and that simply washing the hands failed to completely remove the traces of the drug. Residues of narcotics were detected in a quick screening of book-keeping records of drug dealers. In this approach, the idea is to trap microparticles of the contraband substance on a filter, which is then introduced into the inlet of the IMS, in contrast to the SPME methods described, for which vapors are trapped.

Field experiences described by personnel of the U.S. Coast Guard[28,29] and agents of the Drug Enforcement Administration (DEA)[30] have illustrated the difficulty in collecting good swipe samples in unfavorable conditions. Complications arise from surfaces covered with moisture, oil stains, or paint residues. Tests were made to select the swipe material that gave the best results, and methods were developed for gradual desorption of the suspect material. A handheld explosives monitor was evaluated for extended at-sea deployment. The Sabre 2000 was compared to an IONSCAN 400B for false-negative and false-positive rates for illicit drugs, including heroin and cocaine. The limit of detection (LOD) for heroin was determined from response curves as 20 ng, with close agreements between the analyzers. In contrast, the IONSCAN showed 2-ng limits of detection. The false-positive alarm rate was about 1%, whereas the false-negative rate was about 50% for quantities below 250 ng, about 25% for 250 to 500 ng, and about 10% for more than 500 ng.[28]

Keller et al. extended the subject of bioassays for drug testing of skin as described in the preceding text and developed special procedures for treating hair samples collected from suspects to test them by IMS.[31–36] After rinsing the hair sample to remove external contamination, the sample was digested by a methanolic alkaline solution. A 0.05-mL portion of the digest was applied to a membrane filter and dried before detection of designer drugs[33] and methamphetamines.[31,32] Exotic substances that cause hallucinations, such as psychedelic fungi that contain psilocybin and psilocin (molecular structure shown in Figure 14.1), were identified using mobility spectra.[34] Quantitative analysis was performed by GC–MS after a simple one-step extraction involving homogenization of the dried fruit bodies of fungi in chloroform followed by derivatization.[34] Sweat samples collected from different parts of the body of a drug addict tested positive for cocaine by IMS, while GC–MS confirmed the analysis and identified benzoylecgonine (BE) and ecgoninemethylester (EME).[35] Cocaine, heroin, and 6-acetylmorphine in sweat were detected by ion trap mobility spectrometry, although sweat samples from different parts of the body showed different mobility spectra profiles.[37] SPME has been used to preconcentrate the analytes (narcotic drugs) from headspace vapors, as mentioned.[5]

Differential mobility analyzers and MS identification were used to detect amphetamine and derivatives,[12] morphine, and codeine in human urine.[38] Matz and Hill used an ESI–IMS–MS instrument to study charge competition within the ESI during analysis of amphetamines[39] and to test for benzodiazepines.[40] In real-life situations, cigarette smoke and the nicotine derivatives it contains may interfere with the detection of illicit drugs, particularly amphetamines. Advanced signal-processing methods that use the dynamic data generated by the IMS were developed by Harrington's group to overcome this problem.[41,42]

Cocaine metabolites were detected in urine by solid phase extraction (SPE) IMS with detection limits of 10 and 4 ng/mL for BE and cocaethylene, respectively.[13] The analytes are retained on the SPE cartridge, while salts and polar compounds that may interfere with the analysis remain in solution.[13] Sweat samples have also been examined by ion trap mobility spectrometry (ITMS), and traces of 6-acetylmorphine, heroin, and cocaine were detected.[37] Not only illicit drugs can be detected by IMS, but also food additives or supplements that are undeclared, like the appetite suppressant sibutramine.[43]

The presence of traces of cocaine on banknotes from around the world (mainly euros and U.S. dollars) by different analytical techniques, including GC, liquid chromatography (LC), capillary electrophoresis (CE), tandem MS, and IMS, was reviewed.[44] Drugs were found on almost all U.S. currency banknotes, while in Europe the cocaine was detected with a higher frequency on Spanish euro banknotes than other currencies, and the lowest percentage was found on Swiss banknotes. The amount found varied between a few tenths of a microgram to several dozen micrograms per banknote.[44]

During the last decade, there have been some developments with regard to instrumentation for detection of illicit drugs. In one study, a comparison was made between a corona discharge (CD) IMS and an atmospheric pressure chemical ionization (APCI) MS for detection of cocaine dissolved in acetonitrile.[45] Detection limits of about 0.1 and 0.02 ng were determined for the APCI-MS system and IMS, respectively, when 1 μL of a solution was injected into each of the devices.[45] The negative and positive ion modes were compared for detection of heroin vapors by an ion mobility increment spectrometer (IMIS).[46] It was found that in negative mode there was a strong dependence of the sensitivity on the moisture content in the drift gas.[46] In another study, ultraviolet (UV) irradiation of a solid sample (banknotes, documents, mail items, and hands) contaminated with traces of cocaine hydrochloride or "crack" increased the response of IMIS by a factor of eight, probably due to the release of photolytic degradation products.[47] A further study combined a multicapillary GC column (MCC) with an IMIS detector to improve detection limits for explosives, chemical warfare agents, and drugs, and a detection limit of 0.001 pg/mL was reported for cocaine.[48] Combining a DMS with a handheld MS improved the LOD of diazepam in urine by a factor of four (to 50 ng/mL) compared to the device without the DMS.[49]

A database for the reduced mobilities of prescription and illicit drugs is presented in Table 14.1. It should be noted that the reduced mobility values reported in recent publications are generally in good agreement with the older values given in Table 14.1. For example, the reduced mobility values of several common illicit drugs that were determined by IMS using thermal desorption and a deactivated fused silica transfer line differ by less than 0.02 cm^2 V^{-1} s^{-1} from the values measured two decades earlier.[4] The agreement between the values listed in Table 14.1 and the very accurate values measured at the National Institute for Standards and Technology (NIST)[50] is also good. Differences in the reported reduced mobility values of the same compound by different workers arise mainly from the use of different experimental conditions in the measurement. While agreement is generally good in comparisons of K_o values, differences may arise from experimental conditions in which the product ion is slightly changed. These changes are noticeable in some instances (Table 14.1), although good agreement between measurements made far apart in terms of time and space and by different instruments was also seen.

Screening for a group of illicit drugs, so-called club drugs, rape drugs, or party drugs, was carried out with a dual-mode ITMS.[51] The reduced mobility values of ketamine, GHB (gamma-hydroxybutyrate), ephedrine, flunitrazepam, methamphetamine, MDA (3,4-methylenedioxyamphetamine), amphetamine, Amph-sulfate, and MDMA were reported, and the preferred mode of detection was noted.

In some cases of overlapping peaks, the use of a different drift gas may resolve the overlap and enable the detection of the separated peaks.

14.5 STANDARDS AND CALIBRATION

Traditionally, there are two main techniques for production of standards for quantitative measurement of chemical compounds with low volatility for determination of limits of detection by IMS or other analyzers. One approach is based on preparation of a dilute solution with a known concentration of the compound of interest and injection of a known volume of the solution into the IMS inlet. Thus, multiplying the concentration by the injected volume gives the quantity of the analyte that was introduced into the analyzer. Using solutions that are increasingly dilute, or using smaller volumes, and monitoring the signal intensity will eventually yield the LOD. In another approach, a known volume of the dilute solution is deposited on a suitable material, generally a filter paper, and after the solvent evaporates the filter is inserted in the inlet (which is usually heated), and the vapors emanating from the filter are introduced into the analyzer.

Modern techniques can be used to obtain highly reproducible standards for calibration of analytical instruments. One technique developed by NIST scientists is based on deposition of a controlled amount of analyte using a dot matrix printer that, instead of using printing ink, injects a solution of the analyte, as has been done for explosives, to mention just one example.[52] This technique cannot be used for

TABLE 14.1
Reduced Mobility of Several Common Illicit Drugs and Medications that Can Be Abused (from *Ion Mobility Spectrometry*, 2nd edition)

Name	K_0 (cm²/Vs)	Parameters
Acetaminophen	1.70, 1.76, 1.97	Air, 220°C, skin/cart
N-Acetylamphetamine	1.53	Air, 220°C, sol/wire
Acetylcodeine	1.09, 1.21	Air, 220°C, sol/wire
Alprazolan	1.15	Air, 220°C, sol/wire
Amitryptiline	1.19	Air, 220°C skin/cart
Amobarbital	1.36, 1.53	N_2, 230°C, sol/GC
Amphetamine	1.66	Air, 220°C, sol/wire
Aprobarbital	1.39, 1.56, 1.75	N_2, 230°C, sol/GC
Barbital	0.99, 1.50	N_2, 230°C, sol/GC
Bromazepam	1.24	Air, 220°C, sol
Butabarbital	1.28, 1.35	N_2, 230°C, sol/GC
Cannabinol	1.06	Air, 220°C, sol/wire
CDA	1.18	Air, 220°C, sol/wire
Cocaine	1.16	Air, 220°C, sol/wire
Codeine	1.18, 1.21	Air, 220°C, sol/wire
Diazepam	1.21	Air, 220°C, sol/wire
Flurazepam	1.03	Air, 220°C, sol/wire
Heroin	1.04, 1.14	Air, 220°C, sol/wire
LSD	1.085	
Lorazepam	1.19, 1.22	Air, 220°C, sol/cart
MDA	1.49	Air, 220°C, sol/wire
MDMA	1.47	
Mephobarbital	1.63, 1.81	N_2, 230°C, sol/GC
Methamphetamine	1.63	Air, 220°C, sol/wire
Methyprylan	1.52	Air, 220°C, sol/wire
Morphine	1.22, 1.26	Air, 220°C, sol/wire
Nicotine	1.57	
Nitrazepam	1.22	Air, 220°C, sol/wire
OMAM	1.13, 1.26	Air, 220°C, sol/wire
Opium	1.55	Air, 250°C, wire
Oxazepam	1.23, 1.28	Air, 220°C, sol/cart
PCP	1.27	
Pentobarbital	1.38	N_2, 230°C, sol/GC
Phenobarbital	1.44	N_2, 230°C, sol/GC
Phenylcyclidine	1.27, 1.63, 2.01, 2.23	Air, 220°C, sol/wire
Procaine	1.31	
Secobarbital	1.31, 1.48	N_2,230°C, sol/GC
THC	1.05	Air, 220°C, sol/wire
Thelaine	1.14	Air, 220°C, sol/wire
Triazolam	1.13	Air, 220°C, sol/wire

TABLE 14.1 (continued)
Reduced Mobility of Several Common Illicit Drugs and Medications that Can Be Abused (from *Ion Mobility Spectrometry*, 2nd edition)

Source: DeTulleo-Smith, A.M., Methamphetamine vs. nicotine detection on the Barringer ion mobility spectrometer, IMS meeting, Jackson Hole, WY, August 20–22, 1996; Lawrence, A.H., Detection of drug residues on the hands of subjects by surface sampling and ion mobility spectrometry, *Forens. Sci. Int.* 1987, 34, 73–83.

Air, the drift gas, ionization source flow, and the sample vapors were carried by air; wire, insertion of a metal wire with absorbed vapors or solution (sol).

PCP, phencyclidine; cart, cartridge; MDMA, 3,4-methylenedioxy-*N*-methyl-amphetamine.

---

materials with high vapor pressures but is ideally suitable for most illicit drugs and low-volatility explosives. However, another approach to produce test materials and standards of volatile compounds with long shelf life uses encapsulation techniques by which the analyte is introduced into a gelatin-based material.[53] The emission of an MDMA starting material and odorant, piperonal, was controlled by permeation from a polymer. This was the basis for comparison of the sensitivity of canines (LOD was emission of 1 ng at a rate of 100 ng/s) and SPME-IMS systems (LOD was 2 ng in a static closed system).[54] Validation of IMS-based methods for detection of trace amounts of cocaine and heroin using a vacuum cleaner for sampling of incriminating materials was published.[55] Detection limits of 250 and 1,000 ng were reported for cocaine and heroin on clothes, respectively. A meticulous study of the reliability of qualitative measurements of complex mixtures of eight controlled substances and some other excipients was made, and uncertainties in the reduced mobility values of several common drugs were reduced to less than 0.001 cm^2 V^{-1} s^{-1} according to the report.[50] This made it possible to considerably narrow the detection windows and thus increase the trustworthiness of the identification, as shown in Table 14.2, in which the relative signal intensity for a quantity of 1 ng of the substance is also shown.[50] Nevertheless, this approach does not completely preclude false-positive indications from overlapping peaks from other compounds.

14.6 DATABASE FOR DRUGS

The LODs of the different methods and instruments are difficult to summarize due to the variability of the preconcentration techniques. Several examples were given in Section 14.4. Generally, detection limits of nanograms can be obtained from most systems; however, oversensitive systems could lead to many false-positive responses, especially if one considers that many banknotes may contain micrograms of illicit drugs.

TABLE 14.2
Relative Intensity (in arbitrary units) of 1 ng Controlled Substances and Excipients and the Exact Reduced Mobility Value

Compound (IMS Alarm On)	Intensity at 1 ng (au)		K_0 (cm^{2}V^{-1}s^{-1})		
	Mean	1 SD	Mean	1 SD	n
Meth (1)	496	60	1.6428	0.0009	42
MDMA (2)	352	33	1.4718	0.0004	34
Hydrocodone (3)	138	37	1.1844	0.0004	6
Oxycodone (4)	170	39	1.1709	0.0003	10
Cocaine (5)	151	36	1.1644	0.0006	41
Alprazolam (6)	0[a]		1.1536	0.0003	6
Fentanyl (7)	181	48	1.0550	0.0004	12
Heroin (8)	20	12	1.0463	0.0006	27

Compound (IMS Alarm Off)	Intensity at 1 ng (au)		K_0 (cm^{2}V^{-1}s^{-1})	
	Mean	1 SD	Mean	1 SD
Amphetamine	89	49	1.6753	0.0004
MDA	70	34	1.5018	0.0008
THC	20	9	1.0500	0.0004
Ephedrine	208	61	1.5824	0.0004
Pseudoephedrine	106	41	1.5838	0.0002
Procaine	540	111	1.3117	0.0010
Diphenhydramine	252	54	1.2339	0.0003
Chlorpheniramine	446	124	1.2175	0.0006

Source: Dussy et al., Validation of an ion mobility spectrometry (IMS) method for the detection of heroin and cocaine on incriminated material, *Forens. Sci. Int.* 2008, 177, 105–111.

[a] Intensity at 10 ng = 95 au (1 SD = 20 au).

REFERENCES

1. Karasek, F.W.; Hill, H.H., Jr.; Kim, S.H., Plasma chromatography of heroin and cocaine with mass-identified mobility spectra, *J. Chromatogr.* 1976, 117, 327–336.
2. Ithakissios, D.S., Plasmagram spectra of some barbiturates, *J. Chromatogr. Sci.* 1980, 18, 88–92.
3. Eatherton, R.L.; Siems, W.F.; Hill, H.H., Jr., Fourier transform ion mobility spectrometry of barbiturates after capillary gas chromatography, *J. High Resolut. Chromatogr. Commun.* 1986, 9, 44–48.
4. Kanu, A.B.; Wu, C.; Hill, H.H., Rapid preseparation of interferences for ion mobility spectrometry, *Anal. Chim. Acta* 2008, 610, 125–134.

5. Orzechowska, G.E.; Poziomek, E.J.; Tersol, V., Use of solid-phase micro-extraction (SPME) with ion mobility spectrometry, *Anal. Lett.*, 1997, 30, 1437–1444.

6. Liu, X.; Nacosn, S.; Grigoriev, A.; Lynds, P.; Pawliszyn, J., A new thermal desorption solid-phase microextraction system for hand-held ion mobility spectrometry, *Anal. Chim. Acta* 2006, 559, 159–165.

7. Lai, H.; Guerra, P.; Joshi, M.; Almirall, J.R., Analysis of volatile components of drugs and explosives by solid phase microextraction-ion mobility spectrometry, *J. Sep. Sci.* 2008, 31, 402–412.

8. Guerra, P.; Lai, H.; Almirall, J.R., Analysis of the volatile chemical markers of explosives using novel solid phase microextraction coupled to ion mobility spectrometry, *Sep. Sci.* 2008, 31, 2891–2898.

9. Lai, H.; Corbin, I.; Almirall, J.R., Headspace sampling and detection of cocaine, MDMA, and marijuana via volatile markers in the presence of potential interferences by solid phase microextraction-ion mobility spectrometry (SPME-IMS), *Anal. Bioanal. Chem.* 2008, 392, 105–113.

10. Gura, S.; Guerra-Diaz, P.; Lai, H.; Almirall, J.R., Enhancement in sample collection for the detection of MDMA using a novel planar SPME (PSPME) device coupled to ion mobility spectrometry (IMS), *Drug Test. Anal.* 2009, 1, 355–362.

11. Guerra-Diaz, P.; Gura, S.; Almirall, J.R., Dynamic planar solid phase microextraction-ion mobility spectrometry for rapid field air sampling and analysis of illicit drugs and explosives, *Anal. Chem.* 2010, 82, 2826–2835.

12. McCooeye, M.A.; Mester, Z.; Ells, B.; Barnett, D.A.; Purves, R.W.; Guevremont, R., Quantitation of amphetamine, methamphetamine, and their methylenedioxy derivatives in urine by solid-phase microextraction coupled with electrospray ionization high-field asymmetric ion mobility spectrometry-mass spectrometry, *Anal. Chem.* 2002, 74, 3071–3075.

13. Lu, Y.; O'Donnell, R.M.; Harrington, P.B., Detection of cocaine and its metabolites in urine using solid phase extraction-ion mobility spectrometry with alternating least squares, *Forensic Sci. Int.* 2009, 189, 54–59.

14. Lokhnauth, J.K.; Snow, N.H., Solid phase micro-extraction coupled with ion mobility spectrometry for the analysis of ephedrine in urine, *J. Sep. Sci.* 2005, 28, 612–618.

15. Alizadeh, N.; Mohammadi, A.; Tabrizchi, M., Rapid screening of methamphetamines in human serum by headspace solid-phase microextraction using a dodecylsulfate-doped polypyrrole film coupled to ion mobility spectrometry, *J. Chromatogr. A* 2008, 1183, 21–28.

16. Ochoa, M.L.; Harrington, P.B., Detection of methamphetamine in the presence of nicotine using in situ chemical derivatization and ion mobility spectrometry, *Anal. Chem.* 2004, 76, 985–991.

17. Kanu, A.B.; Hill, H.H., Identity confirmation of drugs and explosives in ion mobility spectrometry using a secondary drift gas, *Talanta* 2007, 73, 692–699.

18. Lawrence, A.H.; Elias, L., Application of air sampling and ion-mobility spectrometry to narcotics detection: a feasibility study, *Bull. Narc.* 1985, 37, 3–16.

19. Lawrence, A.H., Ion mobility spectrometry/mass spectrometry of some prescription and illicit drugs, *Anal. Chem.* 1986, 58, 1269–1272.

20. Lawrence, A.H., Detection of drug residues on the hands of subjects by surface sampling and ion mobility spectrometry, *Forensic Sci. Int.* 1987, 34, 73–83.

21. Nanji, A.N.; Lawrence, A.H.; Mikhael, N.Z., Use of skin sampling and ion mobility spectrometry as a preliminary screening method for drug detection in an emergency room, *J. Toxicol. Clin. Toxicol.* 1987, 25, 501–515.

22. Lawrence, A.H.; Nanji, A.A., Ion mobility spectrometry and ion mobility spectrometry/mass spectrometric characterization of dimenhydrinate, *Biomed. Environ. Mass Spectrom.* 1988, 16, 345–347.

23. Lawrence, A.H.; Nanji, A.A.; Taverner, J., Skin-sniffing ion mobility spectrometric analysis: a potential screening method in clinical toxicology, *J. Clin. Lab. Anal.* 1988, 2, 101–107.

24. Lawrence, A.H., Characterization of benzodiazepine drugs by ion mobility spectrometry, *Anal. Chem.* 1989, 61, 343–349.

25. Chauhan, M.; Harnois, J.; Kovar, J.; Pilon, P., Trace analysis of cocaine and heroin in different customs scenarios using a custom-built ion mobility spectrometer, *J. Can. Soc. Forensic* 1991, 24, 43–49.

26. Eiceman, G.A.; Blyth, D.A.; Shoff, D.B.; Snyder, A.P., Screening of solid commercial pharmaceuticals using ion mobility spectrometry, *Anal. Chem.* 1990, 62, 1374–1379.

27. Fetterolf, D.D.; Donnelly, B.; Lasswell, L.D., Detection of heroin and cocaine residues by ion mobility spectrometry, 39th Conference American Society of Mass Spectrometry, Nashville, TN, 1991.

28. Su, C.-W.; Babcock, K.; Rigdon, S., The detection of cocaine on petroleum contaminated samples utilizing ion mobility spectrometry, *Int. J. Ion Mobility Spectrom.* 1998, 1, 15–27.

29. Su, C.-W.; Babcock, K.; deFur, P.; Noble, T.; Rigdon, S., Column-less GC/IMS (II)—a novel on-line separation technique for IONSCAN analysis, *Int. J. Ion Mobility Spectrom.* 2002, 5, 160–174.

30. DeTulleo, A.M.; Galat, P.B.; Gay, M.E., Detecting heroin in the presence of cocaine using ion mobility spectrometry, *Int. J. Ion Mobility Spectrom.* 2000, 3, 38–42.

31. Miki, A.; Keller, T.; Regenscheit, P.; Bernhard, W.; Tatsuno, M.; Katagi, M., Determination of internal and external methylamphetamine in human hair by ion mobility spectrometry, *Jpn. J. Toxicol. Environ. Health* 1997, 43, 15–24.

32. Miki, A.; Keller, T.; Regenscheit, P.; Dirnhofer, M.; Tatsuno, M.; Katagi, M. Application of ion mobility spectrometry to the rapid screening of methylamphetamine incorporated in hair, *J. Chromatogr. B Biomed. Appl.* 1997, 692, 319–328.

33. Keller, T.; Miki, A.; Regenscheit, P.; Dirnhofer, R.; Schneider, A.; Tsuchihashi, H. Detection of designer drugs in human hair by ion mobility spectrometry (IMS), *Forensic Sci. Int.* 1998, 94, 55–63.

34. Keller, T., Schneider, A.; Regenscheit, P.; Dirnhofer, R.; Rucker, T.; Jaspers, J.; Kisser, W., Analysis of psilocybin and psilocin in *Psilocybe subcubensis* Guzman by ion mobility spectrometry and gas chromatography-mass spectrometry, *Forensic Sci. Int.* 1999, 99, 93–105.

35. Keller, T.; Schneider, A.; Tutsch-Bauer, E.; Jaspers, J.; Aderjan, A.; Skopp, G., Ion mobility spectrometry for the detection of drugs in cases of forensic and criminalistic relevance, *Int. J. Ion Mobility Spectrom.* 1999, 2, 22–34.

36. Keller, T.; Miki, A.; Regenscheit, P.; Dirnhofer, R.; Schneider, A.; Tsuchihashi, H., Detection of methamphetamine, MDMA and MDEA in human hair by means of ion mobility spectrometry (IMS), *Int. J. Ion Mobility Spectrom.* 1998, 1, 38–42.

43. Kudriavtseva, S.; Carey, C.; Ribiero, K.; Wu, C., Detection of drugs of abuse in sweat using ion trap mobility spectrometry, *Int. J. Ion Mobility Spectrom.* 2004, 7, 44–51.

38. McCooeye, M.A.; Ells, B.; Barnett, D.A.; Purves, R.W.; Guevremont, R., Quantitation of morphine and codeine in human urine using high-field asymmetric waveform ion mobility spectrometry (FAIMS) with mass spectrometric detection, *J. Anal. Toxicol.* 2001, 25, 81–87.

39. Matz, L.M.; Hill, H.H., Jr. Evaluating the separation of amphetamines by electrospray ionization ion mobility spectrometry/MS and charge competition within the ESI process, *Anal. Chem.* 2002, 74, 420–427.

40. Matz, L.M.; Hill, H.H., Jr., Separation of benzodiazepines by electrospray ionization ion mobility spectrometry-mass spectrometry, *Anal. Chim. Acta* 2002, 457, 235–245.

41. Reese, E.S.; Harrington, P.B., The analysis of methamphetamine hydrochloride by thermal desorption ion mobility spectrometry and SIMPLISMA, *J. Forensic Sci.* 1999, 44, 68–76.

42. Shaw, L.A.; Harrington, P.D., Seeing through the smoke with dynamic data analysis: detection of methamphetamine in forensic samples contaminated with nicotine, *Spectroscopy* 2000, 15, 40–45.

43. Dunn, J.D.; Gryniewicz-Ruzicka, C.M.; Kauffman, J.F.; Westenberger, B.J.; Buhse, L.F., Using a portable ion mobility spectrometer to screen dietary supplements for sibutramine, *J. Pharm. Biomed. Anal.* 2011, 54, 469–474.

44. Armenta, S.; de la Guardia, M., Analytical methods to determine cocaine contamination of banknotes from around the world, *Trends Anal. Chem.* 2008, 27, 344–351.

45. Choi, S.-S.; Kim, Y.-K.; Kim, O.-B.; An, S.G.; Shin, M.-W.; Maeng, S.-J.; Choi, G.S., Comparison of cocaine detections in corona discharge ionization-ion mobility spectrometry and in atmospheric pressure chemical ionization-mass spectrometry, *Bull. Kor. Chem. Soc.* 2010, 31, 2382–2385.

46. Buryakov, I.A.; Baldin, M.N., Comparison of negative and positive modes of ion-mobility increment spectrometry in the detection of heroin vapors, *J. Anal. Chem.* 2008, 63, 787–791.

47. Kolomiets, Y.N.; Pervukhin, V.V., Effect of UV irradiation on detection of cocaine hydrochloride and crack vapors by IMIS and API-MS methods, *Talanta* 2009, 78, 542–547.

48. Buryakov, I.A., Express analysis of explosives, chemical warfare agents and drugs with multicapillary column gas chromatography and ion mobility increment spectrometry, *J. Chromatogr. B: Anal. Technol. Biomed. Life Sci.* 2004, 800, 75–82

49. Tadjimukhamedov, F.K.; Jackson, A.U.; Nazarov, E.G.; Ouyang, Z.; Cooks, R.G., Evaluation of a differential mobility spectrometer/miniature mass spectrometer system, *J. Am. Soc. Mass Spectrom.* 2010, 21, 1477–1481.

50. Verkouteren, J.R.; Staymates, J.L., Reliability of ion mobility spectrometry for qualitative analysis of complex, multicomponent illicit drug samples, *Forensic Sci. Int.* 2011, 206, 190–196

51. Geraghty, E.; Wu, C.; McGann, W., Effective screening for "club drugs" with dual mode ion trap mobility spectrometry, *Int. J. Ion Mobility Spectrom.* 2002, 5, 41–44.

52. Windsor, E.; Najarro, M.; Bloom, A.; Benner, B.; Fletcher, R.; Lareau, R.; Gillen, G., Application of inkjet printing technology to produce test materials of 1,3,5-trinitro-1,3,5 triazcyclohexane for trace explosive analysis, *Anal. Chem.* 2010, 82, 8519–8524.

53. Staymates, J.L.; Gillen, G., Fabrication and characterization of gelatin-based test materials for verification of trace contraband vapor detectors, *Analyst* 2010, 135, 2573–2578.

54. Macias, M.S.; Guerra-Diaz, P.; Almirall, J.R.; Furton, K.G., Detection of piperonal emitted from polymer controlled odor mimic permeation systems utilizing *Canis familiaris* and solid phase microextraction-ion mobility spectrometry, *Forensic Sci. Int.* 2010, 195, 132–138.

55. Dussy, F.E.; Berchtold, C.; Briellmann, T.A.; Lang, C.; Steiger, R.; Bovens, M., Validation of an ion mobility spectrometry (IMS) method for the detection of heroin and cocaine on incriminated material, *Forensic Sci. Int.* 2008, 177, 105–111.

56. DeTulleo-Smith, A.M., Methamphetamine vs. nicotine detection on the Barringer ion mobility spectrometer, IMS Meeting, Jackson Hole, WY, 1996.

15 Pharmaceuticals

15.1 INTRODUCTION

Traditional applications for ion mobility spectrometry (IMS) have primarily revolved around security and safety because of the rapid response time of the instrument coupled with its selectivity and sensitivity. Its speed, sensitivity, and selectivity, however, open opportunities for expansion into other areas of analysis.[1] Pharmaceuticals offer a particularly promising target area for application of IMS because, unlike environmental and biological samples, pharmaceutical mixtures are usually well defined and not too complex. In addition, many pharmaceuticals have basic sites on the molecules that produce high proton affinities; thus, they respond well to the ion-molecule ionization sources used in positive-mode IMS. Analytical methods currently used in the pharmaceutical industry revolve around some type of chromatography, either liquid or gas, that is usually slow and expensive. For routine, repetitive, and rapid analyses of pharmaceutical samples, IMS offers a unique and efficient alternative to chromatography.

15.2 COMPOUND IDENTIFICATION

IMS for illicit drug analysis has been known since the 1970s (see Chapter 14), but expansion to pharmaceuticals developed slowly and is still only in its infancy of being accepted by the pharmaceutical industry. For volatile or semivolatile pharmaceutical compounds, ^{63}Ni or secondary electrospray ionization (SESI) sources may be used, but the availability of electrospray ionization (ESI) IMS has expanded the number of applications possible for the pharmacy industry. While this process of evaluating the response and measuring the mobilities of individual pharmaceutical standards is ongoing, applications for real-world analyses in the pharmaceutical industry are being developed.

IMS has been used as an inexpensive substitute for high-performance liquid chromatography (HPLC) for a simple, repetitive, and rapid analytical method for over-the-counter drugs and beverages.[2] In this work, a number of active ingredients were determined in drug formulations, including acetaminophen, aspartame, bisacodyl, caffeine, dextromethorphan, diphenhydramine, famotidine, glucosamine, guaifenesin, loratadine, niacin, phenylephrine, pyridoxine, thiamin, and tetrahydrozoline. Aspartame and caffeine could rapidly be determined in a variety of beverages, while other ingredients were detected in 14 over-the-counter drugs. The primary advantage of IMS over HPLC is improved resolving power, rapid analysis, and high sensitivity. The disadvantage is the limited range of separation media available to IMS. For example, in HPLC both mobile and stationary phases can be changed to effect optimized separations for a target analyte. In addition, gradient elution of the analytes (at the cost of speed) is available with HPLC.

15.3 FORMULATION VALIDATION

Used as preservatives in many topical pharmaceutical products, paraben esters have been found to exhibit estrogenic effects that have been linked to breast cancer.[3,4] This has led to requirements for their rapid and quantitative detection. Traditional methods of analysis are slow and tedious. For gas chromatographic methods, derivatization as silyl- or fluoroacetyl derivatives followed by extraction and sample cleanup complicate the process, while liquid chromatographic methods require extraction and long chromatographic run times. To increase speed and decrease complexity, solid phase microextraction with ion mobility spectrometry (SPME–IMS) was used for the determination of parabens in pharmaceutical formulations.[5]

For analysis, 50 mg of the topical formulation was dissolved in water and the SPME fiber exposed directly to the sample. After an optimized period of time, the SPME fiber was presented to the IMS interface as depicted in Figure 15.1, and the sample was desorbed into the IMS.

The IMS instrument used in these studies was a Smiths IONSCAN operating in the negative polarity mode with purified air as the buffer gas. The ionization region was doped with hexachloroethane to suppress interferences. 4-Nitrobenzonitrile was used to calibrate the instrument. The desorption and inlet temperatures were 270°C, and the buffer gas temperature was controlled at 115°C. The airflow was 400 mL/min with a shutter grid opening time of 0.2 ms and a scan period of 30 ms. The sample was desorbed into the IMS for 30 s. Figure 15.2 shows the results of a standard, blank, and sample for the determination of parabens by SPME-IMS.

Other SPME-IMS methods that have been reported for application to pharmaceutical or related samples include those for analysis of ephedrine in urine,[6] methamphetamines in human serum,[7] and captopril in human plasma and pharmaceutical preparations.[8] In a method similar to SPME–IMS, testosterone was collected with a molecular imprinted polymer from urine and desorbed into an IMS. The method was validated with HPLC and determined to have a detection limit of 0.9 ng/mL with a linear dynamic range from 10 to 250 ng/mL.[9]

15.4 CLEANING VALIDATION

In the pharmacy industry, pharmaceuticals are produced in batches, and one type of pharmaceutical may be produced after a different type with the same equipment. Thus, it is imperative that the equipment be cleaned between batches. *Cleaning validation* is the formal process by which a company ensures that there is no contamination from one batch to the next. Specifically, it is defined as

> The process of providing documented evidence that the cleaning methods employed within a facility consistently controls potential carryover of product (including intermediates and impurities), cleaning agents and extraneous material into subsequent product to a level which is below predetermined levels.[10]

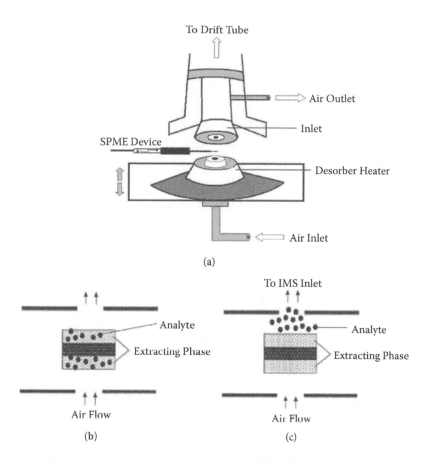

FIGURE 15.1 (a) Schematic diagram of the SPME–IMS interface showing the location of the SPME device. (b) Dynamics of the desorption process: analytes in the fiber coating. (c) Dynamics of the desorption process: analytes evaporating from the coating and entering the flow stream. Note that the fiber is oriented 90° to the flow, unlike traditional SPME–GC (gas chromatography). (From Lokhnauth and Snow, Determination of parabens in pharmaceutical formulations by solid-phase microextraction-ion mobility spectrometry, *Anal. Chem.* 2005, 77, 5938–5946. With permission.)

Many analytical methods are currently used for cleaning validation, and the appropriate analytical tool depends on a variety of factors. Rapid methods of analysis would aid in the efficient preparation when equipment is being switched from one synthesis to the next. Because of the sensitivity to many pharmaceuticals, speed of analysis, and good resolving power, IMS is one of the promising technologies used for cleaning validation.[11] Two sampling methods are commonly used for cleaning evaluation: the swab sampling method and the rinse sampling method.[12] When swab sampling is employed, commercial off-the-shelf (COTS) instruments (mostly used for the thermal desorption and detection of explosives from swabs) can be applied to pharmaceutical cleaning validation. When the rinse method is employed, ESI-IMS can be employed.

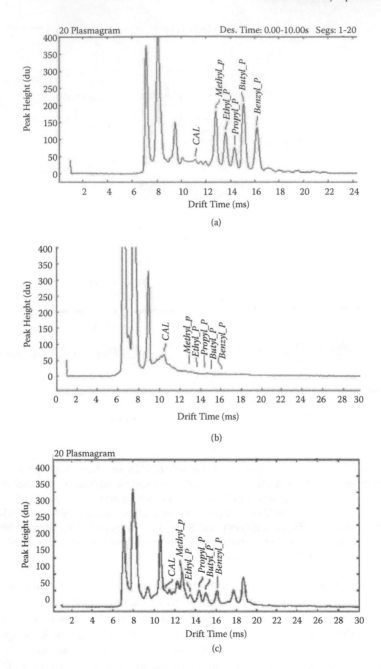

FIGURE 15.2 (a) Ion mobility spectrum of a standard solution of methylparaben, ethyl-paraben, propelparaben, butylparaben, and benzylparaben. (b) IMS spectrum of a sample blank. (C) IMS spectrum of a topical cream showing the presences of parabens at the level of approximately 1 mg/g. (From Lokhnauth and Snow, Determination of parabens in pharma-ceutical formulations by solid-phase microextraction-ion mobility spectrometry, *Anal. Chem.* 2005, 77, 5938–5946. With permission.)

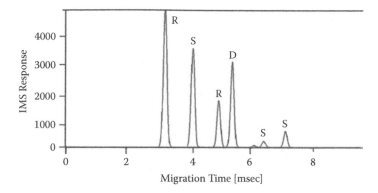

FIGURE 15.3 Ion mobility spectrum of a mixture of 25 µg of duloxetine and 100 µg of proprietary surfactant. The ion peaks labeled with R are the reactant ions, D is the duloxetine, and those labeled S are from the surfactant. (From Strege et al., At-line quantitative ion mobility spectrometry for direct analysis of swabs for pharmaceutical manufacturing equipment cleaning verification, *Anal. Chem.* 2008, 80, 3040–3044. With permission.)

15.4.1 CLEANING EVALUATION WITH THERMAL DESORPTION IMS

One application for the swab procedure for pharmaceutical analysis is the detection of duloxetine hydrochloride, which is an active pharmaceutical ingredient (API) in the drug product Cymbalta. Also, it can evaluate the cleanliness of the equipment from the surfactant components used as the cleaning agent in the manufacturing process.[13] In this approach, an ion trap mobility spectrometer was used with the thermal desorption temperature of 249°C and the IMS drift tube operated at 205°C. A swab wetted with methanol was swiped over the stainless steel equipment and then inserted into the thermal desorber held at 249°C. Vapors generated from the desorber were introduced into the ionization region of an ion trap IMS. Figure 15.3 shows the ion mobility spectrum of a mixture of 25 µg of duloxetine and 100 µg of the proprietary surfactant. The two ion peaks labeled with an R are the reactant ions, the peak labeled D is the duloxetine, and the four peaks labeled S are from the surfactant. This spectrum demonstrates the ability of IMS to detect duloxetine at cleaning validation levels, and that the surfactant used in the cleaning process does not interfere with detection. The dynamic detection range for duloxetine with this method was reported as 5–100 µg per 25-cm² surface area.

Other examples of cleaning validation have been reported, but identities of the compounds were withheld due to proprietary considerations.[14] Nevertheless, quantitative determination of residual APIs and intermediates on equipment surfaces were described in detail. The linear dynamic ranges for the compounds investigated were 0.1–1.0 µg/mL for one compound and 1–10 µg/mL for the others. With IMS technology, they were able to evaluate 30 samples in less than 2 h. In another study, analysis required only 0.5 min/sample compared with 15–30 min/sample using the HPLC method.[15]

The use of swabs for total residue analysis by IMS has been developed for cleaning validation.[16] In this approach, the swabs were 2 × 5 cm strips of polyimide fiber material that were wiped across contaminating surfaces. Figure 15.4 shows the

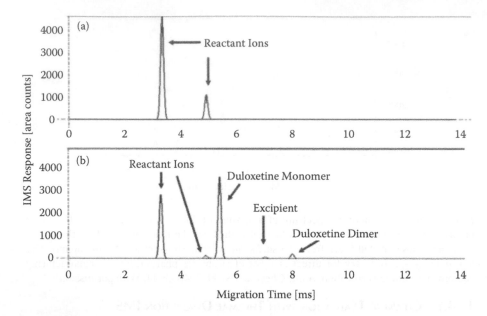

FIGURE 15.4 Ion mobility spectra of 50 µg of Cymbalta drug product thermally desorbed into a ^{63}Ni IMS. (a) The background swab showing on the reactant ions. (b) The swab after swiping it over a contaminated surface. (From Strege, Total residue analysis of swabs by ion mobility spectrometry, *Anal. Chem.* 2009, 81, 4576–4580. With permission.)

thermal desorption ion mobility spectrum of a background swipe compared with a swipe containing 50 µg of Cymbalta drug product. Figure 15.4a is of the blank swab showing only the reactant ions from the ^{63}Ni source. Figure 15.4b shows the detection of the active ingredient, duloxetine, in Cymbalta. Detection limits were determined to be 5 µg of duloxetine. When large quantities of the analyte were present, as was the case for Figure 15.4b, the proton-bound dimer could also be observed in the sample. Quantification of the dimer can serve as a means to extend the dynamic range of the analysis. Although not as selective or as rapid as IMS, the current method for the determination of duloxetine is by ultraviolet (UV) absorbance, which has a dynamic range from 5 to 100 µg.

As demonstrated, thermal desorption IMS systems commonly used in airports and by customs for explosives and drug detection are suitable for cleaning validation provided the target pharmaceuticals or residual APIs are volatile or semivolatile. For ionic species, high molecular weight species, and thermally labile species, thermal desorption is not the best choice. As a rule of thumb, if the target molecule can be analyzed by gas chromatography, then thermal desorption IMS is the method of choice, but if the target molecule requires liquid or ion chromatography, ESI-IMS should be used for cleaning validation.[17,18]

Figure 15.5 shows an ion mobility spectrum of irinotecan, a thermally labile pharmaceutical with a molecular weight of 586.7 Da; a thermal desorption IMS instrument was used for detection. A complex spectrum was obtained containing multiple thermal decomposition products of various mobilities.

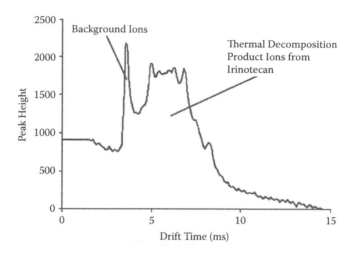

FIGURE 15.5 Ion mobility spectrum of irinotecan using a ^{63}Ni ionization with thermal desorption sample introduction. (From Excellims.)

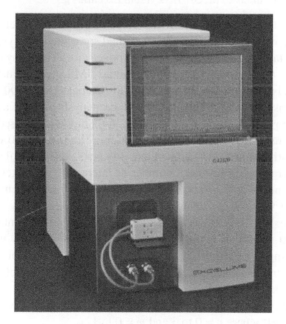

FIGURE 15.6 Commercially available stand-alone ESI-IMS system, Excellims Corporation. (From Excellims.)

15.4.2 CLEANING EVALUATION WITH ESI-IMS

Figure 15.6 shows a commercially available ESI-IMS. It consists of an orthogonal electrospray configuration in which the sample is electrosprayed through a syringe needle of the type used for sample introduction into gas chromatographs. For ESI-IMS, the needle is filled with the sample and placed in the injection port of the

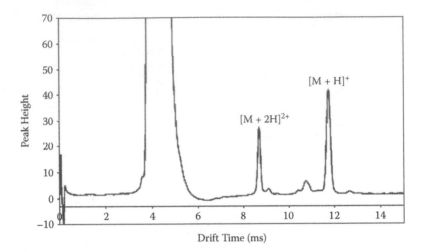

FIGURE 15.7 Electrospray ion mobility spectrum of irinotecan HCl in positive polarity mode using 0.1 μg/μL in 80/20 MeOH/H$_2$O. (From Excellims.)

IMS. The injection port is essentially a syringe pump in which electrical contact is made with the needle to produce the electrified spray, and the syringe plunger is pressed at a constant and controllable rate to produce an electrospray of sample from 0.1 to 10 μL/min. As discussed in previous chapters, IMS spectra can be collected in a few milliseconds, but these spectra are normally averaged from 100 to 3,000 scans, depending on sensitivity requirements, to produce analysis times of a minute or less.

Figure 15.7 shows the determination of the same compounds as used in Figure 15.5 except ESI was used instead of thermal desorption ionization. To obtain a single molecular ion with a sharp, well-defined mobility peak, ESI is required.

The electrospray spectrum has a number of interesting characteristics. First, as with other IMS spectra, the x-axis is the arrival time of an ion swarm given in milliseconds, and the y-axis is the intensity of the signal generated by arriving ions at any point in time. The intensity is quantitatively given as ion current in amperes but is commonly just shown, as is the case in Figure 15.6, as arbitrary units. The large ion swarm arriving at a drift time of about 4.5 ms is that of the solvent ions. In the positive-polarity mode, as in the case for this example, the identity of the solvent ions depends on the composition of the electrospray solvent. In this case, the electrospray solvent was simply water and methanol; thus, the solvent ions would primarily be $(H_2O)_n(CH_3OH)_mH^+$ where $n = 0$ to 3 and $m = 0$ to 1.

Jumping to the end of the spectrum, there is a large, well-defined, sharp peak arriving at the Faraday plate at approximately 11.9 ms. This peak corresponds to the quasi-molecular product ion (MH$^+$) produced from the ionization reaction $M + H_3O^+ \rightarrow MH^+ + H_2O$ occurring in the liquid phase before the sample is electrosprayed into the IMS. In addition to the ion peak at 11.9 ms, there is another sharp, well-defined ion peak at a drift time of about 8.9 ms. In IMS, as in mass spectrometry (MS), dibasic compounds with molecular weights greater than about 500 Da can contain two charges after ESI. Thus, the force exerted by the electric field on the doubly charged ion is twice that acting on a singly charged ion, and the ion travels at a higher velocity, arriving at the Faraday plate sooner than the singly charged ion. The ion swarm that appears at 8.9 ms is the doubly protonated ion of irinotecan (MH_2^{2+}). There are also a few small unidentified ion peaks in the spectrum, which are apparently due to the contaminants on the surfaces of the equipment. Electrospray IMS spectra produce sharp, well-defined, although sometimes multiply charged, ion peaks from high molecular weight or thermal labile compounds.

Whether the application calls for a thermal desorption or electrospray approach, cleaning validation appears to be a perfect match for IMS. IMS is best suited for repetitive analyses of relatively simple mixtures for which high sensitivity and short analysis times are desired.

15.5 REACTION MONITORING

The rapid response of an IMS coupled with the ease that ESI devices can handle liquid samples enable inexpensive, real-time monitoring of pharmaceutical reactions. Many pharmaceutical companies spend significant resources on complex equipment to sample and monitor production processes. Initial work in this area with IMS has demonstrated the possibility of rapid analysis through batch sampling.[19] In this example, a two-step reaction sequence, a Michael addition followed by an intermolecular cyclization, occurred with the addition of cyclohexylethylamine to dimethylitaconate. The products and intermediates of this reaction were monitored over a 72-h time frame. During this process, the ion peak that resulted from the production of the reaction intermediate appeared in the early phase of the reaction and then disappeared, while in the later phase the ion from the reaction product appeared and became dominant, giving a stable peak. In this work, the samples were taken in aliquots from the reaction and immediately electrosprayed into the IMS. In a second example for which IMS was used to monitor the progression of a reaction, sample aliquots were taken at 60, 80, 160, 200, and 24 h and 7 days after the deprotonation of 7-fluor-6-hydroxy-2-methylindole with aqueous sodium hydroxide was initiated.[20]

In a third example, a reaction common in the pharmaceutical industry, reductive amination was used, such as for the production of secondary amines.[21] In addition, this reaction is a challenge to monitor in that the imine intermediates are often susceptible to hydrolysis, making off-line analytical methods such as HPLC difficult.

Figure 15.8 shows a series of spectra taken with an IMS coupled to a time-of-flight (TOF) MS. The spectra presented are of the mass-selected mobility spectra for starting material 1, starting material 2, and the product. Figures 15.8a through 15.8e show the disappearance of starting material 1, monitored at SM1 + H$^+$ (110 Da), and the appearance of the product P + H$^+$ (200 Da). Figures 15.8f through 15.8j show the disappearance of starting material 2, monitored at SM2 + H$^+$ (109 Da), with the product ion at different concentrations. It should be noted that this reaction was too fast to follow with the aliquot methods as has been demonstrated previously. Thus, the spectra in Figure 15.8 demonstrate that the compounds can be measured simultaneously, but they are not under conditions in which they can react. To monitor the reaction, a direct probe from the reaction to the electrospray tip was devised, and the rapid reaction was monitored in real time.

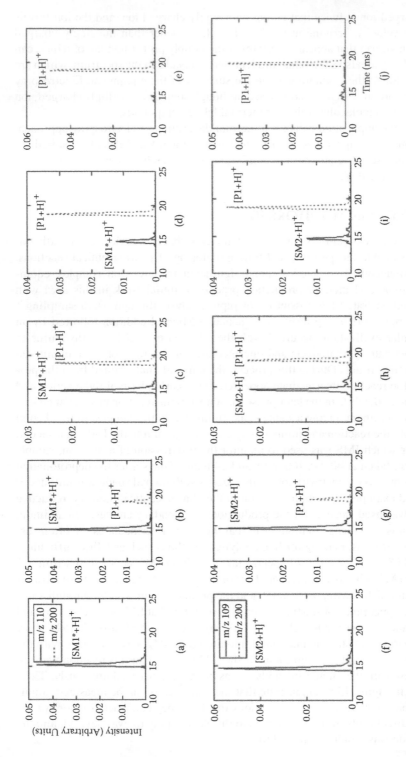

FIGURE 15.8 Selective mass mobility spectra for two components showing the progression of the reaction. (a) through (e) Disappearance of starting material 1 and the appearance of the product. (f) through (j) Disappearance of starting material 2 and the appearance of the product ion. (From Roscioli et al., Real time pharmaceutical reaction monitoring by ion mobility-mass spectrometry, *Anal. Chem.* 2012. With permission.)

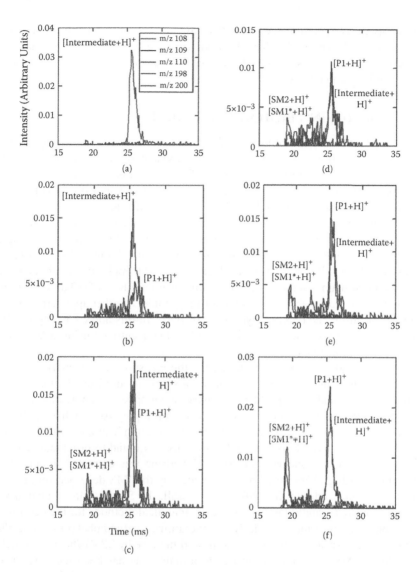

FIGURE 15.9 Example of reductive amination reaction monitored in real time. The reaction was monitored continuously for a period of 300 s. The changes in these spectra occur in the first 50 s. (From Roscioli et al., Real time pharmaceutical reaction monitoring by ion mobility-mass spectrometry, *Anal. Chem.* 2012. With permission.)

Figure 15.9 shows the real-time monitoring of the reaction for the first 300 s of the reaction. While the reaction was monitored continuously, spectra are 50-s averages. For example, IMS spectra for spectrum A were averaged for the first 50 s, spectrum B was averaged for the next 50s, and so on until spectrum F, which is for the time frame 250–300 s. As can be seen, the intermediate species appeared within the first 50 s of the reaction. In the second 50 s, the product appeared and continued to grow as the reaction progressed. This example demonstrated that IMS, especially ion mobility mass–spectrometry (IM–MS), can monitor rapid reactions in a manner that is not possible with HPLC. The detection limit for one of the starting materials, 4-pyridinementhanol, was found to be 1.5 ng, with a linear dynamic range of more than two orders of magnitude.

15.6   MONITORING BIOLOGICAL SAMPLES

For samples that are not too complex or for those for which the target compound has a high ionization efficiency, IMS may provide a viable alternative to HPLC for the rapid, sensitive, and high-resolution analysis of drugs and pharmaceuticals in biological samples. The first example of this was the simultaneous determination of caffeine and theophylline, two compounds with high proton affinities.[22] In this approach, molecular imprinted a polymer solid phase adsorption column was employed to collect and concentrate the compounds from green tea and human serum samples. The target compounds were eluted from the extraction column and electrosprayed into a stand-alone IMS for analysis. Both caffeine and theophylline could be simultaneously quantified from these samples, with detection limits of 0.2 and 0.3 µg/mL and a dynamic range of about two orders of magnitude. Another application of IMS for human serum is the rapid analysis of captropril. In this approach, captropril in the headspace above the serum was collected and concentrated using SPME based on polypyrrole film coupled to IMS.[7,8]

Several methods have been developed for the measurement of pharmaceuticals and similar compounds in urine. Ephedrine, a stimulant used for the treatment of asthma, allergies, and sinus problems, is used as an appetite suppressant and as a (banned) doping agent for athletes. The World Anti-doping Agency has established a concentration of 10 µg/mL of ephedrine in urine as an indication of illegal doping. The use of quantitative SPME coupled with IMS was found to provide a detection limit of 50 ng/mL of ephedrine in urine. Ephedrine was extracted directly from urine samples using SPME; the fiber was heated by the IMS desorber unit, and the analyte was vaporized into the drift tube. Overall, the SPME-IMS compared well with, and was significantly faster than, other analytical methods for the determination of ephedrine in urine.[6] Similarly, amphetamine, methamphetamine, and their methylendioxy derivatives have been monitored in urine by SPME/field asymmetric IMS (FAIMS)/MS.[23] Also, testosterone in human urine has been measured using molecularly imprinted solid phase extraction and corona discharge IMS.[9]

Another interesting application important to the pharmaceutical industry is the measurement of the absorption profile of drugs through the skin.[24] This application is normally accomplished through the use of a Franz diffusion cell (FDC) with quantification by HPLC. IMS methods are more rapid than and compared well with HPLC results when the transdermal analyses of ibuprofen were made. Using IMS, the skin permeability coefficient was found to be 0.013 cm/h, which matched those determined using HPLC.

More complex samples can be analyzed when IMS is coupled to MS. One example is the use of a Synapt traveling wave (TW) IMS for the metabolite profiling of leflunomide (LEF) and acetaminophen (APAP).[25] Compared with quantitative (Q) TOF-MS and Q-TRAP-MS, the ability to provide mobility separation of the MS² and then generate MS^3 fragments demonstrated the potential power of the Synapt TW-IM-MS instrument. IMS coupled with MS^2 and MS^3 experiments were used to identify the microsomal metabolite of acetaminophen.

15.7 SUMMARY AND CONCLUSION

The extension of IMS to pharmaceutical applications was a natural progression from its use for illicit drug detection. Many pharmaceutical compounds are basic and thus have high proton affinities, providing both sensitivity and selectivity in a ^{63}Ni ionization source or a corona discharge ionization source. In addition, the development of stand-alone ESI-IMS provides an efficient introduction and ionization method for nonvolatile pharmaceuticals such as those with high molecular weights, those that are ionic, and those that may decompose at elevated temperatures.

Applications in the pharmaceutical industry are diverse and range from simple characterization of molecular structure to monitoring pharmaceutical reactions in real time. Databases of pharmaceuticals that provide the collision cross section of an ion are being developed and can be used along with molecular modeling to aid in the determination of molecular structure. The rapidity and simplicity of the technique encourage development of protocols for routine process monitoring and for quickly checking the accuracy of drug formulations. IMS has also proven useful for the quantification of routine analyses of targeted pharmaceuticals, metabolites, and APIs in biological matrices such as urine, skin, and blood.

The most successful application has been in cleaning verification, for which analytical methods must ensure that equipment is not contaminated from the previous manufacturing process before a new process can be initiated. Accuracy and speed of the analytical tool are critical for efficient operation of a manufacturing plant. Finally, while the use of IMS for real-time monitoring of pharmaceutical synthesis is in its infancy, the promise of controlling reaction temperatures and adjusting reagent concentrations in real time to optimize reaction rates and yields creates an exciting potential for the continued development of IMS in the pharmacy industry.

The conclusion from the broad array of examples presented in this chapter is similar to that of the other chapters on IMS applications: IMS, in its variety of forms, offers a viable alternate analytical method to traditional chromatographic methods, especially when cost, speed, and sensitivity are important. When samples are complex, however, as in the case of metabolomics for pharmaceutical analyses, chromatographs and MSs can be easily coupled to IMSs to achieve powerful two- and three-dimensional separations.

REFERENCES

1. Cottingham, K., Ion mobility spectrometry rediscovered, *Anal. Chem.* 2003, October, 435A–439A.
2. Fernandez-Maestre, R.; Hill, H.H., Jr., Ion mobility spectrometry for the rapid analysis of over-the-counter drugs and beverages, *Int. J. Ion Mobil. Spectrom.* 2009, 12, 91–102.
3. Routledge, E.J.; Parker, J.; Odum, J.; Ashby, J.; Sumpter, J.P., Some alkyl hydroxy benzoate preservatives (parabens) are estrogenic, *Toxicol. Appl. Pharmacol.* 1998, 153, 12–19.

4. Harvey, P.W., Parabens, oestrogenicity, underarm cosmetics and breast cancer: a perspective on a hypothesis, *J. Appl. Toxicol.* 2003, 23, 285–288.

5. Lokhnauth, J.K.; Snow, N.H., Determination of parabens in pharmaceutical formulations by solid-phase microextraction-ion mobility spectrometry, *Anal. Chem.* 2005, 77, 5938–5946.

6. Lokhnauth, J.K.; Snow, N.H., Solid phase micro-extraction coupled with ion mobility spectrometry for the analysis of ephedrine in urine, *J. Sep. Sci.* 2005, 28, 612–618.

7. Alizadeh, N.; Mohammadi, A.; Tabrizchi, M., Rapid screening of methamphetamines in human serum by headspace solid-phase microextraction using a dodecylsulfate-doped polypyrrole film coupled to ion mobility spectrometry, *J. Chromatogr. A* 2008, 1183, 21–28.

8. Karimi, A.; Alizadeh, N., Rapid analysis of captopril in human plasma and pharmaceutical preparations by headspace solid phase microextraction based on polypyrrole film coupled to ion mobility spectrometry, *Talanta* 2009, 79, 479–485.

9. Mirmahdieh, S.; Mardihallaj, A.; Hashemian, Z.; Razavizadeh, J.; Ghaziaskar, H.; Khayamian, T., Analysis of testosterone in human urine using molecularly imprinted solid-phase extraction and corona discharge ion mobility spectrometry, *J. Sep. Sci.* 2011, 34, 107–112.

10. Active Pharmaceutical Ingredients Committee, Cleaning validation in active pharmaceutical ingredient manufacturing plants, report, APIC, Washington, DC, 1999.

11. Cottengham, K., Ion mobility spectrometry rediscovered, *Anal. Chem.* 2003, October 1, 2003, 439 A.

12. Prabu, S.L.; Suriyaprukash, T.N.K., Cleaning validation and its importance in pharmaceutical industry, *Pharma Times* 2010, 42(7), 21–25.

13. Strege, M.A.; Kozerski, J.; Juarbe, N.; Mahoney, P., At-line quantitative ion mobility spectrometry for direct analysis of swabs for pharmaceutical manufacturing equipment cleaning verification, *Anal. Chem.* 2008, 80, 3040–3044.

14. Qin, C.; Granger, A.; Papov, V.; McCaffrey, J.; Norwood, D.J., Quantitative determination of residual active pharmaceutical ingredients and intermediates on equipment surfaces by ion mobility spectrometry, *J. Pharm. Biomed. Anal.* 2010, 51, 107–113.

15. Walia, G.; Davis, M.; Stefanou, S.; Debono, R., Using ion mobility spectrometry for cleaning verification in pharmaceutical manufacturing, *Pharm. Technol.* April 2003, 72–78.

16. Strege, M.A., Total residue analysis of swabs by ion mobility spectrometry, *Anal. Chem.* 2009, 81, 4576–4580.

17. Krueger, C.A.; Hilton, C.K.; Osgood, M.; Wu, J.; Wu, C., High resolution electrospray ionization ion mobility spectrometry, *Int. J. Ion Mobil. Spectrom.* 2009, 12, 22–37.

18. Wu, C., Electrospray ionization-high performance ion mobility spectrometry for rapid on-site cleaning validation in pharmaceutical manufacturing, Application Notes 2010, IMS-2010-02A, Excellims Corporation, 1–5.

19. Wu, C., Excellims Corporation: Application Notes, 2010.

20. Harry, E.L.; Bristow, A.W.T.; Wilson, I.D.; Creaser, C., Real-time reaction monitoring using ion mobility-mass spectrometry, *Analyst* 2011, 136, 1728.

21. Roscioli, K.M.; Zhang, X.; Shelly, X.L.; Goetz, G.H.; Guilong, C.; Zhang, Z.; Siems, W.F.; Hill, H.H., Jr., Real time pharmaceutical reaction monitoring by ion mobility-mass spectrometry, *Anal. Chem.* 2012.

22. Jafari, M.T.; Rezaei, B.; Javaheri, M., A new method based on electrospray ionization ion mobility spectrometry 1 (ESI-IMS) for simultaneous determination of caffeine and theophylline, *Food Chem.* 2010.

23. McCooeye, M.A.; Mester, Z.; Ells, B.; Barnett, D.A.; Purves, R.W.; Guevremont, R., Quantitation of amphetamine, methamphetamine, and their methylenedioxy derivative in urine by solid-phase microextraction coupled with electrospray ionization-high-field asymmetric waveform ion mobility spectrometry-mass spectrometry, *Anal. Chem.* 2002, 74, 3071–3075.

24. Baert, B.; Van Steelandt, S.; De Spiegeleer, B., Ion mobility spectrometry as a high-throughput technique for in vitro transdermal Franz diffusion cell experiments, *J. Pharm. Biomed. Anal.* 2011, 55, 472–478.
25. Chan, E.C.Y.; New, L.S.; Yap, C.W.; Goh, L.T., Pharmaceutical metabolite profiling using quadrupole/ion mobility spectrometry/time-of-flight mass spectrometry, *Rapid Commun. Mass Spectrom.* 2009, 23, 384–394.

16 Industrial Applications

16.1 INTRODUCTION

Ion mobility spectrometry (IMS) has been utilized in several industrial applications where speed, cost, and specificity of IMS are viewed as strong advantages, particularly in process monitoring. In some instances, the low cost of IMS analyzers and high value added from the convenience of continuous measurement are enough that analyzers are integrated into a component, such as a high-voltage switch in hydroelectric power plants. In other instances, low limits of detection with atmospheric pressure ionization of IMS analyzers have made possible a specific use, such as monitoring air quality in manufacturing rooms where purity is extraordinarily critical. Monitoring industrial emissions, although possibly a category of environmental application, is another example where process-monitoring configurations of IMS instruments were employed for industrial benefit.

The advantage of on-site and high-speed measurements are key motivations to apply IMS in traditional industrial interests, namely, analysis of feedstock purity and product composition, although these are, in this moment, somewhat limited in number and variety. An extension of this idea that has been explored in proof-of-concept applications is monitoring the processes or stages in production of food and beverages. In this application, the IMS measurement is considered objective in measures of odors or flavors, can be automated, and may be integrated into continuous, automated process control.

16.2 INDUSTRIAL PROCESSES

16.2.1 SULFUR HEXAFLUORIDE MONITORING IN HIGH-VOLTAGE SWITCHES

One application for which no other technology is able to complete with mobility spectrometers in cost, size, and analytical response is the continuous monitoring of gas purity of gas insulated switches (GISs) in high-voltage substations.[1–3] Sulfur hexafluoride insulates inside metal surfaces of GISs, and over time or with partial electrical discharges, the gas may undergo progressive chemical degradation with the formation of low levels of corrosive substances. The accumulation of these impurities, in time, may lead eventually to an unpredictable failure of a switch; this would result in damage to the switch and power substation. The economic consequences and inconvenience to power companies and their customers are large.

When the accumulation of impurities can be monitored and maintenance staff can be alerted of an impending failure in a GIS, damage to the overall substation may be prevented by controlled shutdown of turbines. Changes in the concentration of degradation products formed in SF_6 provided such a measure of impurity levels as a trend of drift time versus calendar time. Comparatively simple and inexpensive analyzers permitted each switch to be equipped with a drift tube. In a comparative test, the mobility analyzer performed better than the standard colorimetric detector tubes by a factor of two to four and provided automated, high-speed response with minimum detectable level (MDL) values of 20 ppm, sufficient to detect the impurities formed in a GIS.[2] This application of IMS now has a record of performance of more than 12 years of use.

16.2.2   IMPURITIES IN AMBIENT AIR ATMOSPHERES IN MICROCIRCUIT PRODUCTION

The levels of ammonia in the ambient air inside semiconductor plants, particularly in "clean areas" used for deep ultraviolet (DUV) photoresist and lithographic processes, are important since ammonia and 1-methyl-2-pyrrolidone can affect the acid catalysis of photoresist formulations. Ion mobility spectrometry has provided the necessary analytical capabilities to monitor both compounds at low part-per-billion levels. Compared with gas chromatography (GC), the time required for a measurement with the ion mobility spectrometer was significantly shorter than the 10 min required for thermal desorption/GC.[4,5]

16.2.3   MONITORING EXTENT OF SYNTHETIC REACTIONS

Ion mobility spectrometers continuously monitored the headspace vapors over a polymerization of vinyl acetate, and results provided a measure of reaction progress through unreacted or residual monomer vapors in the headspace.[6] The final quality of a polymer latex is determined by several parameters, including the amount of residual monomer, which is volatile. Thus, odors may be controlled when monomer levels reach a low level, but when is this reached? Currently, samples are analyzed in the laboratory only of the final product, and online monitoring would aid measurements and production efficiency. In a pilot study to demonstrate the suitability of IMS to monitor an emulsion reaction, vinyl acetate was polymerized in a 1-L laboratory reactor. The reaction was followed with gravimetric measurements every 30 min, and an IMS analyzer followed the decrease of the monomer concentration after the injection of the initiator. The monomers vinyl acetate, butyl acetate, and methyl methacrylate were distinguished due to characteristic mobility coefficients even in mixtures of these substances. In this first application, emulsion pumped through a bypass during production. The direct sampling of headspace of the reactor containing the reaction mixture was proposed as an alternative method to monitor the batch reactor.

16.2.4   DISCOVERY OF LEAKS DURING PRODUCTION OF TRANSDERMAL PATCHES

When response to a chemical is strong due to favorable ionization properties of the substance, mobility spectrometers can often provide direct monitoring without effects from potential interferences in the sample matrix. This was seen with the evaluation of nicotine emissions at a production site for transdermal systems (i.e., skin patches) for treating nicotine withdrawal.[7] Nicotine patches are produced by depositing drops of nicotine onto an adsorbent layer that is drawn as a continuous sheet through a machine in which barrier films and adhesive layers are laminated with the nicotine layer and punched to size. Operators of the machine were subjected to inhalation exposures to nicotine vapors during production of the patches, although neither the exact source of the nicotine emissions nor any fluctuations during the manufacturing day were possible to determine using the existing method of air sampling and analysis. The Occupational Safety and Health Administration (OSHA) exposure standard at that time for nicotine was 0.5 mg/m^3 for continuous exposure, and these levels were possible to detect only when sample was preenriched using a sorbent trap and air-sampling pump. The time of sampling caused an integration of sample, losing any spikes in concentrations and preventing rapid surveys of the production site. In one exercise, a chemical agent monitor (CAM) was operated continuously during the workday near an operator's station on the production machine. In another, parts and surfaces of the production machine were scanned rapidly for nicotine vapors by placing the CAM at these locations for several seconds. The entire machine was mapped for nicotine emissions within 15 min.

The CAM exhibited near-instantaneous response, detection limits of 0.006 mg/m^3, and median relative standard deviations of 3.1% for vapor levels of 0.01 to 0.25 mg/m^3. During continuous monitoring of air near the machine, short-lived and elevated concentrations of isopropanol were detected when alcohol-wetted cloths were used to clean the machine after contamination accumulated on particular surfaces, and four places in the machine were identified as sources where nicotine vapors were released into the ambient air. In only a few hours, the sources of nicotine vapors were identified, and this had eluded efforts with traditional methods for months.

16.2.5 Acid Gas Monitoring in Stack Emissions

Acid gases, in particular hydrogen chloride (HCl), hydrogen fluoride (HF), ClO_2, and bromine, are emitted from industrial processes, including aluminum refining, ceramics manufacture, coal combustion, and plastic waste incineration.[8–10] Due to the toxicity and corrosiveness of these acid gases, safety hazards exist within the plant, and pollution of the environment is a concern outside the perimeter. Monitoring of airborne emissions from stacks associated with these processes has been required by federal, state, and often local agencies, and mobility spectrometers have been configured, in the form of process monitors, to meet requirements for monitoring.

Gordon et al. used an IMS analyzer to study the rate of emission and volatilization of ClO_2 and predict exposure to levels of ClO_2 in ambient air.[9] Levels of vapors 2 m from a large mixing tank containing concentrated solutions (<200 to 860 ppm) were 1.4 ppmv as measured with a handheld analyzer. At 4 m, the level was decreased to 0.6 ppmv and fell below the OSHA permissible level at about 7 m. This and the application for emissions of bromine illustrate the best quantitative response with existing IMS instruments.

The determination of HF using IMS should be favored by electron affinity of HF but is complicated by overlap between the product ion peak for HF and the peak for the reactant ion, commonly O_2^-. This was solved using methylsalicylate as a reagent gas; methylsalicylate forms an adduct ion with O_2^- displacing the reactant ion peak to long drift times well displaced from that for HF.[10] The product ion was described as $(HF)_3F^-$ and was well separated from the peak for methylsalicylate adduct to O_2^-.

Interest in monitoring HF arises not only from occupational hygiene and environmental concerns but also from interests in HF as a precursor in the production of chemical warfare agents. Thus, HF is considered a dual-use chemical and important within the sphere of counterterrorism.

The possible chemical reaction between corrosive acid gases and the metallic radioactive foil (^{63}Ni) leads to the formation of nickel halides on the surface of the foil. Such salts are mechanically unstable and could become aerosolized and spread throughout the drift tube. Such salt particles may also be released into ambient environments with unfiltered exhaust flows from the drift tube. To prevent such reactions, the drift tubes for acid gas monitoring are designed to maintain a flow of purified air or nitrogen over the radioactive metal foil, preventing contact between the acid gases and the source. Reactant ions are moved in the clean flow with the voltage gradient into a reaction region containing sample gas. All vapors are vented from the drift tube without entering the drift region or contacting the source. Such instruments should be fitted with interlocks that stop sample flow if purified gas flow fails.

16.2.6 OTHER INDUSTRIAL EMISSIONS

Toluene diisocyanate (TDI) is used to produce flexible polyurethane foam and is available as two positional isomers (2,4-TDI and 2,6-TDI) and a mixture of these known as 80/20 TDI. Exposure to high levels of 2,4-TDI by inhalation causes severe irritation of the skin, eyes, and nose and produces nausea and vomiting. Inhalation exposure at low levels has caused an asthma-like reaction characterized by wheezing, dyspnea, and bronchial constriction in industry workers.

Airborne vapor levels of TDI, in a range of industrial manufacturing conditions found in urethane technology, were monitored using a handheld mobility analyzer with membrane inlet, a derivative of the CAM.[11] Response ranged from 1 to 50 ppb and was little affected by gross excesses of amine and tin catalysts, blowing agents, and surfactants. The CAM was sensitive to 2,4-TDI, 2,6-TDI, and 80/20 TDI at humidity from 0 to 68% relative humidity in sample air. The instrument cleared to original response within seconds after exposure to low vapor levels yet needed a minute with levels of 48 ppb. This can be attributed to material sciences and temperatures of the inlet and membrane. Response was suppressed with 0% humidity tests. Despite attractive analytical behavior, IMS was not incorporated into industrial occupational monitoring. One possible objection to industrial use of a portable IMS analyzer for this application is the regulatory difficulties of moving a sealed radioactive source. This objection is not intrinsic to IMS, and another team demonstrated that a pulsed nonradioactive ion source was suitable for TDI determinations in industrial venues.[12] The pulsed source provided a type of kinetic control over ion formation and ion sampling into the drift tube.

Bromine and some brominated compounds have also been monitored in ambient air at a chemical plant in Israel.[13,14] In the IMS drift tube, bromide ions (Br) are usually the only stable ions in negative polarity and are formed from organobromine compounds. However, Br_3^- was formed from molecular bromine as an adduct ion of Br_2 and Br$^-$ when bromine levels were at elevated concentration. Measurements in storerooms at the production plant showed levels below

10 ppb, and measurements in production areas were detectable and affected by air currents near the reactors. Still, airborne vapors of bromine were below 30 ppb and below the 100 ppb threshold of exposure for an 8-h time-weighted average. This is an example, as described previously with nicotine, when ion chemistry was strongly favored, and interferences such as water were insignificant. Since the spectral patterns for bromine in IMS are concentration dependent with three product ion peaks, neural networks were employed successfully for quantitative vapor determination of bromine in air.[14]

Finally, IMS was applied to the ceramic industry for on-site monitoring of volatile organic compounds (VOCs) produced during tile baking.[15] The instrument was calibrated with a set of reference compounds, including ethyl acetate; ethanol; ethylene glycol; diethylene glycol; acetaldehyde; formaldehyde; 2-methyl-1,3-dioxolane; 2,2-dimethyl-1,3-dioxolane; 1,3-dioxolane; 1,4-dioxane; benzene; toluene; cyclohexane; acetone; and acetic acid using airflow permeation. Testing was made using a laboratory-scale kiln and tiles prepared with selected glycol- and resin-based additives. Results of all experimental measurements were compared to those obtained by solid phase microextraction/gas chromatography/mass spectrometry (SPME/GC/MS). The method was also applied to emissions from two industries in the Modena (Italy) ceramic area as a means of assessing IMS as a real-time monitoring device for emissions from the ceramic industry.[15]

16.3 INDUSTRIAL FEEDSTOCK OR PRODUCTS

16.3.1 PURITY OF GASES

Low levels of ammonia in ethylene, hydrogen, and other light hydrocarbons in industrial processes can cause downstream catalytic poisoning, and an IMS system with a limit of detection of 1 ppb was used to monitor NH_3 in hydrocarbons.[8] A wide variety of process streams was tested without any evident interference from coexisting compounds. In another report, NH_3 in ethylene was measured by an IMS with a silane polymer membrane, ^{63}Ni ion source, $H^+(H_2O)_n$ reactant ion, and nitrogen as the drift and source gas.[16] Because ethylene has no noticeable effect on the analytical results, preconcentration or preseparation were unnecessary. Ethylene's flammability was made inconsequential by the nitrogen atmosphere inside the spectrometer. Response to NH_3 concentrations between 200 and 1,500 ppb was nearly linear, and a calculated minimum detection limit was 25 ppb.

Trace amounts of oxygen were determined in nitrogen using a drift tube operated with argon as the drift gas since argon is chemically transparent with conventional analytical IMS analyzers in both positive and negative polarity.[17] Normally, the detection limit and accuracy for determining O_2 in N_2 by IMS suffers from poor resolution of ion peaks. However, product ions for oxygen formed by ionization chemistry in an argon atmosphere can be resolved and quantitatively determined in mobility spectra. The appearance of any ion peak in the mobility spectrum indicated the presence of impurities in these important industrial gases.[17,18] Impurities from ions that are not present in the pure gases were identified and quantified based on the drift times and peak areas.

16.3.2 Surface Contamination in Semiconductor Manufacturing

In the late 1990s, IMS technology was applied for uses in the semiconductor industry as reported mainly by the group of Budde at Siemens[19–23] and by others.[24–26] In one study, an IMS–MS instrument was used to detect volatile materials that were emitted or outgassed from wafer storage and transport boxes.[19] These vapors were identified as plasticizers and attributed to the polymer additives in the box material. Large variations of up to four orders of magnitude were found in the quantity of vapors emitted by various storage boxes, with clear implications for quality control in manufacturing of electronic components. Furthermore, photoresist solvents and other vapors were found originating from used containers and process media. In follow-up studies, enhanced stress testing was performed at elevated temperatures for polypropylene, polycarbonate, polytetrafluoroethylene, perfluoro alkoxy polymer, polyvinylidene fluoride, and acrylonitrile-butadiene-styrene copolymers.[19] Contamination can arise from several sources in the manufacturing of electronic components, dictating a need for routine measurements. This was demonstrated by Budde's study of organic contamination from wafer boxes, wafer carriers, pods, clean room air filters and filter frames, clean room paper, sealing foils, and other polymeric materials.[21]

Examination of the contamination of surfaces using IMS measurements was also described by Seng, who was concerned with quality control in curing of surface films.[24,25] An IMS analyzer was used to detect trace amounts of coating and ink components that remained unlinked during the ultraviolet (UV) radiation curing process.[26]

16.3.3 Direct Rapid Analysis of Wood for Preservatives and Diseases

16.3.3.1 Preservatives

Three to five chlorine-substituted phenols have been used as fungicides and herbicides since the 1930s, and pentachlorophenol (PCP, not to be confused with a drug of abuse) is used globally as a wood preservative to control mold, mildew, and termites. Roughly 5–10% of PCP is tetrachlorophenol as a contaminant, although tetrachlorophenol is also used alone and sometimes as a replacement for PCP. These preservatives are applied either by a pressure/vacuum process to ensure long-term lumber preservation or by surface applications with spraying, dipping, or brushing. One of the properties of PCP is persistence, and another is vapor pressure, albeit low. Nonetheless, the presence of chlorinated phenols in wood products is a concern because of their accumulation as contaminants in indoor air. These preservatives, when in wood, also complicate safe recycling or combustion of wood. IMS shows distinctive response to chlorinated phenols, and detection limits are very low; IMS was placed in a technology runoff with GC with electron capture detection (ECD) and GC–MS.[27] The IMS method with simple thermal desorption of a sliver of wood was chosen largely based on speed (<1 min), although quantitative response was strong and equal in reliability to GC–MS, a much more complicated measurement method.

Other preservatives, propiconazole and tebuconazole, were also screened using IMS, specifically an IONSCAN with direct solid phase desorption inlet to rapidly detect these substances in Scotch pine (*Pinus sylvestris*).[28] The findings proved that direct thermal desorption with IMS is capable of detecting and distinguishing wood preservatives directly from treated wood shavings. The method required no additional sample preparation or extraction of the wood sample and was easily applied to small samples of wood.

16.3.3.2 Diseases in Wood

At the opposite extreme of wood treated with chemicals as insecticides is wood that has fungal growth. This is an industrial concern since fungi constitute, in enclosed spaces such as homes, a medical concern. Fungi can produce metabolic vapor emissions with characteristic VOCs, which are useful as indicators of fungal growth even when such growth is not visible. Headspace vapors of building materials, specimens of pine sapwood on agar media colonized by the dry-rot fungus *Serpula lacrymans,* and a mixture of six molds were examined over 6 months using IMS.[29] Mobility spectra from an IMS analyzer were computationally processed using principal component analysis (PCA) of the fungi changed during cultivation. Infected wood provided distinctive results and could be distinguished from uncontaminated wood. Also, changes in the emission profiles with time of growth of the fungi were observed in headspace vapors. IMS was recommended as a rapid and sensitive on-site method to indicate actively growing fungi within wood.

Bacterially infected wood known as wetwood in the lumber industry is a severe processing problem that causes defects in lumber drying. These defects appear as excessive and deep surface checking, honeycombs, collapse, and ring separation and result in loss; for oak lumber alone, it is as much as 500 million board feet wasted, more than $25 million per year. Rapid identification of wetwood in northern red oak (*Quercus rubra* L.) could allow the identification and rerouting of infected wood before processing was begun. A method with IMS was developed for this screening and was based on rapid heating to 200° of a sliver of wood.[30] Ion mobility spectra formed from vapors emitted from the sliver allowed classification of normal red oak heartwood from red oak wetwood. The basis for classification was attributed to pyrogallol in the wetwood and its absence in normal heartwood. Resorcinol was detected in the wetwood but not in the normal heartwood. This application illustrates a benefit of high-speed and inexpensive targeted measurements in industry, here a nonchemical industry.

16.4 FOOD PRODUCTION

Foods and beverages share a common certainty: the combination of taste and odor affect the acceptability or preference in human palates, and this can be accompanied by large financial consequences. The food industry has approached quality and control using headspace vapor analysis of foodstuffs from strawberries to spices, and chemical measurements are an integral component of the modern enterprise of food production and research. The historic choices for technology have been GC with a MS as detector, and this will provide the most comprehensive results or thorough accounting of composition of VOCs in headspace vapors. Such measurement capability comes with a cost of reduced sample throughput and comparatively high cost per measurement. For lesser demands for comprehensive knowledge, when a target compound or family of substances governs food quality or control of production, mobility spectrometers have been demonstrated to provide reliable analytical results for a fraction of the cost of a GC–MS instrument. In some instances, a GC inlet is needed owing to the chemical complexity of headspace vapors and the improved reliability of an IMS measurement from a prefractionated sample. High-speed variants of GC are usually sufficient to provide necessary resolving power before the IMS detector.

16.4.1 BEVERAGES

16.4.1.1 Beer

Headspace concentrations of diacetyl and 2,3-pentanedione over beer during fermentation were monitored and used in a brewery as an indication that the beer should be moved to a next step in production. These volatile diketones were indicators of the development of off-tastes or odors in beer, and normally a determination is made intermittently by conventional methods of GC–MS.[31] Since IMS could provide nearly continuous measurements, the onset of these substances past a threshold informed brewmasters that fermentation could be stopped, increasing productivity and saving expenses for fermentation. Monitoring online was implemented using a GC–IMS instrument with a photodischarge lamp as an ion source. A continuous flow of a small portion of wort from the fermentation tank was drawn into a sample vessel, which was heated for a few minutes before the headspace was sampled and analyzed. Concentrations of these analytes were automatically determined every 10 to 15 min, although intervals could be shorter. The method was validated by comparison to the analyses of the same samples using the standard method of a brewery (i.e., GC–MS).

While only these two chemicals were targeted and associated with overbrewing, the GC–IMS could also provide a complete measure of VOCs in plots of ion intensity, retention time, and drift time. The low cost of IMS technology and improved production efficiency were thought meritorious so that the entire production process for beer could benefit from process monitoring. In addition, various types of beer could be assayed using a similar method with regard to alcohol content and precursors of diacetyl and 2,3-pentanedione. The need to draw liquid sample from the fermentation tank was recognized as a complication, and another team used a membrane inlet to isolate headspace vapors from the mobility spectrometer.[32] Their target chemical was ethanol for online measurement of yeast fermentation.

16.4.1.2 Wine

The quality of wine produced, or delivered to consumers, and the control of inventory are two ongoing challenges within the wine industry and food services internationally. While senses and flavors are important for a wine of good quality and flaws can be recognized easily, human judgment of wine quality is also recognized as deficient and complicated by visual cues. This is unsurprising bearing in mind the chemical complexity of wines and variability of human scent; mobility spectrometers were considered for headspace analysis of wine. Specifically sought was ethyl acetate, which adds "depth of body, richness, and sweetness to wine," using an

Owlstone Nanotech field asymmetric IMS (FAIMS) analyzer.[33] Excessive levels of ethyl acetate, greater than 100 to 200 mg/L in the headspace, produced an unpleasant aroma, and the wine was regarded as spoiled and was discarded. Although the FAIMS instrument showed a response for ethyl acetate 30 times better than human thresholds, ethanol vapors in authentic wines saturated response of the instrument, and a GC was necessary to isolate the ethyl acetate from ethanol.

One method to introduce selectivity into an IMS method is to replace ion sources, exchanging the proton-based ion chemistry of a radioactive ion source with another principle of selectivity in response. This was done with UV ionization, for which presumably ketones and aldehydes were favored in ionization, and sampling was made with a continuous flow system, including a gas phase separator.[34] This combination of sample preparation and vapor ionization with IMS was used to analyze various white wines. Characteristic profiles for each type of wine were obtained, and classification was made using chemometric tools. Four types of wines were selected for study, and classification performance was obtained using PCA to reduce data dimensionality, linear discriminant analysis (LDA), and finally a k-nearest neighbor (kNN) classifier. This provided a classification rate with an independent validation set of 92.0% at a 95% confidence level. The same white wines were analyzed using GC with a flame ionization detector (GC-FID), which, with similar data processing, showed similar results to the IMS method, yet with more complex technology and longer analysis times.

Online monitoring of fermentation and other microorganism-based processes by headspace sampling and IMS analysis was combined with unsupervised neurocomputing.[34] Tests were made using a strain of brewer's yeast (*Saccharomyces cerevisiae*) and a benchtop fermenter from which vented gases were analyzed. Results showed the presence of five phases of yeast growth based on vapor analysis. Ion mobility spectra with distinctive profiles were also able to alert to the presence of contaminated cultivations in yeast fermentation, suggesting a value to this method even only in a small-scale demonstration.[35]

A chemical marker for the off-flavor of wine, in bottles with cork stoppers, has been proposed as 2,4,6-trichloroanisole (2,4,6-TCA). In one approach,[36] the analyte was extracted from liquid sample using a single drop of an imidazolium-based ionic liquid. The ionic liquid was heated, releasing the 2,4,6-TCA into an IMS in negative polarity. The limits of detection and quantification were 0.2 and 0.66 ng/L, respectively, and precision was reported for 10 ng/L as 1.4% (repeatability, $n = 5$) and 2.2% (reproducibility, $n = 5$ during 3 days).

In a related study, the ionization chemistry of 2,4,6-TCA was described for the atmospheric pressure gas phase ion source with an IMS.[37] The ion of greatest intensity in positive polarity was a protonated monomer ion with reduced mobility values K_o of 1.58 cm²/Vs; minor amounts of a proton-bound dimer at K_o of 1.20 cm²/Vs were reported. In negative polarity, two product ions were tentatively assigned identities as a trichlorophenoxide with $K_o = 1.64$ cm²/Vs and chloride attachment adducts of a TCA $K_o = 1.48$ and 1.13 cm²/Vs for MCl^- and M_2Cl^-, respectively. The limit of detection for 2,4,6-TCA dissolved in dichloromethane and deposited on a filter paper was 2.1 µg, or 1.7 ppm in the gas phase. Preconcentration and preseparation were recommended for actual wine samples.

16.4.1.3 Olive Oils

High-speed, reliable analytical methods are sought when high value or expensive products can be adulterated or misrepresented and chemical analyses may clarify quality or origins. Such is the subject of olive oils and their origins. Headspace vapors over extravirgin olive oil, olive oil, and pumice olive oil were analyzed with an IMS equipped with a photodischarge (UV) lamp and with a GC–IMS with a tritium source.[38] The three grades of olive oil were characterized using plots of ion intensity, retention time, and drift time, with various chemometric tools to classify samples. The rates with independent validation sets were 86.1% with a UV-based photo ionization source and IMS drift tube and 100% for a GC–IMS.

In a related work with similar technology, three categories of virgin olive oil were examined from headspace vapors.[39] Two of these, virgin and extravirgin olive oil, have similar characteristics and are difficult to distinguish by a standard analytical method. Duplicate analysis was made of 98 samples of the three categories using chemometric methods and classified according to sensory quality. The success of classification was 97%.

Finally, improvements in headspace measurements over olive oils was possible with the addition of a Tenax TA sorbent trap to sample volatile aldehydes with carbon numbers between 3 and 6.[40] The GC, used in methods described previously, was eliminated, simplifying the method and technology, and the measurement was based on thermal desorption of the sorbent traps and vapor analysis by IMS alone. Sensitivity was increased with the preconcentration step, and selectivity was introduced using temperature programmed thermal desorption of the trap. The limits of detection were below 0.3 mg/kg, and the relative standard deviation was better than 10%. One-way analysis of variance (ANOVA) of the chemical measurements (i.e., peak heights for the target aldehydes) showed significant differences between olive oil grades.

16.4.2 FOOD SPOILAGE AND BIOGENIC AMINES

Foods undergo spoilage after preparation or during storage through enzymatic and microbial processes, with the formation of volatile substances, notably biogenic amines formed by degradation of amino acids through decarboxylation reactions. These amines are found in a range of foods, including fish and seafood, meat products, milk products, and others. Since amines have favorable ionization chemistry with an atmospheric pressure ionization source and mobility spectra for amines are distinctive, headspace vapor monitoring of muscle foods for volatile biogenic amines could disclose the extent of spoilage.[41] Sample preparation was minimal, the addition of a few drops of an alkaline solution. The limit of detection for trimethylamine (TMA) was 2 ng, the time of analysis was less than 2 min, and short- and long-term reproducibility were 15% and 25%, respectively. As expected, spoilage increased with time at room temperature, as shown in Figure 16.1, and the IMS results were correlated with microorganism populations. Biogenic amines were proposed as indicators of food spoilage or freshness.

Chemometric methods were refined for the IMS-based quantitative determination of TMA in chicken meat juice to gauge food spoilage.[42] The lowest amount of TMA

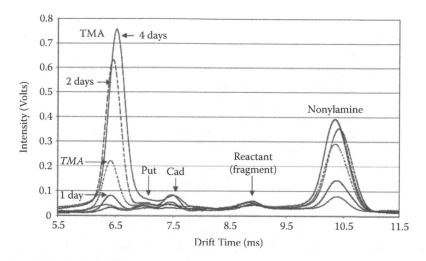

FIGURE 16.1 The progress of chicken spoilage at room temperature is seen in the growth of biogenic amines. A sample of broth from freshly ground chicken meat was diluted 10-fold and measured after 1, 2, and 4 days. These are shown along with the background spectrum and a calibration spectrum for TMA. The amount of TMA increased from 1.1 ng/g in a fresh sample to 41 ng/g after 4 days. Corresponding levels of bacteria increased in count from 10^3 to 10^9.

detected in chicken juice was 0.6 ± 0.2 ng. Data were processed using partial least squares (PLS), and fuzzy rule-building expert system (FuRES) the day of spoilage could be associated with the concentration of TMA. Others sought a direct measure of *Escherichia coli* in cooked and raw meats by IMS.[45] In this method, which was derived from enzyme-based reaction methods, O-nitrophenol (ONP) vapors are produced from extracellular enzyme reactions with a substrate, O-nitrophenyl-β-D-glucuronide. The ONP vapors were detected in negative polarity with an IMS, allowing the detection of *Escherichia coli* within 9 h.

16.4.3 CONTAMINATION IN FOODS

Concerns with the quality of food have arisen in IMS studies in two categories: pesticide residues on surfaces of fruit and mycotoxins in nuts. In both instances, IMS was seen as a method suitable for targeted compound determinations rather than comprehensive chemical characterizations. Residues of pesticides on fruits, specifically oranges, were recognized as a potential hazard, particularly the use of pesticides banned in North America, yet arriving on imported supplied. The surface of samples were wiped using a patch of Teflon or fiberglass, which were then analyzed using a GC IONSCAN, a derivative instrument of the well-known commercial instrument for explosive/narcotic detection.[44] Negative polarity was used in the determination for most pesticides. The method of wiping the exterior surface was not an efficient extraction method, with 1% for parathion and 5.9% for ß-hexachlorocyclohexane with a Teflon patch or 4.1% and 2.8%, respectively, with a fiberglass filter; however, the method was effective by metrics of speed, cost, simplicity, and finally efficacy. Residues of γ-BHC were found on the skin of unspiked oranges purchased in local supermarkets, and limits for pesticides analyzed ranged from 10 to 300 pg, with a linear range of 10 to 6,000 pg.

In companion publications,[45,46] pesticides were sampled with a laser; desorbed vapors were passed into a differential mobility spectrometer (DMS). Fast detection of pesticides was made using apples, grapes, tomatoes, and peppers, with detection limits in the nanogram range. The detection of pesticides was improved for DMS with an atmospheric pressure photoionization ion source modified with dopants such as benzene, anisole, and chlorobenzene. Improvements of detection limits up to two orders of magnitude were observed, and peaks were displaced on the compensation voltage (CV), axis as expected with modified gas atmospheres in DMS.

A last example of IMS methods applied to food quality is the determination of aflatoxins B1 and B2 on pistachio samples; methanol extracts of samples were introduced into an IMS analyzer equipped with an IMS equipped with a nonradioactive, corona discharge ion source.[47] The response to aflatoxins was acceptable for screening purposes, with a linear range of roughly 100, relative standard deviations 10% or lower, and a limit of detection of 0.25 ng for both aflatoxins. Ammonia could be used as a dopant, and detection limits were improved by a factor of 2.5. Authentic pistachio samples were analyzed without difficulties.

16.5   CONCLUSIONS

In comparison to previous summaries of industrial applications of IMS, the discussion here illustrated a wide breadth of applications in the past decade. In addition, significant depth is seen, with methods applied in actual online process control, monitoring of stacks, ambient air monitoring, and measurements with authentic samples, not only standard laboratory solutions. The versatility and capability of IMS are now proven in the range of chemicals monitored, the complex matrices for some samples, and the demands on instruments. The advantages of IMS include reliability and quantitative performance, speed of response, durability in accepting sample with little preparation, and low detection limits. When samples are complex enough to create complications from competitive reactions in the ion source, remedies have been found with fast GC or temperature-ramped desorption of sorbent traps.

REFERENCES

1. Pilzecker, P.; Baumbach, J.I.; Kurte, R., Detection of decomposition products in SF$_6$: a comparison of colorimetric detector tubes and ion mobility spectrometry, *Proceedings of the Conference on Electrical Insulation and Dielectric Phenomena,* 2002, 865–868; DOI 10.1109/CEIDP.2002.1048932.
2. Baumbach, J.I.; Pilzecker, P.; Trindade, E., Monitoring of circuit breakers using ion mobility spectrometry to detect SF$_6$-decomposition, *Int. J. Ion Mobil. Spectrom.* 1999, 2, 35–39.
3. Soppart, O.; Pilzecker, P.; Baumbach, J.I.; Klockow, D.; Trindade, E., Ion mobility spectrometry for on-site sensing of SF$_6$ decomposition, *IEEE Trans. Dielectr. Electr. Insul.* 2000, 7, 229–233.

4. Dean, K.R.; Carpio, R.A., Real-time detection of airborne contaminants in DUV lithographic processing environments, *Proc. IES* 1995, 41, 9–17.
5. Vigil, J.C.; Barrick, M.W.; Grafe, T.H., Contamination control for processing DUV chemically amplified photoresists, *Proc. SPIE Int. Soc. Opt. Eng.* 1995, 2438, 626–643, Adv. Resist Technol. Proc. XII, Editor Allen, R.D.
6. Vautz, W.; Mauntz, W.; Engell, S.; Baumbach, J.I., Monitoring of emulsion polymerisation processes using ion mobility spectrometry—a pilot study, *Macromol. React. Eng.* 2009, 3(2–3), 85–90.
7. Eiceman, G.A.; Sowa, S.; Lin, S.; Bell, S.E. Ion mobility spectrometry for continuous on-site monitoring of nicotine vapors in air during the manufacture of transdermal systems, *J. Hazard. Mater.* 1995, 43, 13–30.
8. Bacon, T.; Weber, K., PPB level process monitoring by ion mobility spectroscopy (IMS), and hydrogen chloride and hydrogen fluoride continuous emission monitoring, brochure, Molecular Analytics, Sparks, MD, http://www.ionpro.com.
9. Gordon, G.; Pacey, G.; Bubnis, B.; Laszewski, S.; Gaines, J., Safety in the workplace: ambient chlorine dioxide measurements in the presence of chlorine, *Chem. Oxidation* 1997, 4, 23–30.
10. Spangler G.E.; Epstein, J., Detection of HF using atmospheric pressure ionization (API) and ion mobility spectrometry (IMS), 38th ASMS Conference on Mass Spectrometry and Allied Topics, Tucson, AZ, June 1990.
11. Brokenshire, J.L.; Dharmarajan, V.; Coyne, L.B.; Keller, J., Near real time monitoring of TDI vapour using ion mobility spectrometry (IMS), *J. Cell. Plastics* 1990, 26, 123–142
12. Baether, W.; Zimmermann, S.; Gunzer, F., Pulsed ion mobility spectrometer for the detection of toluene 2,4-diisocyanate (TDI) in ambient, *Sensors J. IEEE* 2012, 12, 1748–1754.
13. Karpas, Z.; Pollevoy, Y.; Melloul, S., Determination of bromine in air by ion mobility spectrometry, *Anal. Chim. Acta* 1991, 249, 503–507.
14. Boger, Z.; Karpas, Z., Use of neural networks for quantitative measurements in ion mobility spectrometry (IMS), *J. Chem. Inf. Comp. Sci.* 1994, 34(3), 576–580.
15. Pozzi, R.; Bocchini, P.; Pinelli, F.; Galletti, G.C., Rapid analysis of tile industry gaseous emissions by ion mobility spectrometry and comparison with solid phase micro-extraction/gas chromatography/mass spectrometry, *J. Environ. Monit.* 2006, 8(12), 1219–1226.
16. Cross, J.H.; Limero, T.F.; Lane, J.L.; Wang, F., Determination of ammonia in ethylene using ion mobility spectrometry, *Talanta* 1997, 45, 19–23.
17. Dheandhanoo, S.; Ketkar, S.N., Improvement in analysis of O_2 in N_2 by using Ar drift gas in an ion mobility spectrometer, *Anal. Chem.* 2003, 75, 698–700.
18. Pusterla, L.; Succi, M.; Bonucci, A.; Stimac, R., A method for measuring the concentration of impurities in helium by ion mobility spectrometry, PCT Int. Appl. 2002. Application number: 10/601,383. Publication number: US 2004/0053420A1. Filing date: Jun 23, 2003.
19. Budde, K.J., Determination of organic contamination from polymeric construction materials for semiconductor technology, *Mater. Res. Soc. Symp., Proc. Ultraclean Semiconductor Processing Technology and Surface Chemical Cleaning and Passivation*, 1995, 386, 165–176.
20. Budde, K.J.; Holzapfel, W.J.; Beyer, M.M., Detection of volatile organic contaminants in semiconductor technology—a comparison of investigations by gas chromatography and by ion mobility, *Proc. 39th Annu. Tech. Meeting IES, Las Vegas, NV* April 1993, p. 366.
21. Budde, K.J., Organic surface analysis in semiconductor technology by ion mobility spectrometry, *Proc. Electrochem. Soc.* 1995, 95–30, 281–296.
22. Budde, K.J.; Holzapfel, W.J.; Beyer, M.M., Application of ion mobility spectrometry to semiconductor technology: outgassings of advanced polymers under thermal stress, *J. Electrochem. Soc.* 1995, 142, 888–897.

23. Budde, K.J.; Holzapfel, W.J., Organic contamination analysis in semiconductor silicon technology detrimental "cleanliness" in cleanrooms, in H.R. Huff, U. Gösele, and H. Tsuya (Eds.) *Semiconductor Silicon*, Volume 2, Electrochemical Society, Pennington, NJ, 1998, pp. 1496–1510.

24. Carr, T.W. Analysis of surface contaminants by plasma chromatography-mass spectroscopy, *Thin Solid Films* 1977, 45(1), 115–122.

25. Seng, H.P., Controlling method for UV-curing processes. A method for the end user, *Eur. Coat. J.* 1998, 11, 838–841.

26. Dean, K.R.; Miller, D.A.; Carpio, R.A.; Petersen, J.S.; Rich, G.K., Effects of airborne molecular contamination on DUV photoresists, *Photopolym. Sci. Technol.* 1997, 10, 425–444.

27. Schröoder, W.; Matz, G.; Kübler, J., Fast detection of preservatives on waste wood with GC/MS, GC-ECD and ion mobility spectrometry, *Field Anal Chem. Tech.* 1998, 2(5), 287–297.

28. Rasmussen, J.S.; Felby, C.; Prasad S.; Schmidt H.; Eiceman, G.A., Rapid detection of propiconazole and tebuconazole in wood by solid phase desorption ion mobility spectrometry, *Wood Sci. Technol.* 2001, 45(2), 205–214.

29. Hübert, T.; Tiebe, C.; Stephan, I., Detection of fungal infestations of wood by ion mobility spectrometry *Int. Biodeterior. Biodegradation* 2011, 65(5), 675–681

30. Pettersen, R.J.; Ward, J.C.; Lawrence, A.H., Detection of northern red oak wetwood by fast heating and ion mobility spectrometric analysis, *Holzforschung* 1993, 47, 513–552.

31. Vautz, W.; Baumbach, J.I.; Jung, J., Beer fermentation control using ion mobility spectrometry – results of a pilot study, *J. Institute Brewing*, 2006, 112(2), 157–164.

32. Tarkiainen, V.; Kotiaho, T.; Mattila, I.; Virkajärvi, I.; Aristidou, A.; Ketola, R.A., On-line monitoring of continuous beer fermentation process using automatic membrane inlet mass spectrometric system, *Talanta* 2005, 65, 1254–1263.

33. Morris, A.K.R.; Rush, M.; Parris, R.; Sheridan, S.; Ringrose, T.; Wright, I.P.; Morgan, G.H., Quantification of ethyl acetate using FAIMS, Analytical Research Forum, Loughborough, UK, July 26–28 2010.

34. Garrido-Delgado, R.; Arce, L.; Guamán, A.V.; Pardo, A.; Marco, S.; Valcárcel, M., Direct coupling of a gas-liquid separator to an ion mobility spectrometer for the classification of different white wines using chemometrics tools, *Talanta* 2011, 84, 471–479.

35. Kolehmainen, M.; Rönkkö, P.; Raatikainen, O., Monitoring of yeast fermentation by ion mobility spectrometry measurement and data visualisation with self-organizing maps, *Anal. Chim. Acta* 484(1), 93–100.

36. Márquez-Sillero, I.; Aguilera-Herrador, E.; Cárdenas, S.; Valcárcel, M., Determination of 2,4,6-tricholoroanisole in water and wine samples by ionic liquid-based single-drop microextraction and ion mobility spectrometry, *Chim. Acta* 2011, 702(2), 199–204.

37. Karpas, Z.; Guamán, A.V.; Calvob, D.; Pardo, A.; Marco, S., The potential of ion mobility spectrometry (IMS) for detection of 2,4,6-trichloroanisole (2,4,6-TCA) in wine, *Talanta* 2012, 93, 200–205.

38. Garrido-Delgado, R.; Mercader-Trejo, F.; Sielemann, S.; de Bruyn, W.; Arce, L.; Valcárcel, M., Direct classification of olive oils by using two types of ion mobility spectrometers, *Anal. Chim. Acta* 2011, 696(1–2), 108–115.

39. Garrido-Delgado, R.; Arce, L.; Valcárcel, M., Multi-capillary column-ion mobility spectrometry: a potential screening system to differentiate virgin olive oils, *Anal. Bioanal. Chem.* 2011; 402(1), 489–498.

40. Garrido-Delgado, R.; Mercader-Trejo, F.; Arce, L.; Valcárcel, M., Enhancing sensitivity and selectivity in the determination of aldehydes in olive oil by use of a Tenax TA trap coupled to a UV-ion mobility spectrometer, *J. Chromatogr. A* 2011, 42, 7543–7549.

41. Karpas, Z.; Tilman, B.; Gdalevsky, R.; Lorber, A., Determination of volatile biogenic amines in muscle food by ion mobility spectrometry (IMS), *Anal. Chim. Acta* 2002, 463, 155–163.
42. Bota, G.M.; Harrington, P.B., Direct detection of trimethylamine in meat food products using ion mobility spectrometry, *Talanta* 2006, 68, 629–635.
43. Ogden, I.D.; Strachan, N.J.C., Enumeration of *Escherichia coli* in cooked and raw meats by ion mobility spectrometry, *J. Appl. Microbiol.* 1993, 74, 402–405.
44. DeBono, R.; Grigoriev, A.; Jackson, R.; James, R.; Kuja, F.; Loveless, A.; Le, T.; Nacson, S.; Rudolph, A.; Yin, S., Rapid analysis of pesticides on imported fruits by GC-IONSCAN, *Int. J. Ion Mobil. Spectrom.* 2001, 4, 16–19.
45. Borsdorf, H.; Roetering, S.; Nazarov, E.G.; Weickhardt, C., Rapid screening of pesticides from fruit surfaces: preliminary examinations using a laser desorption-differential mobility spectrometry coupling, *Int. J. Ion Mobil. Spectrom.* 2009, 12, 15–22.
46. Roetering, S.; Nazarov, E.G.; Borsdorf, H.; Weickhardt, C., Effect of dopants on the analysis of pesticides by means of differential mobility spectrometry with atmospheric pressure photoionization, *Int. J. Ion Mobil. Spectrom.* 2010, 13, 47–54,
47. Tabrizchi, M.; Ghaziaskar, H.S.; Sheibani, A., Determination of aflatoxins B1 and B2 using ion mobility spectrometry, *Talanta* 2008, 75, 233–238.

17 Environmental Monitoring

17.1 INTRODUCTION

A trend in measurements for environmental sciences beginning in the 1980s was to locate analyzers on-site and as close as possible to the test location. There is no need with on-site instruments to obtain and transport samples to a central laboratory, resulting in fast results, reduced costs, and possibly improved accuracy and precision. This is especially attractive when technology can be transferred from laboratory benches to actual on-site venues (some of which are harsh environments) without compromises in analytical performance. Unfortunately, the demands on weight, power, and size necessitate design compromises when adapting laboratory instruments for portable or transportable uses. Since ion mobility spectrometry (IMS) analyzers emerged from a strong tradition in engineering for the demanding requirements of military and security uses, existing technologies are highly compatible with this philosophic trend toward on-site environmental measurements. The very selection of ion mobility methods by the military arose from considerations of simplicity and reliability available in ambient-pressure-based instruments without liquid reagents or solutions.

An emphasis in applications of IMS or of gas chromatography (GC) with ion mobility dectors decidedly has existed toward volatile or semivolatile substances, and instruments today encompass a range from the small, simple aspirator IMS (aIMS) designs of Environics Oy to large, sophisticated, integrated instruments, such as the Volatile Organic Analyzer. A tendency today is to apply whatever handheld, rugged IMS is available to a user even if not optimally configured for a specific use. While this can sometimes frustrate users, the benefits of military hardening of technology has value with users minimally trained in handling analytical instruments. In this chapter, priority is given to those reports of actual in-field or on-site monitoring, thus distinguishing environmental monitoring strictly from industrial or other applications. Division of the discussion in the chapter is by media, prominent monitoring studies, and then chemical families.

There is value in noting that interest in environmental monitoring and the thrust toward on-site or in-situ analyzers followed a pattern in which an energetic marketplace developed from 1985 to 1995, and after the mid-1990s, interest in such measurements plummeted. A nationwide contraction in companies making environmental measurements was mirrored by loss in private and government funding in detection technologies. Unlike medical, clinical, and even industrial applications of IMS, advances in environmental applications have lagged in both number and ambition.

Surprisingly, a review of IMS in environmental analyses has recently appeared and suggested that interest in environmental analyses worldwide has persisted, and that IMS has been involved or explored as a tool in a large range of studies.[1]

17.2   AIRBORNE VAPORS

17.2.1   MONITORING AMBIENT AIR AND AIR SAMPLING

A significant record of air sampling or monitoring studies might have been expected from the relatively long history of development of handheld IMS analyzers for vapor monitoring, beginning with the chemical agent monitor (CAM), RAID, automatic chemical agent detector and alarm (ACADA), JUNO, and lightweight chemical detector (LCD). Such a record can be found, perhaps owing to the availability of hardware, only within military establishments. Nonetheless, an early description of vapor plume drift from a point source over a meadow near the Chesapeake Bay demonstrated the convenience and ease of continuous monitoring by IMS.[2] The study showed the complexity of plume dynamics in real-time monitoring of vapor concentrations in ambient atmospheres and demonstrated the vulnerability of point sensors to location in reference to the direction and speed of wind and the proximity to a vapor source.

An exercise in continuous monitoring of ambient air a decade later with a differential mobility spectrometer (DMS) equipped with a photodischarge ion source disclosed the diurnal excursions of vapor levels for benzene, toluene, ethylbenzene, and xylenes (BETXs) in an urban setting.[3] Patterns of concentration changes inside a building were associated with traffic on a nearby thoroughfare. When vapors were monitored near an interstate highway, vapor measurements could be quantitatively associated with estimates of volumetric displacements of combustion engines in cars and trucks. This study, along with the work mentioned previously, illustrates the resolution of quantitative information available with continuous monitoring of air inside buildings and in open landscapes. The detection of vapors in each instance was unchallenging owing to abundance of source materials. In these instances, sampling was continuous and uncomplicated because levels of contaminants in the airborne vapors were significantly greater than the detection limits of the instrument, levels at parts per billion by volume to parts per million by volume. Otherwise, preconcentrators are needed, and some were designed specifically for use in series with an IMS analyzer.[4]

17.2.2   SPECIFIC CHEMICAL STUDIES

At the end of the 1970s, IMS instruments were known to provide response at trace concentration levels to a broad range of compounds with a beta-emission source (Chapter 2). Depending on the selection of an appropriate ion source and instrument parameters, response could be obtained for practically any volatile substance, in either positive or negative polarities; however, the number of actual monitoring episodes is relatively small. Unsurprisingly, the chemicals at the center of such studies are those with strong environmental or toxicological impact.

17.2.2.1 Chlorocarbons

Vinyl chloride was examined under laboratory conditions only using GC–IMS, which was shown to have 2-ppbv detection levels, and these could be lowered to the 0.02-ppbv level when 1 L of an air sample was concentrated on an adsorbent trap.[5] High temperatures of the mobility detector increased response to vinyl chloride, which exhibited product ions in both negative and positive polarity. In negative ion spectra, the chloride ion was observed and arose from electron capture and dissociation mechanisms. Another measure of response with the GC–IMS was given for vinyl chloride as 7 pg/s. The GC–IMS also provided chloride-specific and bromide-specific detection through selected-ion monitoring in IMS. Retention times added analytical value to the measurements and contributed largely by providing simplified ionization chemistry. Eighteen Environmental Protection Agency (EPA) priority pollutants and 34 other volatile organic compounds were reported, and the concept of nonselective, positive ion detection of 30 of the 34 compounds was demonstrated along with selective, electron-capture-type detection of 29 others.

A novel photoemissive electron source that serves as a selective ionization source for compounds that can capture electrons with quasi-thermal energy was developed for use in IMS.[6] Pulsed ultraviolet (UV) irradiation (from a flash lamp or laser) impinging on a metal layer causes emission of electrons in distinct packages, so gating is not required. These electrons are captured first by oxygen molecules in air, forming O_2^- ions that subsequently can transfer an electron to molecules with high electron affinity, like halocarbons and nitro compounds. The negative ions were characterized by the IMS, and real-time detection limits of low parts per million by volume were obtained for chlorinated species. Two feasibility studies were made using DMS[7] and IMS[8] with chlorocarbons.

17.2.2.2 Pesticides

A team in Iran has examined the response of IMS in positive polarity to several classes of chemicals in two reports and given limits of detection as 5.3×10^{-10}, 5.8×10^{-10}, and 4.5×10^{-10} g for sevin, amitraz, and metalaxyl, respectively.[9] The working range of these compounds was about three orders of magnitude, and the relative standard deviation (RSD) of repeatability at the 5-μg/mL level were all below 14%. Similar analytical response studies were made with organophosphorus pesticides such, as malathion, ethion, and dichlorovos.[10] The limits of quantification (LOQs) were 1.0×10^{-9}, 1.0×10^{-9}, and 5.0×10^{-9} g for malathion, ethion, and dichlorovos, respectively. The working range of these compounds was about three orders of magnitude, and the RSDs of repeatability at the 5-μg/ml level were all below 15%. No actual environmental studies were made. The findings in this last work were consistent with an earlier report[11] in which a molecule with a trivalent phosphorus atom had higher mobilities than the phosphonate isomers. Substitution of an alkoxy group by chlorine had little effect on the mass mobility correlation. A final citation here is the work for developing an aIMS for pesticide monitoring[12] for which the sensitivity for detection of pesticides decreased in the order diazinon, aldicarb, dimethoate, and parathion. The main advantages of the aspiration mobility method were identified as "fast response, high sensitivity, real time vapor monitoring, straightforward maintenance and low cost. In addition, the cell tolerates high chemical concentrations and still recovers quickly."[12]

An actual field study was made with pyrethroids during inhalation exposures; indoor air was monitored as several commercial insecticides were sprayed as aerosols for 10- to 15-s intervals into a 100-m³ room.[13] Air was trapped and enriched during 1 min through volumetric sampling with a Teflon filter and a 3-L/min low-volume sampler. Concentrations of permethrin in the Teflon patches ranged from 16.8 to 196.4 ng cm^{-2} or 2.0 to 31.1 µg/cm²/h, and those regions of a body receiving high levels were the exposed areas of chest, arm, and head (Figure 17.1). Airborne vapor levels after use of household insecticides were 25 to 51 µg/m³ for permethrin and 63 to 81 µg/m³ for tetramethrin. In contrast, levels of deltamethrin were found from 15 to 25 µg m^{-3}. The authors concluded that "the fast detection time, the low levels of detection, and the ability for continually holding a high throughput make IMS one of the fastest and most sensitive analytical methods available for dermal and respiratory occupational exposure assessment."[13]

17.2.2.3 Formaldehyde

Formaldehyde has been a persistent analytical challenge owing to reactivity and surface adsorption of vapors. Nonetheless, a need exists for airborne vapor monitoring in buildings and, for one group, during manned space flights. Efforts were made to derivatize formaldehyde, making a determination technically feasible with the sorbent trap/thermal desorption inlet of the GC–IMS.[14] Previously, this approach had been shown plausible using GC–MS;[15] a denuder tube was coated with 2-hydroxymethylpiperidine and connected to an adsorbent sampler packed with Tenax TA A volatile derivative. Hexahydrooxazolo[3,4-a]pyridine was formed and retained on the adsorbent trap, and a measurement was made using thermal desorption coupled to GC mass spectrometry (MS).

17.2.2.4 Aerosols Containing Microorganisms

A demanding and unusual air-monitoring application of IMS was the program of Snyder et al.,[16–19] who combined an aerosol sampler with pyrolysis-gas chromatography–ion mobility spectrometry (Py-GC–IMS) for stand-alone continuous analysis of air for biological aerosols. Field studies were made at the Defense Research Establishment Suffield, Alberta, Canada, where biological aerosols were dispersed in 42 trials. Simple data analysis correlated with the bioaerosol challenge in 30 trials (71%), and an aerosol of biological origin was determined in another 7 trials. Two additional trials with sprays of blank water had no discernible, unambiguous biological response. Limits of detection were below 0.5 bacterial analyte-containing particles per liter of air using an aerosol concentrator for 2,000 liters of sample in 2.2 min. Snyder concluded "that the Py-GC–IMS can provide information more specific than a biological or nonbiological analysis to an aerosol when the time of dissemination is unknown to the operator."[17] In addition, the Py-GC–IMS measurement permitted discrimination between aerosols of a Gram-positive spore, a Gram-negative bacterium, and a protein.

In a series of investigations based on the pyrolysis of bacteria with vapor characterizations by GC–DMS,[20–23] mobility methods were developed that could provide at

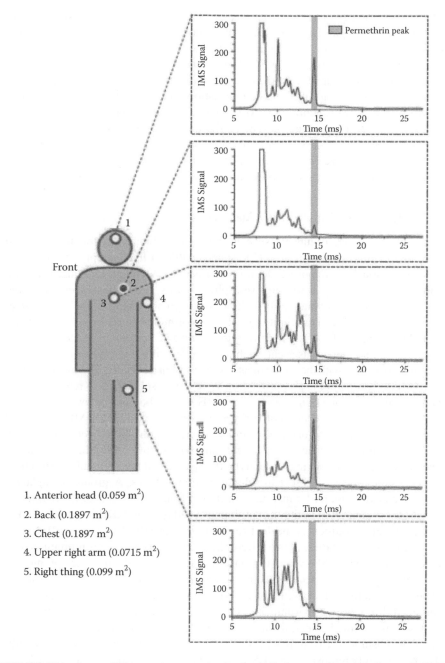

FIGURE 17.1 Ion mobility results were obtained rapidly from patches placed on an individual using hand spray application in a 100-m³ closed room, as might be in an actual pesticide application. Bursts of spray were for 15–30 s regularly over a period of 10 min. The IMS analyzer provided direct results without sample preparation. (From Armenta and Blanco, Ion mobility spectrometry as a high-throughput analytical tool in occupational pyrethroid exposure, *Anal. Bioanal. Chem.* 2012, 404(3), 635–648.)

the very least a biological trigger and at best an identification of bacterial species from an air or surface sample. The dependence of results on bacteria growth cycle, nutrient source, and temperature of growth is a limitation once thought fundamental to pyrolysis and reported for comparable GC–MS investigations. This was surmounted using computational processing of redundant chemical information in the topographic plots of ion intensity, retention time, and compensation voltage. The soft ionization with an atmospheric pressure ionization (API) source was also considered important in the successive development by preventing the data sets from becoming extremely complex, as seen in GC–MS results with electron impact sources. The method was unable to distinguish bacteria at the strain level, and this was attributed to insufficient intensity of chemical information from the highly inefficient pyrolysis step.

17.2.3 Quality of Recirculated Air Atmospheres

In recirculated air environments such as found in submarines, the International Space Station, and the now-retired U.S. space shuttle, a fixed supply of air is processed continuously to remove CO_2 and replenish O_2. Although IMS analyzers are not ideal for monitoring these light atmospheric gases, other substances, including metabolites, may accumulate in air without thorough scrubbing in the air purification system. In addition, a capability for on-site continuous monitoring of air is valuable and potentially lifesaving, following the release of poisons from a contingency or unexpected and dangerous event, such as a fire or leaking torpedo.

Vapors associated with the air purification can present a concern, and monoethanolamine (MEA) is used in the air purification system on a submarine to remove carbon dioxide from the ambient air. The level of MEA must be kept below well-defined exposure limits, and continuous monitoring of vapor concentrations of MEA is needed and must be convenient to operate, calibrate, and maintain. Moreover, demands on operations from consumables, size, weight, and cost should be minimized. Finally, instrumentation must resist, in both analytical response and reliability of construction, the high humidity levels present in submarine atmospheres. A group in England[24] developed an IMS monitor for measuring airborne levels of MEA near the maximum permissible concentration limit for a continuous 90-day period. Ketones were used as reagent gases and modifiers of the drift gas[25,26] to improve the detection of MEA in an atmosphere containing diesel fumes, freon 22, and ammonia. The minimum detectable level was 5 ppb of MEA, and response was selective in the presence of interferences. This approach to detecting alkanolamines in airborne vapors was confirmed.[27]

The atmosphere of the International Space Station is recycled, and air quality is maintained through the Environmental Control and Life Support System (ECLSS), which scrubs nominal contaminants from the recirculated atmosphere. Concern exists that substances arising from leaks, spills, and small fires cannot be immediately or completely removed by ECLSS, and continuous exposure, even to low levels of some pollutants, is regarded as a health hazard. Thus, a GC–IMS system known as the Volatile Organic Analyzer (VOA) was developed in the 1990s to provide identification and quantitative determinations for a suite of about 25 volatile organic compounds, including alcohols, aldehydes, aromatic compounds, and halocarbons. The VOA was comprised of two separate GC–IMS analyzers with integrated electronics and flow control.[28-30] A sorbent trap inlet was integrated into the design, and the GC columns were programmable in temperature. At the installation of the VOA on the International Space Station in September 2001, this was the only industry-built GC–IMS instrument in recent decades and the only integrated design ever made. The VOA was retired in spring 2010.

17.3 WATER

Water samples may be introduced into ion mobility analyzers indirectly from headspace vapors or directly using electrospray ion sources. Alternatively, membrane interfaces may be used to extract or isolate into a sample gas flow dissolved volatile organic compounds from aqueous solution. A sequential method by which substances are extracted and concentrated on solid phase microextraction (SPME) has also been demonstrated for water analysis, although to date there has been little interest in joining mobility spectrometers to liquid extractions. Perhaps the application of strongest interest for water-based applications of IMS is in groundwater monitoring for chlorocarbons, gasoline, and others for which the robustness and size of mobility spectrometers are attractive for either on-site continuous monitoring or down-hole monitoring in test wells.

17.3.1 Groundwater Monitoring

17.3.1.1 Gasoline-Related Contamination

Leaking storage tanks for fuel placed into the ground without a liner received intense attention a decade ago from the threat to groundwater from benzene, toluene, ethyl benzene, and xylene (BTEX). These substances were detected in water using a tubular silicone membrane interface with a portable IMS.[31] Vapors of toluene transferred from the water through the membrane and to the gas phase were detected at 0.101 mg/L in purified water, equivalent to 2.75 µg/m^3 with static sampling. At high levels of moisture experienced with a membrane interface, toluene was not detected at these levels. In field studies, trace concentrations of gasoline components in river water were determined with a response time of several seconds.

One constituent of gasoline with strong environmental interest, methyl-tert-butyl-ether (MTBE), which is a gasoline additive and as an oxygenate is used to raise the octane number of gasoline. This and benzene, toluene, and m-xylene (BTX) were characterized in water samples[32] using two ionization sources, a radioactive ^{63}Ni source and a UV photoionization source with a silicone membrane interface and GC inlet. These substances were clearly separated by differences in retention times and in drift times. Detection limits for MTBE from the complete method were 2 µg/L with the UV ion source and 30 pg/L with a ^{63}Ni ion source. Results with water samples were 20 mg/L (UV) and 1 µg/L (^{63}Ni), with RSDs of between 2.9% and 9%. The method was proven to be suitable for near-real-time monitoring as the total analysis time was less than 90 s.

On-site detection of MTBE in water without a GC column was based on headspace vapor sampling with an adsorbent (EXtrelut) column to isolate the MTBE.[33]

The peak of the MTBE proton bound dimer, with a reduced mobility value of 1.50 cm^2/Vs, was used to determine MTBE near more than 30 μg/L in water. Interferences from hydrocarbons (alkanes, alkenes, and cycloalkanes) were negligible due to their low proton affinity relative to that of MTBE. The time required for the determination of MTBE, including sample preparation, was approximately 5 min. While the limits of detection were much improved over that described previously, the step of isolating MTBE from the major constituents of the matrix introduced material costs and delays. An alternative to this, with the advantage of detection of multiple components, is the use of a multicapillary column for preseparation and improved ionization chemistry with a UV discharge source.[34] Several types of sample preparation, other than headspace vapors and membrane extraction, were described and based on extraction or enrichment methods. SPME is an accepted and now widely employed method to extract organic compounds from various gas and liquid matrices. Extraction by SPME with dodecylsulfate-doped polypyrrole-coated fiber and 30-min extraction of headspace vapors over water were combined with IMS methods to assay MTBE.[35] In one calibration curve, the linear range was 2 to 17 ng/mL, with a detection limit of 0.7 ng/mL. RSDs for three replicates in water samples were less than 10%. The method was applied to the analysis of MTBE in three groundwater samples and regular unleaded gasoline from a gas station in the Tehran central district, Iran.

Another approach to extraction was the use of a single drop of an ionic liquid, 1-methyl-3-octyl-imidazolium hexaflurophosphate, to extract BTEX from water.[36] The limits of detection were 20 ng/L for benzene and 91 ng/L for o-xylene. The repeatability of this method for $n = 5$ was 3.0% for o-xylene and 5.2% for toluene.

17.3.1.2 Chlorocarbon-Related Contamination

Chlorocarbons are a particular concern owing to their persistence in the environment and the lack of metabolic cleansing pathways in organisms and high fat solubility. Consequently, these substances bioaccumulate in the food chain, with unclear or deleterious effects on health and environmental quality. Chlorocarbons are considered separate from pesticides (see separate discussion) in this discussion and are chlorinated benzenes or phenols and chlorinated alkanes or alkenes. Volatile compounds in water were purged for determination by IMS with a corona discharge (CD) ion source.[37] Chlorobenzene at 3 to 30 mg/L was determined in 5 min, which was considered suitable for on-site measurements at a restoration site.

Subsurface soil-gas sampling in test wells was made using an IMS encased in a 51-mm diameter stainless steel probe along with supporting electronics.[38] Resolving power was 38 for DtBP and 31 for tetrachloroethylene or perchloroethylene (PCE). This instrument was placed into service at a site contaminated with PCE. The presence of PCE was confirmed by GC–MS analysis of a gas sample at a laboratory certified by the Environmental Protection Agency (EPA). This demonstrated the viability of a down-well IMS-based analyzer.

Water sampling with a spiral hollow polydimethylsiloxane (PDMS) membrane was used with IMS for in situ sampling and analysis of trace chlorinated hydrocarbons in water in a single procedure.[39] Aqueous contaminants permeate across the membrane into a gas flow, through the membrane tube, and into a specially made

IMS analyzer. The IMS analyzer exhibited a resolving power of $R = 33$ and 41, respectively, for negative- and positive-mode reactant ions, and the detection limits in negative polarity were 80 µg/L for PCE and 74 µg/L for trichloroethylene (TCE). The time-dependent response to TCE was experimentally and theoretically studied for various concentrations, membrane lengths, and flow rates. Findings supported the use of membrane extraction for the continuous monitoring of chlorinated hydrocarbons in water with an IMS analyzer.

17.3.2 SPECIFIC CHEMICAL STUDIES

17.3.2.1 Ammonia in Water

Ion mobility spectrometers with hydrated proton ion chemistry is exquisitely sensitive to ammonia, and two teams have sought to engineer IMS-based analyzers for ammonia in water.[40,41] In the earlier development,[40] ammonia was thermally purged through a silicone membrane, and the limit of detection in this approach was 1.2 mg/L. The membrane was free of memory effects, and pH was controllable and permitted determination of ammonium ion concentrations. The system was engineered to avoid biofouling.

In a later study,[41] an ion mobility spectrometer equipped with a corona discharge ion source was used with pyridine as an alternate reagent gas to enhance selectivity and sensitivity for the determination of ammoniacal nitrogen in river and tap water samples. The limit of detection was about 9.2×10^{-3} µg/mL, and the linear dynamic range was obtained from 0.03 to 2.00 µg/mL. The RSD was about 11%, and analytical results on actual environmental samples compared favorably with the Nessler method.

17.3.2.2 Pesticides

While pesticides show favorable response in IMS analyzers with a conventional radioactive ion source, high levels of moisture from a wet vapor sample might affect response, and electrospray ionization (ESI) with a high-resolving-power IMS drift tube was developed as a direct field analytical method for the detection and identification of mixtures of sulfonylurea (SU) herbicides in aqueous samples.[42] These herbicides have been used increasingly and show persistence and the potential for crop damage. Thus, a fast environmental method of analysis is necessary or desirable. Eight herbicides were chosen based on availability and extent of use; these were evaluated with ESI ion mobility–mass spectrometry (IM–MS). Product ions for these showed characteristic reduced mobility values; response was favorable to mixtures of rimsulfuron, metsulfuron-methyl, prosulfuron, sulfometuron-methyl, tribenuron-methyl, and primisulfuron-methyl.

17.3.2.3 Anions from Disinfection

An electrospray ion source enlarges the range of analytes accessible to IMS, field asymmetric IMS (FAIMS), or DMS measurements, and ESI-FAIMS-MS was used in the determination of aqueous-based haloacetic acids, a class of disinfection by-products regulated by the U.S. EPA.[43] The method of ESI-FAIMS effectively discriminated against background ions resulting from the electrospray of tap water solutions containing the haloacetic acids, and this simplified mass spectral response. The selectivity of ESI-MS was improved, and the limits of detection were lowered for six haloacetic acids to 0.5 and 4 ng/mL in 9:1 methanol/tap water (5 and 40 ng/mL in the original tap water samples, respectively). This was achieved without preconcentration, derivatization, or chromatographic separation prior to analysis.

The same ESI-FAIMS-MS was employed also for the determination of perchlorate at low nanomolar levels, relatively free from common interferences.[44] For example, FAIMS isolation of ions eliminated isobaric overlaps of bisulfate and dihydrogen phosphate with perchlorate. The ESI-FAIMS-MS showed a detection limit for perchlorate of 1 nM ($\approx$0.1 ppb) in a solvent of 9:1 methanol-water with 0.2 mM ammonium acetate and 10 μM sulfate. Finally, this same method was extended to the determination of a mixture of nine chlorinated and brominated haloacetic acids.[45] Haloacetate anions of the mono- and dihalogenated acids and the decarboxylated anions of three of the trihalogenated acids were detected in a nitrogen gas atmosphere. While signal was not observed for bromodichloroacetic acid (BDCAA) at a dispersion voltage of −3400 V, addition of low levels of carbon dioxide into the nitrogen gas provided detection of the pseudomolecular trihaloacetate anions, including BDCAA, and significant increases in sensitivities for the trihalogenated species. There was little effect from the addition of carbon dioxide on the determination of mono- and dihalogenated anions. Flow injection analysis of the nine haloacetic acids showed detection limits between 5 and 36 ppt in 9/1 methanol/water (v/v) containing 0.2 mM ammonium acetate.

Anions have also been determined using conventional IMS with an ESI ion source and included arsenate, phosphate, sulfate, nitrate, nitrite, chloride, formate, and acetate.[46] Distinct peak patterns and reduced mobility constants were observed for respective anions. Application to authentic water samples for the determination of nitrate and nitrite demonstrated the feasibility of using ESI-IMS as a rapid analytical method for monitoring nitrate and nitrite in water systems. The method was used for on-site measurement by exchanging air for nitrogen as the drift gas without complications. The linear dynamic range was 1,000, and detection limits were 10 ppb for nitrate and 40 ppb for nitrite.

17.3.2.4   Metals in Solution

Just as ESI enlarged possibilities for measurements to anion determinations, cations were also now easily accessible, reminiscent of the early studies of mobility (1900 to 1935) when metal ions, often alkali cations, were used in mobility investigations. No analytical work with metal ions had ever been published, only a few articles on organometallic compounds. A first report of this in 2001 demonstrated that inorganic cations in aqueous solutions could be measured using ESI-IMS.[47,48] Substances producing a single mobility peak included aluminum sulfate, lanthanum chloride, strontium chloride, uranyl acetate, uranyl nitrate, and zinc sulfate. Those producing multiple peaks included aluminum nitrate and zinc acetate. Cations were detected at lower limits of 0.16 to 13 ng/L, and although the ion species were not mass identified, positively charged cation-solvent or cation-solvent-anion complexes were suggested from the MS literature. One facet of these studies was the influence of the counterion or anion on the drift times of ion peaks, and this suggests that the ion species are forms of $M_nX_m^+$ species for which values for n and m are not yet determined.

17.4   SOIL

The least examined environmental media in the development of IMS methods or applications is soil; 5 to 10 journal articles have been released over the past 40 years. The compounds or families of substances determined include 2,4-dichlorohenoxy-acetic acid (2,4-D),[49,50] aromatic hydrocarbons,[51] polycyclic aromatic hydrocarbons (PAHs),[52] polychlorinated biphenyls (PCBs),[53] and explosives.[54]

The first article in IMS involving soils was the determination of 2,4-D residues in soils. In this method, samples were extracted, esterified with methanol in the presence of a boron trifluoride catalyst, and then analyzed by capillary GC with an IMS as detector.[49] The procedure depended on the use of an IMS analyzer, for which mobility selection of an ion was made using a dual-shutter design. Selective monitoring of the product ion formed with the methyl ester of 2,4-D permitted direct analysis of the derivatized extract without further preparation. Recovery was 93% of 2,4-D for 50 ppb in spiked soils. Later, supercritical fluid chromatography was substituted for the GC,[50] allowing measurements without derivatization of the soil extract. The detection limit for 2,4-D was estimated as 500 ppb.

Soils contaminated with gasoline paralleled the environmental complications of leaking underground storage tanks, and an IMS analyzer fitted with a photodis-charge lamp was used to distinguish among common petrochemical fuels, including leaded gasoline, unleaded gasoline, kerosene, and diesel fuel.[51] In this, headspace vapors over samples were measured, and positive ion mobility spectra obtained in air at ambient pressure were comprised of three to five peaks, which were tentatively identified as benzene, alkylated benzenes, naphthalene, and alkylated naphthalenes. Discrimination between unleaded gasoline and diesel fuel in fuel mixtures across a broad range of gas phase concentration ratios was possible using IMS with photo-ionization through a discharge lamp as the ion source. Reproducibility in prep-aration and analysis of soil samples was 10–60% RSD for individual components, while instrumental reproducibility with toluene alone was 5% RSD. Water vapor had no effect on response for fuel vapors alone, but increased moisture content in soils caused an enhanced response. Drilling mud from a natural gas reserve pit was used to demonstrate environmental application of IMS in sensing a water-saturated sample. Effects on sensor sensitivity from weathering and other environmental alter-ations of fuels in soils and groundwater were not explored.

Where thermal desorption is inadequate to remove an analyte from the surface, a laser beam can be directed, focused or unfocused, against a solid, and compounds on surfaces of solids can be vaporized and ionized at ambient pressure in air. In one application of laser-based IMS to environmental analyses, soils contaminated with petroleum products were assayed for PAHs.[52] In this, a laser was used to irradiate soil, vaporizing PAHs into the gas phase. This provided a direct, fast, extraction-free method for soil analyses.

Polychlorinated biphenyls, used mainly in transformer oil, are environmentally important due to their suspected toxicity. Ritchie et al.[53] used an IMS detector in the negative polarity to determine PCB isomers (congeners) having five or more chlo-rine atoms at detection levels of 35 ng by extracting them into an iso-octane solu-tion. Mixtures of up to four PCB congeners showed characteristic multiple peaks, although detection of Aroclors in transformer oil was suppressed by the presence of the antioxidant BHT (2,6-di-t-butyl-4-methylphenol) in the oil.

Last, explosives in soils can undergo degradation over time by photochemical and biochemical pathways, notably trinitrotoluene (TNT) undergoes biochemical degradation to 2-amino-4,6-dinitrotoluene (2-ADNT) and photochemical degradation to 1,3,5-trinitrobenzene (TNB). These explosives and degradation products are concerns at environmental remediation sites worldwide. The simple mixing of dry soil with a filter paper produces enough transfer for rapid IMS surveying of soils in contamination sites. Some environmental residues, such as 2,4-D and 2,4-dichlorophenol (DCP) were found to have IMS responses that overlapped those of the TNT degradation products. Moreover, the chloride anion as a reactant ion was not always successful in resolving peak overlap of analytes and interferents. The exchange of Cl⁻ with Br⁻ permitted resolution of analyte peaks from those of interferences.[54] Also, stability of adduct ions was enhanced by lowering the temperature of the IMS analyzer.

17.5 CONCLUSIONS

Mobility spectrometers have been directed toward in-field measurements from the inception as a modern analytical method, but this was for use in battlefields not brownfields, estuaries, or water-processing plants. All of the necessary advantages for in-field environmental analyses exist along with one large, until recently, disadvantage, a radioactive source. The common sealed source in historic designs of IMS analyzers severely limits legal portability or transportability of the analyzers. Will the surge in nonradioactive sources be seen as a fresh phase in IMS technology and lead to renewed use in environmental measurement? Will environmental monitoring grow in importance in the next years or decades? Will speedy, on-site analyzers be seen again as a viable alternative to laboratory-based measurement? Answers to these questions, more than those for the now-proven capabilities of IMS instruments, will decide if IMS ever has a significant presence in environmental analyses.

REFERENCES

1. Márquez-Sillero, I.; Aguilera-Herrador, E.; Cárdenas, S.; Valcárcel, M., Ion-mobility spectrometry for environmental analysis, *Trends Anal. Chem.* 2011, 30, 677–690.
2. Eiceman, G.A.; Snyder, A.P.; Blyth, D.A., Monitoring of airborne organic vapors using ion mobility spectrometry, *Int. J. Environ. Anal. Chem.* 1990, 38, 415–425.
3. Eiceman, G.A.; Nazarov, E.G.; Tadjikov, B.; Miller, R.A., Monitoring volatile organic compounds in ambient air inside and outside buildings with the use of a radio-frequency-based ion-mobility analyzer with a micromachined drift tube, *Field Anal. Chem. Tech.* 2000, 4, 297–308.
4. Martin, M.; Crain, M.; Walsh, K.; McGill, R.A.; Houser, E.; Stepnowski, J.; Stepnowski, S.; Wu, H.D.: Ross, S., Microfabricated vapor preconcentrator for portable ion mobility spectroscopy, *Sens. Actuators B Chem.* 2007, 126, 447–454.
5. Simpson, G.; Klasmeier, M.; Hill, H.; Atkinson, D.; Radolovich, G.; Lopez Avila, V.; Jones, T.L., Evaluation of gas chromatography coupled with ion mobility spectrometry for monitoring vinyl chloride and other chlorinated and aromatic compounds in air samples, *J. High Resolut. Chromatogr.* 1996, 19, 301–312.

6. Walls, C.J.; Swenson, O.F.; Gillispie, G.D., Real-time monitoring of chlorinated aliphatic compounds in air using ion mobility spectrometry with photoemissive electron sources, *Proc. SPIE Int. Soc. Optical Eng.* 1999, 3534, 290–298.

7. Eiceman, G.A.; Krylov, E.V.; Tadjikov, B.; Ewing, R.G.; Nazarov, E.G.; Miller, R.A., Differential mobility spectrometry of chlorocarbons with a micro-fabricated drift tube, *Analyst* 2004, 129, 297–304.

8. Sielemann, S.; Baumbach, J.I.; Pilzecker, P.; Walendzik, G., Detection of trans-1,2-dichloroethene, trichloroethene and tetrachloroethene using multi-capillary columns coupled to ion mobility spectrometers with UV-ionisation sources, *Int. J. Ion Mobil. Spectrom.* 1999, 2, 15–21.

9. Jafari, M.T.; Azimi, M., Analysis of sevin, amitraz, and metalaxyl pesticides using ion mobility spectrometry, *Anal. Lett.* 2006, 39, 2061–2071.

10. Jafari, M.T., Determination and identification of malathion, ethion and dichlorovos using ion mobility spectrometry, *Talanta* 2006, 69, 1054–1058.

11. Karpas, K.; Pollevoy, Y., Ion mobility spectrometric studies of organophosphorus compounds, *Anal. Chim. Acta* 1992, 259, 333–338.

12. Tuovinen, K.; Paakkanen, H.; Hänninen, O., Detection of pesticides from liquid matrices by ion mobility spectrometry, *Anal. Chim. Acta* 2000, 404, 7–17.

13. Armenta, S.; Blanco, M., Ion mobility spectrometry as a high-throughput analytical tool in occupational pyrethroid exposure, *Anal. Bioanal. Chem.* 2012, 404(3), 635–648.

14. Veasey, C.; Thomas, C.P.; Limero, T., The determination of formaldehyde using thermal desorption-ion mobility spectrometry, SAE Technical Paper 2001-01-2197, SAE International, Warrendale, PA, 2001.

15. Thomas, C.L.P.; McGill, C.D.; Towill, R., Determination of formaldehyde by conversion to hexahydrooxazolo[3,4-a]pyridine in a denuder tube with recovery by thermal desorption, and analysis by gas chromatography–mass spectrometry, *Analyst* 1997, 122, 1471–1476.

16. Snyder, A.P.; Dworzanski, J.P.; Tripathi, A.; Maswadeh, W.M.; Wick, C.H., Correlation of mass spectrometry identified bacterial biomarkers from a fielded pyrolysis-gas chromatography-ion mobility spectrometry biodetector with the microbiological gram stain classification scheme, *Anal. Chem.* 2004, 76, 6492–6499.

17. Snyder, A.P.; Maswadeh, W.M.; Parsons, J.A.; Tripathi, A.; Meuzelaar, H.L.C.; Dworzanski, J.P.; Kim, M.-G., Field detection of bacillus spore aerosols with stand-alone pyrolysis–gas chromatography–ion mobility spectrometry, *Field Anal. Chem. Technol.* 1999, 3, 315–326.

18. Snyder, A.P.; Tripathi, A.; Maswadeh, W.M.; Ho, J.; Spence, M., Field detection and identification of a bioaerosol suite by pyrolysis-gas chromatography-ion mobility spectrometry, *Field Anal. Chem. Technol.* 2001, 5, 190–204.

19. Snyder, A.P.; Tripathi, A.; Maswadeh, W.M.; Eversole, J.; Ho, J.; Spence, M., Orthogonal analysis of mass and spectral based technologies for the field detection of bioaerosols, *Anal. Chim. Acta* 2004, 513, 365–377.

20. Prasad, S.; Schmidt, H.; Lampen, P.; Wang, M.; Gűth, R.; Rao, J.V.; Smith, G.B.; Eiceman, G.A., Analysis of bacterial strains with pyrolysis-gas chromatography/differential mobility spectrometry, *Analyst* 2006, 131, 1216–1225.

21. Prasad, S.; Pierce, K.M.; Schmidt, H.; Rao, J.V.; Gűth, R.; Bader, S.; Synovec, R.E.; Smith, G.B.; Eiceman, G.A., Analysis of bacteria by pyrolysis gas chromatography–differential mobility spectrometry and isolation of chemical components with a dependence on growth temperature, *Analyst* 2007, 132, 1031–1039.

22. Prasad, S.; Pierce, K.M.; Schmidt, H.; Rao, J.V.; Gűth, R.; Bader, S.; Synovec, R.E.; Smith, G.B.; Eiceman, G.A., Constituents with independence from growth temperature for bacteria using pyrolysis-gas chromatography/differential mobility spectrometry with analysis of variance and principal component analysis, *Analyst* 2008, 133, 760–767.

23. Prasad, S., Detection and classification of bacteria through gas chromatography differential mobility spectrometry, PhD thesis, New Mexico State University, Las Cruces, May 2008.

24. Bollan, H.R.; West, D.J.; Brokenshire, J.L., Assessment of ion mobility spectrometry for monitoring monoethanolamine in recycled atmospheres, *Int. J. Ion Mobil. Spectrom.* 1998, 1, 48–53.

25. Bollan, H.; Eiceman, G.A.; Brokenshire, J.L.; Rodriguez, J.E.; Stone, J., Ion chemistry of hydrazines with ketone reagent chemicals in ion mobility spectrometry, *J. Am. Soc. Mass Spectrom.* 2007, 18(5), 940–951.

26. Bollan, H.R., The detection of hydrazine and related materials by ion mobility spectrometry, DPHL thesis, Sheffield Hallam University, Sheffield, UK, March 1998.

27. Gan, T.H.; Corino, G., Selective detection of alkanolamine vapors by ion mobility spectrometry with ketone reagent gases, *Anal. Chem.* 2000, 72, 807–815.

28. Limero, T.; Brokenshire, J.; Cummings, C.; Overton, E.; Carney, K.; Cross, J.; Eiceman, G.; James, J., A volatile organic analyzer for space station: description and evaluation of a gas chromatography/ion mobility spectrometer, *International Conference on Environmental Systems 921385*, July 1, 1992, Seattle, WA.

29. Limero, T.; Martin, M.; Reese, E., Validation of the Volatile Organic Analyzer (VOS) for ISS operations, *Int. J. Ion Mobil. Spectrom.* 2003, 6, 5–10.

30. Limero, T.; Reese, E., First operational use of the ISS VOA in a potential contingency event, *Int. J. Ion Mobil. Spectrom.* 2002, 5(3), 27–30.

31. Wan, C.; Harrington, P.B.; Davis, D.M., Trace analysis of BTEX compounds in water with a membrane interfaced ion mobility spectrometer, *Talanta* 1998, 46, 1169–1179.

32. Baumbach, J.I.; Sielemann, S.; Xie, Z.; Schmidt, H., Detection of the gasoline components methyl tert-butyl ether, benzene, toluene, and m-xylene using ion mobility spectrometers with a radioactive and UV ionization source, *Anal. Chem.* 2003, 75, 1483–1490.

33. Stach, J.; Arthen-Engeland, T.; Flachowsky, J.; Borsdorf, H., A simple field method for determination of MTBE in water using hand held ion mobility (IMS), *Int. J. Ion Mobil. Spectrom.* 2002, 5, 82–86.

34. Sielemann, S.; Baumbach, J.I.; Schmidt, H.; Pilzecker, P., Quantitative analysis of benzene, toluene, and m-xylene with the use of a UV-ion mobility spectrometer, *Field Anal. Chem. Technol.* 2000, 4, 157–169.

35. Alizadeh, N.; Jafari, M.; Mohammadi, A. Headspace-solid-phase microextraction using a dodecylsulfate-doped polypyrrole film coupled to ion mobility spectrometry for analysis methyl tert-butyl ether in water and gasoline, *J. Hazard. Mater.* 2009, 169, 861–867.

36. Aguilera-Herrador, E.; Lucena, R.; Cárdenas, S.; Valcárcel, M., Ionic liquid-based single-drop microextraction/gas chromatographic/mass spectrometric determination of benzene, toluene, ethylbenzene and xylene isomers in waters, *J. Chromatogr. A* 2008, 1201, 106–111.

37. Borsdorf, H.; Rammler, A.; Schulze, D.; Boadu, K.O.; Feist, B.; Weiss, H., Rapid on-site determination of chlorobenzene in water samples using ion mobility spectrometry, *Anal. Chim. Acta* 2001, 440, 63–70.

38. Kanu, A.B.; Hill, H.H.; Gribb, M.M.; Walters, R.N., A small subsurface ion mobility spectrometer sensor for detecting environmental soil-gas contaminants, *J. Environ. Monitor.* 2007, 9, 51–60.

39. Du, Y.Z.; Zhang, W.; Whitten, W.; Li, H.Y.; Watson, D.B.; Xu, J., Membrane-extraction ion mobility spectrometry for in situ detection of chlorinated hydrocarbons in water, *Anal. Chem.* 2010, 82, 4089–4096.

40. Przybylko, A.R.M.; Thomas, C.L.P.; Anstice, P.J.; Fielden, P.R.; Brokenshire, J.; Irons, F., The determination of aqueous ammonia by ion mobility spectrometry, *Anal. Chim. Acta* 1995, 311, 77–83.

41. Jafari, M.T.; Khayamian, T., Direct determination of ammoniacal nitrogen in water samples using corona discharge ion mobility spectrometry, *Talanta* 2008, 76, 1189–1193.

42. Clowers, B.H.; Steiner, W.E.; Dion, H.M.; Matz, L.M.; Tam, M.; Tarver, E.E.; Hill, H.H., Evaluation of sulfonylurea herbicides using high resolution electrospray ionization ion mobility quadrupole mass spectrometry, *Field Anal. Chem. Technol.* 2001, 5, 302–312.

43. Ells, B.; Barnett, D.A.; Froese, K.; Purves, R.W.; Hrudey, S.; Guevremont, R., Detection of chlorinated and brominated byproducts of drinking water disinfection using electrospray ionization-high-field asymmetric waveform ion mobility spectrometry-mass spectrometry, *Anal. Chem.* 1999, 71, 4747–4752.

44. Handy, R.; Barnett, D.A.; Purves, R.W.; Horlick, G.; Guevremont, R., Determination of nanomolar levels of perchlorate in water using ESI-FAIMS-MS, *J. Anal. At. Spectrom.* 2000, 15, 907–911.

45. Ells, B.; Barnett, D.A.; Purves, R.W.; Guevremont, R., Detection of nine chlorinated and brominated haloacetic acids at part-per-trillion levels using ESI-FAIMS-MS, *Anal. Chem.* 2000, 72, 4555–4559.

46. Dwivedi, P.; Matz, L.M.; Atkinson, D.A.; Hill, H.H, Jr., Electrospray ionization ion mobility spectrometry: a rapid analytical method for aqueous nitrate and nitrite analysis, *Analyst* 2004, 129, 139–144.

47. Dion, H.M.; Ackerman, L.K.; Hill, H.H., Initial study of electrospray ionization-ion mobility spectrometry for the detection of metal cations, *Int. J. Ion Mobil. Spectrom.* 2001, 4, 31–33.

48. Dion, H.M.; Ackerman, L.K.; Hill, H.H., Detection of inorganic ions from water by electrospray ionization-ion mobility spectrometry, *Talanta* 2002, 57, 1161–1171.

49. Baim, M.A.; Hill, H.H., Jr., Determination of 2,4-dichlorohenoxyacetic acid in soils by capillary gas chromatography with ion mobility detection, *J. Chromatogr.* 1983, 279, 631–642.

50. Morrissey, M.A.; Hill, H.H., Jr., Selective detection of underivatized 2,4-dichlorophenoxyacetic acid in soil by supercritical fluid chromatography with ion mobility detection, *J. Chromatogr. Sci.* 1988, 27, 529–533.

51. Eiceman, G.A., Fleischer, M.E.; Leasure, C.S., Sensing of petrochemical fuels in soils using headspace analysis with photoionization-ion mobility spectrometry, *Int. J. Environ. Anal. Chem.* 1987, 28, 279–296.

52. Roch, T.; Baumbach, J.I., Laser-based ion mobility spectrometry as an analytical tool for soil analysis, *Int. J. Ion Mobil. Spectrom.* 1998, 1, 43–47.

53. Ritchie, R.K.; Rudolph, A. Environmental applications for ion mobility spectrometry, Third International Workshop on Ion Mobility Spectrometry; Galveston, TX, October 16–19, 1994.

54. Daum, K.A.; Atkinson, D.A.; Ewing, R.G.; Knighton, W.B.; Grimsrud, E.P., Resolving interferences in negative mode ion mobility spectrometry using selective reactant ion chemistry, *Talanta* 2001, 54, 299–306.

18 Biological and Medical Applications of IMS

18.1 INTRODUCTION AND GENERAL COMMENTS ON BIOLOGICAL AND MEDICAL APPLICATIONS

Advances in methods to introduce samples into ion mobility spectrometry (IMS) drift tubes as described in Chapter 3 have enabled the range of applications of IMS to be expanded beyond the measurements of gases, vapors, and volatile and semi-volatile compounds normally associated with IMS. Molecules and samples, wholly new to IMS measurements, such as peptides, proteins, and carbohydrates, have been explored during the past decade and may have importance in medical research or clinical use. Such studies have been made possible in large part through the adaptation of methods of sample delivery and ionization that have been successfully pioneered in mass spectrometry (MS) during the past 20 years. These include electrospray ionization (ESI) and nano-ESI for aqueous samples or matrix-assisted laser desorption ionization (MALDI) for solid samples as discussed in Chapter 3. The importance of these methods in chemical sciences was recognized with the award of the 2002 Nobel Prize in Chemistry to John B. Fenn[1] and Koichi Tanaka[2] for developing ESI and MALDI, respectively. In some cases, IMSs may replace MSs in simple biological investigations; in other cases, the mobility spectrometer may serve as a filter for removing interferences in MS measurements or even afford some mode of separating isomeric ions and obtaining structural information. These systems are sometimes referred to as ion mobility–mass spectrometry (IM–MS), and several variations have been used for laboratory studies of macromolecules (see Chapter 9). Compared with MS, mobility spectrometers are inexpensive, portable, and sensitive analyzers that use low power. What may be compromised in density of information in comparison to a MS is rewarded with savings in cost and convenience.

Several reviews that surveyed the role that IMS technology plays in biological systems,[3–6] food safety and bioprocess control,[7,8] and detection of chemical and biological warfare agents[9] have been published during the last decade. Apart from the applications that are likely to remain as laboratory or research tools, IMS analyzers have been examined or developed for clinical uses, such as the diagnosis of diseases or measurements of exposure to anesthetic gases. Bacteria can be determined using IMS by a method of fast sample preparation that produces volatile or semivolatile compounds. In another approach, bacteria are thermally decomposed to produce volatile compounds, and these may be used to determine bacteria in air. A few examples of these types of IMS studies are described in this chapter to illustrate certain applications that may alter both the perceptions and the value of IMS during the next decades. This is not an exhaustive review, but an effort was made to offer a balanced presentation of the wide range of methods or applications. In general, practical applications with a commercial potential that are based on IMS drift tubes operated at ambient pressure can be distinguished from laboratory or research drift tubes operated at reduced pressure with an inert gas and usually as an IM–MS instrument (see Chapter 9).

The chemical principles underlying these biological applications of IMS technology are quite similar to those described in detail in other chapters of this monograph. Namely, proton transfer reactions are the major pathway for production of positive ions, but unlike small molecules that usually have a single site for forming a stable protonated species in air at atmospheric pressure, studies of macromolecules of biologic importance, particularly in helium at a pressure of a few millibars, may have several such sites, which could lead to the formation of multiply charged ionic species. Furthermore, under these conditions some of the macromolecules would also accept an alkali ion, like lithium or sodium, which would lead to formation of lithiated or sodiated ions. Formation of clustered ions, in positive or negative ion modes, is also a suitable ion formation mechanism in biological studies, although in most cases only positive ion mode is of practical interest.

In general, the ability of IMS in separating ions according to their mobility, rather than according to mass alone, allows the distinction between isomeric ions based on their conformation or three-dimensional (3D) structure. This is useful in biologic studies, in which the activity of a macromolecule is governed to a large extent by its structural properties, for instance, whether a protein is folded or unfolded. Another aspect of using ion mobility measurements to study complex biological systems is that the information is usually reduced to a small number of peaks. Thus, the response of an IMS provides at best only a partial view of the biological system but can thus avoid a lot of the clutter that may mask the important facets of the system. For example, when diagnosing a spoiled food product, with proper control of the ionization process, only biogenic amines that indicate the extent of food spoilage will be observed; many other compounds that are present in the sample do not interfere with the measurements.

Thus, IMS can serve as a stand-alone instrument in medical and biological applications for which only limited information from a small number of compounds is required. Ion mobility separation can be used to preseparate or filter interfering compounds prior to MS analysis or serve as a means of distinguishing between isomers based on structural differences. As mentioned, mobility data can also be used to determine the stereoscopic conformation of macromolecules and can thus serve as a means for assessing their biologic activity.

18.2 MEDICAL DIAGNOSTICS USING IMS

Since ancient times, odors or vapors from breath or urine have been understood to reflect illnesses in humans. For example, a sweet fruity odor could be associated with diabetes or starvation or an ammoniacal odor could indicate uremia.[10] The use of advanced analytical instruments, including gas chromatograph/mass spectrometers (GC-MSs), has been explored since the 1960s to replace human senses with refined chemical measurements.[10] Extensions of this concept to GC–MS or to mobility spectrometers alone have been made in the past decade and are not restricted to only physiological vapors but also include vapors from molds, bacteria, and certain medically useful gases, such as anesthetics.

18.2.1 Breath Analysis

The potential for using IMS to detect, identify, and monitor volatile compounds such as halothane, enflurane, and isoflurane, which are used as anesthetic gases in operating theaters, was investigated in the late 1980s.[11,12] The presence and concentration of these compounds following inhalation of a small dosage was monitored in the respired air exhaled directly into a handheld IMS analyzer. In the negative-ion mode, the dominant ions were the chloride adduct ion and, with increases in anesthetic gas concentrations, the chloride-bound dimer. In general, residual low levels of these anesthetic gases were detectable even more than 1 h after the end of treatment. Real-time monitoring of breath for these substances could make possible an objective measure of the depth of anesthesia to complement existing clinical observations. In addition, minutes after the onset of treatment, vapors sampled from the skin surface provided evidence of an exposure event, as shown in Figure 18.1.[11]

Diagnosis of pathological conditions through breath analysis by IMS or GC–IMS is rapidly growing, mainly through the efforts of research groups in Germany (Vautz and Baumbach) and the United Kingdom (Thomas and Creaser). In principle, volatile chemicals inhaled by a subject could be observed in respired air or from skin

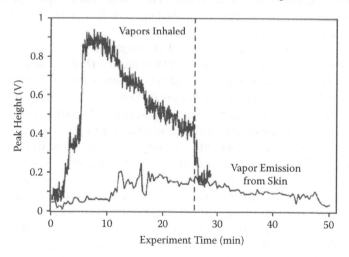

FIGURE 18.1 Plots of peak intensity for the product ion of isoflurane in IMS determination of vapors containing isoflurane and inhaled by a rabbit and vapor concentrations of enflurane obtained from gases respired from skin. The source of isoflurane was removed from the inhalation chamber at a time marked by the dashed line. Both the inhaled vapors and the emissions from the skin were monitored using military-grade chemical agent monitors in negative ion polarity. (From Eiceman et al., Ion mobility spectrometry of halothane, enflurane and isoflurane anesthetics in air and respired gases, *Anal. Chem.* 1989, 61, 1093–1099. With permission.)

emissions and could provide a noninvasive assessment of the extent of exposure. Lung diseases may be recognized through changes in the chemical composition of volatile organic compounds (VOCs) in the air exhaled by an ill subject. Preliminary studies indicated that the mobility spectra from GC-IMS analysis, especially with multicapillary columns (MCCs), of the breath of those with lung damage of the type that may be incurred by cancer differs from that of healthy subjects.[13-18] The underlying assumption was that the respired VOCs reflected their concentration in blood through exchange of gases at the lung interface.[13] In that study, 11 substances (mainly ketones, alkanes, and diones) with room temperature reduced mobility values of 1.08 to 1.97 cm^2 V^{-1} s^{-1} served as markers of illness. A photodischarge lamp (10.6 eV) afforded a degree of selectivity, although some kind of preseparation technique is preferred to resolve mutual interferences between the components. In another study by the same group, a high-speed capillary column was coupled to an IMS drift tube with an ultraviolet (UV) ionization source to determine VOCs that may arise through diseases or certain occupational exposures. These compounds included some common ketones, benzene, and some substituted benzenes.[14] The importance of eliminating interferences by background chemicals that may be present in the room air where biomarkers exhaled in breath are measured was also investigated, with and without human personnel.[16]

Another investigation examined the presence of acetone in breath using a membrane extraction module, a sorbent trap, and a GC with dual detectors: a flame ionization detector and a mobility spectrometer. The last quarter liter portion of the stream of exhaled breath, which better reflects the content of volatile compounds in the lung tissue, was analyzed. The membrane removed much of the respired moisture, blocking interference with the analyses.[17]

In recent years, breath analysis focused on diagnosis of specific diseases, like chronic obstructive pulmonary disease (COPD), with GC–DMS (differential mobility spectrometry) using an absorbent trap to collect the sample.[18] COPD and lung cancer were diagnosed by MCC–IMS in exhaled breath in a large-scale study that included 132 persons with COPD, persons with lung cancer, and healthy volunteers.[19] Principal component analysis was used to classify the spectra, with a positive predictive value of 95%, and cyclohexane was identified as playing a major role in differentiating between the groups.[19] Lung cancer and airway infections were also investigated with MCC–IMS, and volatile metabolites were identified.[18,20,21] The feasibility of diagnosing sarcoidosis by MCC–IMS was examined, and the group with the disease could be differentiated from the group with mediastinal lymph node enlargement.[22-24] Another disease that was tested by breath analysis as a means for diagnosis was *Aspergillus* fumigates.[25] Several analytical techniques were studied, including IMS, to validate that 2-pentylfuran could be considered as a marker for the disease.[25] The level of propofol was monitored in exhaled breath of 13 patients undergoing anesthesia using MCC–IMS, and the results compared with GC–MS analysis of the serum; a bias of −10.5% was found.[26]

In a way, objective scientifically based analytical techniques for analysis of organic compounds in breath or respired air are consistent with, and supported by, centuries of experience in traditional medicine. Examples of diagnostic vapors include acetone from diabetes, isoprene from lung damage, and styrene, 2-methylheptane, propylbenzene, decane, and undecane from lung cancer.[27]

The growing interest in breath analysis by IMS, DMS, or GC is reflected in the increasing number of investigators and reports mentioned previously and others that focused on improvements in instrumentation,[28] analytical methodology,[29–34] or signal processing.[35,36] Hence, given the sensitivity and specificity of IMS analyzers to such compounds, commercial development of VOCs for clinical diagnosis of illness through analysis of exhaled air is indeed occurring. However, this is dependent on a suitable instrument configuration, probably a fast GC-IMS analyzer.

18.2.2 Diagnosis of Vaginal Infections

The degradation of proteins, peptides, and amino acids leads to the formation of several low molecular weight compounds, including biogenic amines. Biogenic amines are a family of compounds that includes monoamines, diamines, triamines, and tetramines, such as trimethylamine (TMA), putrescine and cadaverine, spermidine and spermine, respectively. These and several other compounds, such as histamine, tyrosine, skatole, and other amines, are particularly suitable for IMS determinations due to the high proton affinities of amines. The vapors may arise through several degradation pathways, including chemical reactions, enzymatic reactions, and microbial processes. For example, decarboxylation of lysine produces cadaverine, whereas loss of CO_2 from histidine produces histamine, as shown in Equations 18.1 and 18.2:

$$lysine \rightarrow cadaverine + CO_2 \tag{18.1}$$

$$histidine \rightarrow histamine + CO_2 \tag{18.2}$$

An example of these amines being indicative of illness is bacterial vaginosis (BV), which is the most common form of vaginal infections that afflict females of almost all ages, races, and societies.[37] The fishy odor, which is a typical symptom of BV, arises from the presence of TMA.[38] BV may occur when the delicate balance between lactobacilli (microorganisms that excrete lactic acid and peroxides and maintain a low pH, from 3.8 to 4.2) and pathogenic microorganisms is disturbed in the vaginal discharge fluid. This may be due to several external and internal effects, such as use of medication (antibiotics, in particular, reduce the lactobacilli population); habits of personal hygiene (rinsing too frequently or not enough); allergy; and more. One of the manifestations of BV is the enhanced production of biogenic amines, TMA and putrescine particularly. This has been extensively investigated in Israel and the United States by Q-Scent (now 3QBD).[39–41]

In practice, a swab of vaginal discharge fluid is collected by the gynecologist during a routine visit or after complaints by the patient and placed in a vial and capped. The addition of an alkaline solution to the sample enhances the volatilization of amines even at room temperature, and headspace vapors containing volatile amines, if present, may be transferred to an IMS analyzer for quantitative determination. After about 10 s, the swab can be heated rapidly, and in some cases also treated with a few drops of a dilute acid solution, and headspace vapors can be sampled for other

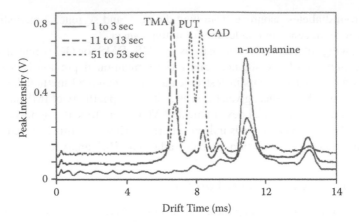

FIGURE 18.2 Mobility spectra measured at delays after introduction of sample. Each spectrum is the average of three scans in a sample collected from a patient with bacterial vaginosis. Note the changes in the background spectrum (solid) as volatile trimethylamine reached the drift tube (dashed) and later as the semivolatile amines became dominant (dotted line). The reagent gas was n-nonylamine, and the amines were trimethylamine (TMA), putrescine (PUT), and cadaverine (CAD). (From Amsel et al., Nonspecific vaginitis: Diagnostic criteria and microbial and epidemiologic associations, *Am. J. Med.* 1983, 74(1): 14–22. With permission.)

volatile amines. To minimize interferences, *n*-nonylamine (or other compounds with high proton affinity) may be used as the dopant gas.

In a sample from a BV-infected patient, TMA will appear initially, and putrescine will be observed later when the sample is heated. During the measurement, the dynamic trends for a sample collected from a patient with BV will show first an increase in the TMA peak area and concomitant decrease in the reactant ion peak area; this is followed by a decrease in the TMA peak area and increase in the putrescine peak area (Figure 18.2). Samples from a healthy subject will exhibit only the reactant ion peak without peaks for biogenic amines. Other common vaginal infections, such as candidiasis (yeast infection) and trichomoniasis, are also detectable as elevated levels of other biogenic amines, cadaverine, and putrescine, with little or no TMA. This analysis is fast (1 min per test), sensitive, specific (overall above 95% accuracy with low false negative and false positive), simple in methodology, and inexpensive compared to other screening methods for BV. Thus, an IMS-based test could replace the conventional diagnostic procedure (i.e., the Amsel test[42]) and diagnostic methods for other infections, as suggested by the favorable classifications shown in Figure 18.3. This method is soon to be commercially available and could become one of the first successes in clinical applications of stand-alone mobility spectrometers.

18.3   FOOD FRESHNESS, MOLDS, AND ODOR DETECTION

18.3.1   Muscle Food Freshness

All foods that contain muscle tissue (meat, poultry, and fish) can form biogenic amines through enzymatic and bacterial action on proteins and amino acids, as described

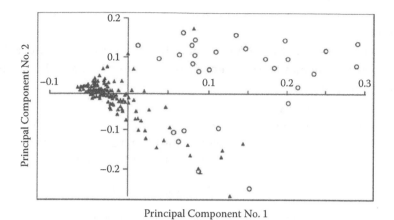

FIGURE 18.3 Summary of the measurements, based on principal component analysis, of biogenic amines in vaginal discharge fluid. The circles and triangles denote samples that were diagnosed as BV positive and BV negative, respectively, in the Amsel test. (From Amsel et al., Nonspecific vaginitis: Diagnostic criteria and microbial and epidemiologic associations, *Am. J. Med.* 1983, 74(1): 14–22. With permission.)

previously for vaginal infections.[43] For example, TMA is formed by a stepwise degradation of choline to betaine and then to TMA, according to Equation 18.3:

$$Choline \rightarrow Betaine + TMA \qquad (18.3)$$

The unpleasant odor of decaying fish (the same as in BV) has been attributed mainly to TMA,[38] although other compounds may also contribute to this odor. A handheld GC-IMS analyzer was used to determine the odors released from aging fish, and TMA was observed in the odors.[44] However, this was not pursued further until a systematic investigation of the spoilage of meat, poultry, and fish was undertaken.[45] The content of biogenic amines was determined as a function of the storage temperature and time in different types of muscle food. Amine concentrations were correlated to the levels of six types of microorganisms counted by standard culture growth techniques (Table 18.1). To enhance the sensitivity toward semivolatile compounds, the sample introduction method described previously for diagnosis of vaginal infections was used. In Figure 18.4, mobility spectra are shown for different types of muscle food after a few days at room temperature. The amount of biogenic amines formed depends on the type of meat, the storage period, and the temperature of storage. Storage of food in a deep freeze (–18°C) slowed the degradation processes and practically arrested decomposition for periods of several months. In meat at room temperature, biogenic amines were produced rapidly, and large concentrations of these amines were evident within 24 h. The results showed that IMS detectors can be used as a diagnostic tool for food freshness or for the determination of food spoilage. Another IMS study, using multivariate modeling as a freshness index, was also carried out.[46]

Another approach to screening bacteria is based on extracellular enzymes that are secreted by bacteria and found on the outside surface of the cell wall. When these

TABLE 18.1
IMS Results for TMA and Total Amines and the Microorganism Culture Count for Samples of Ground Chicken When Fresh and after 1, 2, and 4 Days at Room Temperature

Sample	TMA (ng/g)	Total Amines (arbitrary units)	1	2	3	4	5	6
Fresh	1.1 ± 0.3	1.2 ± 0.1	3.8E3	8.4E3	2.3E2	2.3E2	4.5E2	4.5E2
1 days	3.7 ± 1.0	2.1 ± 0.5	7.4E6	1.0E7	1.3E5	8.6E3	1.5E5	1.9E5
2 days	32.8 ± 8	11.3 ± 3	3.9E8	4.7E8	6.6E5	4.2E6	4.7E6	6.3E6
4 days	41.0 ± 8	$14\,3 \pm 4$	1.4E9	1.9E9	5.9E4	1.0E7	7.8E6	7.5E6

Source: From Karpas et al., Determination of volatile biogenic amines in muscle food products by ion mobility spectrometry, *Anal. Chim. Acta* 2002, 463(2): 155–163. With permission.

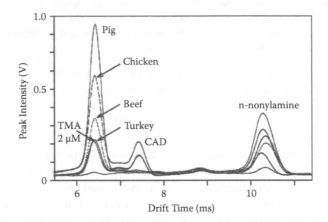

FIGURE 18.4 Mobility spectra showing the formation of volatile amines from the spoilage of pork, turkey, beef, and chicken during storage at room temperature for 1 day. Calibration was with 2 ng of TMA (trimethylamine). TMA and cadaverine (CAD) are apparent in the mobility spectrum for each muscle food. The reagent gas was n-nonylamine. (From Karpas et al., Determination of volatile biogenic amines in muscle food products by ion mobility spectrometry, *Anal. Chim. Acta* 2002, 463(2): 155–163. With permission.)

enzymes are allowed to react with a suitable substrate, a volatile substance is produced and is able to be detected using a mobility spectrometer. This approach was tested and adapted with a sampling system to automatically screen food for bacteria associated with food safety.[47,48] Although this method is effective, fast, and sensitive, the level of extracellular enzymes is governed in part by the health and recent history of the bacteria. A related method with improved reliability and specificity is based on immunoassay methods as discussed in Section 18.5.

Direct detection by IMS of TMA in chicken meat juice with excellent detection limits of 0.6 ± 0.2 ng using partial least squares (PLS) and fuzzy rule-building expert system (FuRES) was also reported.[49] A combination of thermal desorption with GC separation and differential mobility spectrometry detection (TD-GC-TDS) has been used to determine the level of putrescine and cadaverine in chicken meat by sampling headspace vapor.[50]

The clearance of TMA from saliva after consumption of tuna chunks was studied by IMS with triethylphosphate (TEP) used as the dopant.[51] Two types of canned tuna were studied: tuna in oil and tuna in water. As biogenic amines, including TMA, are hydrophilic, the level of TMA in the water of tuna canned in water was high (about 0.5 mg mL^{-1}) due to extraction from the tuna meat, while for tuna canned in oil the TMA was retained by the tuna chunks. Subsequently, the level in saliva after consumption of tuna chunks canned in oil was significantly higher than after consuming tuna meat that was conserved in water. However, in both cases after about 20 min the level of TMA in saliva returned to the level before consuming the tuna.[51]

18.3.2 Molds and Mycotoxins

Mycotoxins are toxic metabolites produced and released by molds (fungi or yeasts), which usually thrive in an enclosed atmosphere, particularly in dark, moist locations. Benign molds can be present on food products and add to their flavor, as is the case of some types of cheese, but these organisms or the chemicals released by their metabolism (mycotoxins) may cause headaches, allergic reactions, irritation of the respiratory tract and skin, as well as other health problems, principally among asthmatic or sensitive subjects.[52,53] Some of the chemicals that have been considered as metabolic by-products of molds and fungi, mainly ketones and alcohols, have been characterized by IMS devices with a photodischarge UV lamp and with a ^{63}Ni ion source.[52] Minimum detection levels with the ^{63}Ni ion source were about an order of magnitude better than those with the UV discharge lamp. It was demonstrated that an IMS analyzer was capable of monitoring the level of these chemicals in indoor air at subclinical concentrations. The chemicals emitted from bread mold cultures were directly measured with a ^{63}Ni-based IMS analyzer and then with a GC-UV-IMS instrument after solid phase microextraction (SPME). The direct measurements of the mixture by a stand-alone IMS analyzer were difficult to interpret as several chemicals were present in the sample, and the mobility spectrum was further complicated by the extensive fragmentation of the alcohol ions. The use of SPME for sampling headspace vapors and the thermal desorption for chromatographic prefractionation with an IMS detector provided improved measurements.[53]

Aflatoxins are a type of mycotoxins associated with plants, such as peanuts and corn, and they cause undesirable health effects and in some cases may be carcinogenic. Methanol extracts of pistachio nuts containing spikes of B1 and B2 aflatoxins were measured by corona discharge IMS, and with the use of ammonia as a dopant detection limits of 0.1 ng were achieved.[54] In a subsequent work by the same group, ochratoxin A on licorice root was determined by inverse IMS, and under optimal conditions, limits of detection (LODs) of 0.01 ng were reported.[55] The mycotoxin zearalenone and its metabolites in cornmeal were analyzed using high-field asymmetric waveform IMS–MS, and LODs of 0.4–3 ng mL^{-1} were reported.[56] Gaseous metabolites that serve as indicators for the presence of molds were investigated by IMS, and their identity was confirmed by GC–MS.[57] Characteristic spectra of headspace vapor were recorded for different microbial species under different conditions.[57] In another study, an aspirator IMS and semiconductor sensors were used to detect and monitor VOCs emitted by microbial specimens, and the results were compared with sterile samples.[58]

18.3.3 Other Applications of IMS in the Food Industry

The applications of IMS technology in food quality and safety[8] and in bioprocesses were reviewed[7] recently. The fermentation process of beer was continuously monitored with a GC–IMS to determine the concentrations of diacetyl and 2,3-pentanedione and stop the fermentation process before their level increased above the odor threshold.[59] This technique can replace the traditional off-line analysis that is carried out in a laboratory once a day and ensure the quality of the brew. Mobility spectra of headspace volatile compounds emitted from heated samples of Iberian pig fat were used to differentiate between free-range pigs and confined pigs and to authenticate the feeding regime.[60] With the use of chemometrics, only 2.3% of the 65 samples were misclassified. IM–MS was used to detect traces of chemical warfare agents and their hydrolysis products in spiked food and beverage samples; detection limits were in the part-per-billion range (nanograms/gram in food or nanograms/milliliter in beverages).[61] Residues of three veterinary drugs (furazolidone, chloramphenicol, and enrofloxacin) in chicken meat were measured by corona discharge IMS after solid phase extraction (SPE), and detection limits of nanograms per gram were established with the use of spiked samples for calibration.[62] TMA was quantitatively measured by IMS in samples of chicken meat juice as a means of assessing the degradation of the product.[63] Two chemometric approaches, PLS and FuRES, were used to classify the extent of spoilage.

18.4 MACROMOLECULES: BIOMOLECULES AND BIOPOLYMERS

Until 2005, the separation, detection, or identification of proteins, peptides, and amino acids using IMS had been explored only by a handful of research groups in the United States, namely, those of Hill, Bowers, Russell, Jarrold, Clemmer, and Guvremont (in Canada). Common to all these groups is the method of sample handling; samples were introduced into a drift tube as liquids via ESI or as solids with MALDI. In most cases, a MS was used for detecting and identifying the ions, and a mobility spectrometer was used to preseparate components and to obtain reduced mobility coefficients (see Chapter 9 for details on IM–MS instruments). There is increasing interest in using ion-mobility-based methods to investigate complex biological systems; this can be observed in Figure 18.5, which shows (based on SciFinder search) the number of publications involving proteins and IMS in the years 2001–2010.

These studies involved mainly the estimation of the conformation of the gas phase ions formed by these macromolecules and were derived from molecular models and the collision cross sections obtained from K_o values. One facet of forming gas phase ions from large molecules with several functional groups is that multiply charged ions are commonly observed at ambient pressure with ESI. Thus, mobility spectra for a single compound from substances such as proteins and oligonucleotides (and synthetic polymers) will exhibit peaks for these various charge states. Even a

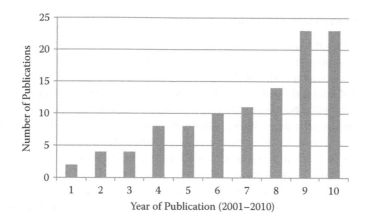

FIGURE 18.5 The number of publications involving proteins and ion mobility spectrometry in the years 2001–2010 (based on SciFinder search).

simple mixture of biological molecules will be very complex with poorly resolved or unresolved peaks. Mass spectra of such mixtures will also be too complex for interpretation. Examples of the instrumentation used in these research projects include high-field asymmetric waveform IM–MS[64–66] and an IMS with time-of-flight (TOF) MS for multidimensional separation of complex mixtures.[67–73] Among the compounds studied are those that are formed in a mixture from tryptic digestion of peptides[74] and peptide libraries,[67,72,75–79] bradykinin (BK),[65,80] polyglycine and polyalanine,[81] carbohydrates,[69] ubiquitin,[82] chemically modified DNA oligonucleotides with up to eight bases in length,[71] proteins,[81] valinomycin,[83] and others.[84]

The use of mobility spectrometers for studies of macromolecules can be categorized into three groups: single-compound structural studies, the use of an IM–MS instrument for characterizing ion mixtures, and the use of a mobility spectrometer as an ion filter before MS. In some of these, the mobility spectrometers are operated at reduced pressures of less than 100 torr (usually 3 torr) with a drift gas of helium or argon. Although these are not strictly analytical applications of IMS, the use of a mobility measurement in combination with other techniques illustrates an instance when similar information is difficult to obtain by other means. Mobility measurements used to obtain structural information on large biomolecules rely on drift tubes that operate at reduced pressure, usually with helium as the drift gas. Although these are usually not considered "classic" analytical IMS instruments, the principles of ion mobility are central to the studies and represent a type of refined mobility experiment.

A few review papers discussing the use of ion mobility and IM–MS measurements for studying macromolecules have been recently published.[3,5,85,86] Combining MALDI with IM–MS was discussed in view of its applications in solvent-free structural studies, sequencing, and protein identification.[3] The mobility measurements provided a new dimension in the analysis of biomolecules due to the fact that separation is based on the size and shape of the ion and not only on its mass.[85] The role of multidimensional assemblies and aggregations in normal cellular processes and diseases, based on ESI-IM-MS experiments, was analyzed.[86]

18.4.1 Conformation Studies

As an ion travels through the drift gas, there are several types of interactions between the ion and the neutral molecules of the drift gas (see Chapter 10). These interactions depend on the characteristics of the ion (size, total charge, charge distribution within the ion, and shape) and the drift gas molecule (size, dipole and quadrupole moments, and polarizability). The magnitude of these interactions determines the drift velocity of the ion, namely, the mobility. This is the basis for elucidation of ion structure or conformation from mobility measurements and has been applied to elucidate the structure of small ions like anilines[87] and diamines[88] as well as that of large ions like protonated polyglycine and polyalanine.[81]

The relationship between protein crystal structure and their collision cross sections determined from gas phase ion mobility measurements has been discussed in a few recent publications.[89,90] According to one publication, the gas phase conformation of a protein can be related to the structure in liquid phase in many cases.[89] Cross sections were estimated based on mobility measurements in a traveling wave ion mobility spectrometer (TW-IMS) and compared with values obtained through x-ray crystallography and nuclear magnetic resonance (NMR) spectroscopy, and good correlations were found. The gas phase mobility measurements also yield information on the relative stability of the different charge states, as the mobility of the protein, and therefore its cross section, changes with its charge.[89]

The work of Clemmer's group demonstrates the unique advantage of using IMS to examine the influence of solvent composition and capillary temperature on the gas phase conformations of ubiquitin ions (+6 to +13) formed during ESI.[82] Three general conformer types were observed: compact folded forms (favored for the +6 and +7 charge states); partially folded conformers (favored for the +8 and +9 ions); and unfolded conformers (favored for the +10 to +13 charge states). The population distribution of different conformers was highly sensitive to solvent composition and the capillary temperature used for ESI. The differences in mobility, and therefore ion cross section, were associated with the number of charges on the ion; increases in charge state caused increases in coulomb repulsion and unfolding of the ion. Naturally, folded compact conformers will exhibit shorter drift times than the unfolded conformers with large cross sections of collision. Cross sections of ions in the gas phase are difficult to obtain by any other measurement techniques besides IMS.

With the advent TW-IMSs, more measurement of cross sections of macromolecules to obtain structural information were carried out.[89,91–94] The system was calibrated by using relatively small molecules and ions with known cross sections, like oligoglycine peptides, and the results were compared with theoretical calculations and verified by repeating the measurements in two drift gases (helium and nitrogen).[91] Thus, leucine and isoleucine were separated in the mobility spectrometer, and other structural isomers that are not distinguishable by MS could be separated.[91] A large data set of collision cross sections of proteins was acquired with a quadrupole-IMS-TOF-MS hybrid MS.[92] The cross-section measurements of denatured peptides and proteins and native-like proteins and peptides were carried out in helium and nitrogen, and for large macromolecules, they were well correlated, but for smaller ions significant differences were found.[92]

Unfolding of protein ions in the gas phase was induced by collision activation with argon, and the cross section of unfolded conformation was measured and compared with that of the more compact and stable folded conformation.[93] In another publication, TW-IMS-MS was used to measure cross sections to show that a protein retained its solution conformation of both recombinant and solvent-disrupted versions.[94] TW-IMS-MS was also used to study the speciation of different ions bound to bovine carbonic anhydrase, and the conclusion was that ions that are strongly bound to the enzyme (Zn^{+2} and Cu^{+2}) modify its structure more than weakly bound ions.[95]

18.4.2 Alkali Ions of Biomolecules

In addition to protonated species, other ions, such as Na^+ and Li^+, may be attached to macromolecules and can be readily observed in ESI/IMS/MS experiments. Hill's group derived structural information about the most probable location of the charge in protonated and sodiated BK and kemptide (where sodium replaced a proton in singly and doubly charged peptides). The structures were derived from the cross sections determined with a mobility spectrometer.[78] A mobility difference between protonated and sodiated species was observed, and it appeared as if the doubly charged sodiated peptides had a smaller collision cross section than the doubly charged protonated ones, leading to the conclusion that the gas phase conformations of these ions are different with respect to intramolecular interactions. In Figure 18.6, the mobility spectra of mixtures of isomeric peptides are shown. The top frame is for pentapeptides with inverse amino acid sequences, and the bottom frame shows hexapeptides differing by N-terminal amino acid and the fourth amino acid.

Protonated and sodiated ions may also be formed in MALDI experiments, as shown by Bowers's group.[80] They noted that several cationized species were generated in the gas phase in BK, including the protonated form (BKH^+) and the sodiated forms ($BKNa^+$) and ($BK-H+2Na$)$^+$. All three species had similar cross sections of 245 ± 3 Å^2, independent of temperature from 300 to 600 K. They concluded that BK is wrapped around the charge centers in a globular shape where the dynamics of structural variation exhibit little change in time-averaged sizes up to 600 K. In another publication, the collision cross sections of valinomycin–alkali ion complexes (Li, Na, K, Rb, and Cs) were derived from the measured mobility of the ions in 3 torr of helium in a short drift tube with a MALDI source.[83] The systematic increase in cross section with ion size indicated that the backbone folding of the cyclic valinomycin molecule was dependent on the size of the alkali ion.

18.4.3 Further Studies of Macromolecules

Cook's group noticed in 1994 that positively charged horse heart apomyoglobin (molecular weight 16,951 Da) appeared in two forms in an ESI/tandem MS instrument: a high-charge state with distribution centered around $(M + 20H)^{20+}$ and a second distribution centered around $(M + 10H)^{10+}$, which were predominant at low

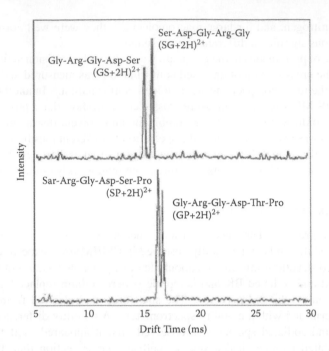

FIGURE 18.6 Ion mobility spectra of mixtures of isomeric peptides. The top frame is for pentapeptides with inverse amino acid sequences, and the bottom frame shows hexapeptides differing by N-terminal amino acid and the fourth amino acid. The spectra were obtained with nitrogen as a drift gas at 250°C and ambient pressure. The mass spectrometer was operated in the single-ion-monitoring mode at mass-to-charge ratios of 246 (top) and 302 (bottom), showing only the doubly charged gas phase ions of the peptides. (From Wu et al., Separation of isomeric peptides using electrospray ionization/high-resolution ion mobility spectrometry, *Anal. Chem.* 2000, 72, 391–395. With permission.)

and high target gas pressures, respectively.[84] These two distinct charge-state distributions were interpreted as an open conformational form for the high-charge state and a folded form for the low-charge state. Preferential charge selection was dependent on the nature and pressure of the target gas as well as the nature of the protein, and conformers could be selected by control of the collision gas pressure, favoring one form over the other. It was noted that bimodal distributions were observed at intermediate pressures, but that charge states between the two distributions were not effectively populated under most of the conditions examined. Hard-sphere collision calculations showed large differences in collision frequencies and in the corresponding kinetic energy losses of the two conformational states and demonstrated that the observed charge-state selectivity could be explained through elastic collisions.[84]

Several studies were concerned with the structural characterization of peptides using different combinations of mobility spectrometry and MS.[67,72,74–77,96–99] A novel approach is deployment of ion-mobility-based techniques to image proteins in tissues[100] and mapping proteomes.[101] DMS, combined with MS, is also gaining popularity in studies of macromolecules.[56,61,64–66,102,103]

18.5 DETECTION AND DETERMINATION OF BACTERIA

18.5.1 PYROLYSIS GC–IMS METHODS

The widespread use and acceptance of mobility spectrometers in the armed forces created interest in exploring the possibility of using handheld chemical agent monitors (CAMs) to detect bacteria with extracellular enzymes[104] and later by the immunoassay methods[105] described in the following text. These methods were not suitable for continuous monitoring of air without dispensable reagents, so alternate approaches were sought using methods of pyrolysis that effectively convert biological substances into chemical information. Nonetheless, the concept of retaining a common analytical method for detection of both chemical and biological agents was appealing, and a method of pyrolysis coupled to a GC–IMS has been pioneered by the U.S. Army team of A. P. Snyder and colleagues. The technology was a derivative of the CAM drift tube as described in Chapter 6[106] and was modified further with a pyrolysis inlet[107] that could accept a sample of air containing bacteria. To detect bacterial aerosols in ambient air, a large volume of air was passed through a glass tube. After the collection step, the sample was rapidly heated to 300°C (i.e., pyrolyzed). The pyrolysis products were characterized using a high-speed GC column with IMS detector (Py-GC–IMS).[108–112] The findings were supported by experiments made in parallel using a GC–MS instrument.[113] The presence of compounds that arise from the pyrolysis of spores and bacteria was determined at specific GC retention times and drift times. Certain compounds are characteristic of *Bacillus* spores and are regarded as biomarkers, that is, chemicals that are always present in an organism regardless of history or life stage of the bacteria. For example, Gram-positive spores such as *Bacillus subtilis* var. *globigii* (BG) spores contain 5–18% by weight of calcium dipicolinate that can be pyrolyzed to dipicolinic acid (DPA), picolinic acid (PA), and pyridine. PA has a high proton affinity, and it is detected in a sensitive fashion and identified by a conventional IMS analyzer. PA occupies a unique region in the GC–IMS data domain with respect to other bacterial pyrolysis products. A 1,000-to-1, air-to-air aerosol concentrator was interfaced to the Py-GC–IMS instrument to test controlled aerial releases of the spores. In the 21 BG trials, the Py-GC–IMS instrument experienced two true negatives and no false positives and developed a software failure in one trial. The remaining 18 trials gave true positive results for the presence of BG aerosol in ambient air after a biorelease. The LOD for the Py-GC–IMS instrument was estimated at approximately 3,300 BG spore-containing particles.[108]

In a later publication, the application of the Py-GC–IMS method was shown to be able to discriminate between aerosols of a Gram-positive spore (BG), a Gram-negative bacterium (*Erwinia herbicola*, EH), and a protein (ovalbumin), shown in Figure 18.7.[109] The kinetics of thermal decomposition of biological substances (*Bacillus* Gram-positive spores) were investigated by thermogravimetric analysis (TGA) and differential thermogravimetry (DTG), and the emitted gases were monitored by Py-GC–IMS for comparison of decomposition profiles.[110] The biomarkers released from a pyrolized sample containing *Bacillus anthracis* were characterized by Py-GC–MS, and 2-pyridinecarboxyamide that was traced back to compounds in the cell wall that were identified.[111]

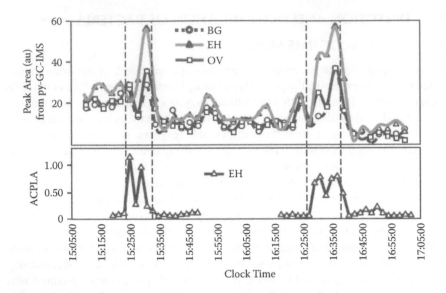

FIGURE 18.7 Results from monitoring by Py-GC–IMS of ambient air following a release of biological aerosols. The upper frame shows instrument response to spores of gram-positive *Bacillus subtilis* var. *globigii* (BG), gram-negative *Erwinia herbicola* (EH), and ovalbumin protein (OV). The dashed vertical lines represent the time boundaries of the aerosol releases. The lower frame shows results from samples collected and analyzed by the agar petri dish bacterial growth in units of ACPLA, agent containing particles per liter of air (viable bacteria component of the aerosol). (From Snyder et al., Field detection and identification of a bio-aerosol suite by pyrolysis-gas chromatography-ion mobility spectrometry, *Field Anal. Chem. Technol.* 2001, 5, 190–204.)

Recent extensions of Py-GC–IMS have been made using pyrolysis with a GC–DMS, employing methods that generally paralleled those of Snyder et al. In this work by Eiceman et al.,[114] the DMS detector provided differential mobility spectra simultaneously in positive and negative polarity. Three-dimensional profiles of ion intensity, compensation voltage, and retention time as shown in Figure 18.8 exhibited distinctive patterns that allowed the categorization of Gram-positive, Gram-negative, and spore forms of bacteria. The attraction of the Py-GC–DMS instrument was the advantage of the microfabricated drift tube, including practical aspects of size, weight, and power, and the fundamental measurement facet of continuous ion analysis. Quantitative studies were made of the method, with direct application of bacteria in solution to a pyrolysis ribbon using a microliter syringe; 6,000 bacteria were detected with an unoptimized inlet, and quantitative precision was about 10% relative standard deviation. In a series of investigations, data sets from Py-GC–DMS were computationally processed to allow classification of bacteria regardless of age and temperature of growth, which had historically frustrated pyrolysis methods for bacterial classification. In another study, three strains belonging to the genus *Bacillus* were investigated with a Py-GC–DMS, and the long-term reproducibility of the instrument was evaluated over a period of 60 days using a Scotch whisky quality control.[115] The data were preprocessed by two approaches: correlation optimized warping (COW) and

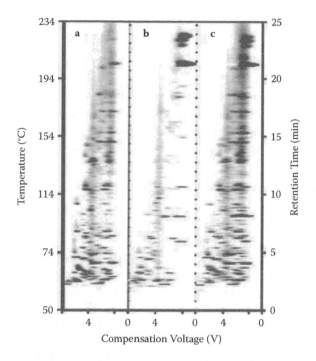

FIGURE 18.8 Topographic plots of compensation voltage versus retention time from the Py-GC-DMS characterization of positive ions for *E. coli* (a), *Micrococcus luteus* (b), and *B. megaterium* (c). The intensity scale ranges from 0.9 V (white) to 2.5 V (black) in equal steps of 0.1 V. (From Schmidt et al., Microfabricated differential mobility spectrometry with pyrolysis gas chromatography for chemical characterization of bacteria, *Anal. Chem. 2004*, 76, 5208–5217. With permission.)

asymmetric least squares (ALS). The separation between *B. subtilis* and *B. megaterium* was readily observed by principle components analysis, but supervised learning was required to separate the two strains of *B. subtilis*.[115]

Another approach to thermal processing of bacteria was demonstrated using a commercial mobility spectrometer; microgram quantities of whole bacterial cells were thermally desorbed in a heated anvil, producing complex patterns of positive and negative ion mobility spectra in a handheld IMS analyzer.[116] The spectra differed reproducibly for different strains and species and for different conditions of growth and can be used for the classification and differentiation of specific strains and species of bacteria, including pathogens. This provided a means to detect specific components of bacterial cells and to identify and classify bacteria within a minute without specialized test kits or reagents. Methods for improved ion peak detection were also described for sequential sample desorption at stepped increases in temperature (programmed temperature ramping).

Thermal desorption with an Itemiser IMS detection and chemometric modeling were deployed to characterize and differentiate between whole-cell bacteria. In situ hydrolysis and methylation were required to differentiate *Escherichia coli* strains that could not be distinguished by their mobility spectra.[117] The metabolic profile of

Escherichia coli was characterized by IM(TOF)MS with a MALDI source and with an ESI source.[118,119]. The preseparation afforded by the mobility spectrometer made it possible to separate metabolites in the mass range below 1,800 Da. ESI-IMS was also used to characterize intact virus particles, and it was found that icosahedral virus particles retain their structure in the gas phase.[120]

The ability of a DMS to detect pyrolized spores of *Bacillus subtilis* (a stimulant for *Bacillus anthracis*) was demonstrated, even when the spores were suspended in water.[121] The spores had to heated to quite high temperatures (above 550°C for at least 10 s) to produce fragment particles considerably below 10 kDa.

18.5.2 Enzyme-Based Immunoassay IMS

IMSs have been employed as detectors for well-established methods for the determination of bacteria and enzyme-linked immunosorbent assays (ELISA). In ELISA methods, primary antibodies attach to epitopes on the bacterial wall. Each antibody has a structure containing numerous epitopes that can be associated with a secondary antibody. The secondary antibody also has a region with enzymatic activity. This enzymatic region is able to react with a substrate to cleave a product that either is colored and can be determined by a spectrophotometer or is volatile and can be determined by headspace analysis. In the method developed by Snyder et al.[105] and quantitatively explored by Smith et al.,[117] the final product exhibited a distinctive negative product ion peak, as observed with a mobility spectrometer. This was accomplished with the widely deployed military-grade CAM, which was used without modification and suggested that it could serve as a potential bacteria analyzer, provided reagent kits and an inlet adaptor were also distributed.

In practice, a sample containing *Bacillus cereus* was placed in a vial where bacteria would adhere to the inner walls of the vial and the sample could be washed to remove impurities. In stages, reagents were added and sample washed to build the primary and secondary antibody structure. In the last step, ortho-nitrophenyl-β-D-galactoside (ONPG) was added to the sample; the ONPG reacted with a β-galactosidase of the secondary antibody. The product of this reaction was orthonitrophenol or o-nitrophenol (ONP), a volatile compound that can be thermally desorbed into an IMS analyzer. Responses were compared to those of the conventional spectrophotometric assay. Both detection techniques produced a sigmoid-shape curve characteristic of immunoassay experiments. The bacterial detection limit with the IMS technique was estimated at below 1,000 cells for an 8-min assay time. The only advantage of the ELISA-coupled IMS method is that the mobility spectrometer is capable of better detection limits than an optical spectroscopic method. This will have the benefit of reduced assay times or improved detection limits at fixed reaction times. The method has not been accepted by ELISA suppliers, perhaps because the existing ELISA methods are seen as sufficient to meet existing needs.

18.6   OTHER BIOLOGICAL APPLICATIONS

The concept of using chemical instrumentation to replace olfactometry expert panels, employed in the perfume and flavors industries, has long been sought, and simple, inexpensive sensor arrays have recently been promoted as "artificial noses" or "electronic noses" (e-noses). However, these devices lack specificity in response because the signal is based on principles of solubility of vapors in polymers. IMSs have been used to detect odors (pleasant like perfume or unpleasant like decaying meat) since the advent of the technique in the late 1960s. Perfumes, flowers, and sagebrush species have been differentiated using mobility spectra combined with principal component analysis (PCA) for interpretation of the spectra.[122] In another test, the vapors emanating from aluminum films that were coated with white and orange color prints were analyzed using a GC–IMS analyzer equipped with a photo-discharge lamp as the ion source.[123] Distinct patterns were observed for each coating. Comparison of the mobility spectra of coated and uncoated transparent films (one burdened with malodor) used in food wrapping showed differences in the response, and the malodorous film was readily detected. The use of GC for prese-paration was important as the mobility constants of many of the compounds studied had similar values. Once again, the potential of IMS methods and particularly GC–IMS instruments for such applications is only beginning to emerge.

Efforts to assist search-and-rescue teams in locating people buried under collapsed structures by using handheld IMS instruments have been proposed.[124,125] One study focused on detection of VOCs emitted in expired air of fasting monks in Greece, with the objective to simulate the breath exhaled by people who could be trapped under collapsed buildings in an event such as an earthquake; indeed, a 30-fold increase in acetone level was found using a portable GC–IMS.[124] In another study, the objective was to detect vapors emanating from human urine, mainly acetone, to indicate the presence of people under debris.[125]

Animal studies also deploy IMSs for various purposes.[126–128] A feasibility study was carried out to demonstrate the potential of IMS for animal breath analysis, even for mice.[126] In another IMS analysis of the breath exhaled by 20 Sprague–Dawley rats, a difference in the spectral pattern was observed between the healthy rats and those in which sepsis was induced, and the results were confirmed by GC–MS analysis.[127] In addition to diagnosing vaginitis in women by measuring biogenic amines (Section 18.2.2), a similar approach was quite successfully adopted for diagnosing vaginal infections in domestic animals like sows and cows.[128] Due to their potential medical applications, detection, identification, and monitoring of biogenic amines has received some attention recently.[129–131] The reduced mobility values of biogenic amines were measured by an IMS drift tube with a corona discharge source with n-nonylamine serving as a calibrant for the mobility scale, but due to erroneous peak assignment, these were overestimated.[129] The reduced mobility values of TMA, putrescine, cadaverine, spermidine, and spermine published in the literature were remeasured, reevaluated, and reported.[130] The effect of humidity on measurements of amines and its effect on sensitivity were studied.[131] The reduced mobility values of several amino acids and other small biomolecules were determined with IM(TOF) MS with a MALDI source.[132]

Finally, studies of alcohols present in human saliva using selective membrane extraction and a GC–DMS detector[133] and a method for rapid analysis of hair based on ESI–IM–MS were published.[134]

18.7   CONCLUSION

Although mobility spectrometers are often defined as vapor analyzers, the record of the past decade demonstrates that macromolecules of high molecular mass and low volatility can also be characterized for gas phase mobility. Such measurements do not simply replace a MS but provide details about the molecule, now an ion, in ways not possible using any other methods or principles of measurement. Thus, a mobility spectrometer can uniquely offer insights into the structure and properties of the macromolecule. These methods are currently research tools only. In contrast, the possibility of using a mobility spectrometer, specifically a field-dependent mobility analyzer, as a filter before a MS to reduce chemical noise in MS determinations of macromolecules with an electrospray ion source is being offered commercially as an inlet for a MS. Whether this method or any of the biological initiatives with mobility spectrometry will endure or grow is wholly speculative. Practical analytical methods such as pyrolysis GC–IMS or GC–DMS and ELISA-coupled IMS have been demonstrated and await further developments by manufacturers or those needing such tools. The clinical uses of IMS analyzers for diagnosing bacterial infections such as BV may become a large market for use, unlike any previous use of IMS technology in medicine. This remains to be developed commercially and will succeed or fail, probably in the next decade.

REFERENCES

1. Fenn, J.B., Electrospray wings for molecular elephants (Nobel lecture), *Angew. Chem.* 2003, 42, 3871–3894.
2. Tanaka, K., The origin of macromolecule ionization by laser irradiation (Nobel lecture), *Angew. Chem.* 2003, 42, 3861–3870.
3. McLean, J.A.; Ruotolo, B.T.; Gillig, K.J.; Russell, D.H., Ion mobility-mass spectrometry: a new paradigm for proteomics, *Int. J. Mass Spectrom.* 2005, 240, 301–318.
4. Bohrer, B.C.; Merenbloom, S. I.; Koeniger, S.L.; Hilderbrand, A.E.; Clemmer, D. E., Biomolecule analysis by ion mobility spectrometry, *Annu. Rev. Anal. Chem.* 2008, 1, 10.1–10.35.
5. Guharay, S.K.; Dwivedi, P.; Hill, H.H., Ion mobility spectrometry: ion source development and applications in physical and biological sciences, *IEEE Trans. Plasma Sci.* 2008, 36(4, Pt. 2), 1458–1470.
6. Uetrecht, C.; Rose, R.J.; van Duijn, E.; Lorenzen, K.; Heck, A.J.R., Ion mobility mass spectrometry of proteins and protein assemblies, *Chem. Soc. Rev.* 2010, 39, 1633–1655.
7. Vautz, W.; Baumbach, J.I., Analysis of bio-processes using ion mobility spectrometry, *Eng. Life Sci.* 2008, 8, 19–25.
8. Vautz, W.; Zimmermann, D.; Hartmann, M.; Baumbach, J.I.; Nolte, J.; Jung, J., Ion mobility spectrometry for food quality and safety, *Food Addit. Contam.* 2006, 23, 1064–1073.
9. Seto, Y.; Kanamori-Kataoka, M.; Tsuge, K.; Ohsawa, I.; Maruko, H.; Sekiguchi, H.; Sano, Y.; Yamashiro, S.; Matsushita, K.; Sekiguchi, H., Development of an on-site detection method for chemical and biological warfare agents, *Toxin Rev.* 2007, 26, 299–312.
10. Manoli, A., The diagnostic potential of breath analysis, *Clin. Chem.* 1983, 29, 5–18.
11. Eiceman, G.A.; Shoff, D.B.; Harden, C.S.; Snyder, A.P.; Martinez, P.M.; Fleischer, M.E; Watkins, M.L., Ion mobility spectrometry of halothane, enflurane and isoflurane anesthetics in air and respired gases, *Anal. Chem.* 1989, 61, 1093–1099.

12. Martinez-Sandoval, P., Atmospheric pressure ion chemistry and physiological respiration of halogenated anesthetic gases, MS thesis, New Mexico State University, Las Cruces, May 1991.

13. Ruzsanyi, V.; Sielemann, S.; Baumbach, J.I., Determination of VOCs in human breath using IMS, *Int. J. Ion Mobil. Spectrom.* 2002, 5, 45–48.

14. Xie, Z.; Sielemann, S.; Schmidt, H.; Li, F.; Baumbach, J.I., Determination of acetone, 2-butanone, diethyl ketone and BTX using HSCC-UV-IMS, *Anal. Bioanal. Chem.* 2002, 372, 606–610.

15. Baumbach, J.I., Ion mobility spectrometry coupled with multi-capillary columns for metabolic profiling of human breath, *J. Breath Res.* 2009, 3, 034001/1–034001/16.

16. Bödeker, B.; Davies, A.N.; Maddula, S.; Baumbach, J.I., Biomarker validation—room air variation during human breath investigations, *Int. J. Ion Mobil. Spectrom.* 2010, 13, 177–184.

17. Lord, H.; Yu, Y.F.; Segal, A.; Pawliszyn, J., Breath analysis and monitoring by membrane extraction with sorbent interface, *Anal. Chem.* 2002, 74, 5650–5657.

18. Basanta, M.; Jarvis, R.M.; Xu, Y.; Blackburn, G.; Tal-Singer, R.; Woodcock, A.; Singh, D.; Goodacre, R.; Thomas, C.L.P.; Fowler, S.J., Non-invasive metabolomic analysis of breath using differential mobility spectrometry in patients with chronic obstructive pulmonary disease and healthy smokers, *Analyst,* 2010, 135, 318–320.

19. Westhoff, M.; Litterst, P.; Maddula, S.; Boedeker, B.; Rahmann, S.; Davies, A.; Baumbach, J.I., Differentiation of chronic obstructive pulmonary disease (COPD) including lung cancer from healthy control group by breath analysis using ion mobility spectrometry, *Int. J. Ion Mobil. Spectrom.* 2010, 13, 131–139.

20. Baumbach, J.I.; Westhoff, M., Ion mobility spectrometry to detect lung cancer and airway infections, *Spectrosc. Eur.* 2006, 18, 22–27.

21. Westhoff, M.; Litterst, P.; Freitag, L.; Urfer, W.; Bader, S.; Baumbach, J.I., Ion mobility spectrometry for the detection of volatile organic compounds in exhaled breath of patients with lung cancer: results of a pilot study, *Thorax* 2009, 64, 744748.

22. Bunkowski, A.; Boedeker, B.; Bader, S.; Westhoff, M.; Litterst, P.; Baumbach, J.I., MCC/IMS signals in human breath related to sarcoidosis-results of a feasibility study using an automated peak finding procedure, *J. Breath Res.* 2009, 3, 046001/1–046001/10.

23. Bunkowski, A.; Boedeker, B.; Bader, S.; Westhoff, M.; Litterst, P.; Baumbach, J.I., Signals in human breath related to Sarcoidosis. Results of a feasibility study using MCC/IMS, *Int. J. Ion Mobil. Spectrom.* 2009, 12, 73–79.

24. Westhoff, M.; Litterst, P.; Freitag, L.; Baumbach, J.I., Ion mobility spectrometry in the diagnosis of sarcoidosis: results of a feasibility study, *J. Physiol. Pharmacol.* 2007, 58 Suppl 5(Pt 2), 739–751.

25. Chambers, S. T.; Bhandari, S.; Scott-Thomas, A.; Syhre, M., Novel diagnostics: progress toward a breath test for invasive *Aspergillus* fumigates, *Med. Mycol.* 2011, 49, S54–S61.

26. Perl, T.; Carstens, E.; Hirn, A.; Quintel, M.; Vautz, W.; Nolte, J.; Juenger, M. Determination of serum propofol concentrations by breath analysis using ion mobility spectrometry, *Br. J. Anesth.* 2009, 103, 822–827.

27. Miekisch, W.; Schubert, J.K.; Vagts, D.A.; Geiger, K., Analysis of volatile disease markers in blood, *Clin. Chem.* 2001, 47, 1053–1060.

28. Molina, M.A.; Zhao, W.; Sankaran, S.; Schivo, M.; Kenyon, N.J.; Davis, C. E., Design-of-experiment optimization of exhaled breath condensate analysis using a miniature differential mobility spectrometry (DMS), *Anal. Chim. Acta* 2008, 628, 185–161.

29. Bunkowski, A.; Maddula, S.; Davies, A.N.; Westhoff, M.; Litterst, P.; Boedeker, B.; Baumbach, J.I., One-year time series of investigations of analytes within human breath using ion mobility spectrometry, *Int. J. Ion Mobil. Spectrom.* 2010, 13, 141–148.

30. Ruzsanyi, V.; Baumbach, J.I.; Sielemann, S.; Litterst, P.; Westhoff, M.; Freitag, L., Detection of human metabolites using multi-capillary columns coupled to ion mobility spectrometers, *J. Chromatogr. A* 2005, 1084, 145–181.
31. Ulanowska, A.; Ligor, M.; Amann, A.; Buszewski, B., Determination of volatile organic compounds in exhaled breath by ion mobility spectrometry, *Chem. Analityczna* 2008, 53, 953–965.
32. Vautz, W.; Baumbach, J.I., Exemplar application of multi-capillary column ion mobility spectrometry for biological and medical purpose, *Int. J. Ion Mobil. Spectrom.* 2008, 11, 35–41.
33. Vautz, W.; Baumbach, J.I.; Westhoff, M.; Zuechner, K.; Carstens, E.T.H.; Perl, T., Breath sampling control for medical application, *Int. J. Ion Mobil. Spectrom.* 2010, 13, 41–46.
34. Vautz, W.; Nolte, J.; Fobbe, R.; Baumbach, J.I., Breath analysis—performance and potential of ion mobility spectrometry, *J. Breath Res.* 2009, 3, 036004/1–036004/8.
35. Boedeker, B.; Vautz, W.; Baumbach, J.I.; Peak comparison in MCC/IMS-data-searching for potential biomarkers in human breath data, *Int. J. Ion Mobil. Spectrom.* 2008, 11, 89–93.
36. Perl, T.; Boedeker, B.; Juenger, M.; Nolte, J.; Vautz, W., Alignment of retention time obtained from multicapillary column gas chromatography used for VOC analysis with ion mobility spectrometry, *Anal. Bioanal. Chem.* 2010, 397, 2385–2394.
37. Sobel, J.D., Vaginitis, *N. Engl. J. Med.* 1997, 37, 1896–1903.
38. Brand, J.M.; Galask, R.P., Trimethylamine: the substance mainly responsible for the fishy odor often associated with bacterial vaginosis, *Obstet. Gynecol.* 1986, 63, 682–685.
39. Karpas, Z.; Chaim, W.; Tilman, B.; Gdalevsky, R.; Lorber, A., Diagnosis of vaginal infections by ion mobility spectrometry, *Int. J. Ion Mobil. Spectrom.* 2002, 5, 49–54.
40. Karpas, Z.; Chaim, W.; Gdalevsky, R.; Tilman, B.; Lorber, A., A novel application for ion mobility spectrometry: diagnosing vaginal infections, *Anal. Chim. Acta* 2002, 474, 118–123.
41. Chaim, W.; Karpas, Z., Lorber, A., New technology for diagnosis of bacterial vaginosis, *Eur. J. Obstet. Gynecol. Reprod. Biol.* 2003, 111, 83–87.
42. Amsel, R., Nonspecific vaginitis: diagnostic criteria and microbial and epidemiological associations, *Am. J. Med.* 1983, 74, 14–22.
43. Karovicova, J., Kohajdova, Z., Biogenic amines in food, *Chem. Pap.* 2005, 59, 70–79.
44. Snyder, A.P.; Harden, C.S.; Davis, D.M.; Shoff, D.B.; Maswadeh, W.M., Hand portable gas-chromatography ion mobility spectrometer for the determination of the freshness of fish, 3rd International Workshop on Ion Mobility Spectrometry, Galveston, TX, October 16–19, 1994, pp. 146–166.
45. Karpas, Z.; Tilman, B.; Gdalevsky, R.; Lorber, A., Determination of volatile biogenic amines in muscle food by ion mobility spectrometry (IMS), *Anal. Chim. Acta* 2002, 463, 185–163.
46. Raatikainen, O.; Reinikainen, V.; Minkkinen, P.; Ritvanen, T.; Muje, P.; Pursiainen, J.; Hiltunen, T.; Hyvoenen, P.; von Wright, A.; Reinikainen, S., Multivariate modeling of fish freshness index based on ion mobility spectrometry measurements, *Anal. Chim Acta* 2005, 544, 128–134.
47. Strachan, N.J.C.; Nicholson, F.J.; Ogden, I.D., An automated sampling system using ion mobility spectrometry for the rapid detection of bacteria, *Anal. Chim. Acta* 1995, 313, 63–67.
48. Ogden, I.D.; Strachan, N.J.C., Applications of ion mobility spectrometry for food analysis, Special publication—Royal Society of Chemistry, *Biosensors for Food Analysis*, 1998, 167–162.
49. Bota, G.M.; Harrington, P.B., Direct detection of trimethylamine in meat food products using ion mobility spectrometry, *Talanta* 2006, 68, 629–635.

50. Awana, M.A.; Fleet, I.; Thomas, C.L.P., Optimising cell temperature and dispersion field strength for the screening for putrescine and cadaverine with thermal desorption-gas chromatography–differential mobility spectrometry, *Anal. Chim. Acta* 2008, 611, 226–232.

51. Barnard, G.; Atweh, E.; Cohen, G.; Golan, M.; Karpas, Z., Clearance of biogenic amines from saliva following the consumption of tuna in water and in oil, *Int. J. Ion Mobil. Spectrom.* 2011, 14, 207–211.

52. Ruzsanyi, V.; Sielemann, S.; Baumbach, J.I., Determination of microbial volatile organic compounds MVOC using IMS with different ionization sources, *Int. J. Ion Mobil. Spectrom.* 2002, 5, 138–142.

53. Ruzsanyi, V.; Baumbach, J.I.; Eiceman, G.A., Detection of the mold markers using ion mobility spectrometry, *Int. J. Ion Mobil. Spectrom.* 2003, 6, 53–57.

54. Sheibani A.; Tabrizchi, M.; Ghaziaskar, H.S., Determination of aflatoxins B1 and B2 using ion mobility spectrometry, *Talanta* 2008, 18, 233–238.

55. Khales, M.; Zeinoddin, M.S.; Tabrizchi, M., Determination of ochratoxin A in licorice root using inverse ion mobility spectrometry, *Talanta* 2011, 83, 988–993.

56. McCooeye, M.; Kolakowski, B.; Boison, J.; Mester, Z., Evaluation of high-field asymmetric waveform ion mobility spectrometry mass spectrometry for the analysis of the mycotoxin zearalenone, *Anal. Chim. Acta* 2008, 627, 112–116.

57. Tiebe, C.; Hubert, T.; Koch, B.; Ritter, U.; Stephan, I., Investigation of gaseous metabolites from moulds by ion mobility spectrometry (IMS) with gas chromatography-Mass spectrometry (GC-MS), *Int. J. Ion Mobil. Spectrom.* 2010, 13, 17–24.

58. Rasanen, R.M.; Hakansson, M.; Viljanen, M., Differentiation of air samples with and without microbial volatile organic compounds by aspiration ion mobility spectrometry and semiconductor sensors, *Build. Environ.* 2010, 45, 2184–2191.

59. Vautz, W.; Baumbach, J.I.; Jung, J., Continuous monitoring of the fermentation of beer by ion mobility spectrometry, *Int. J. Ion Mobil. Spectrom.* 2004, 7, 1–3.

60. Alonso, R.; Rodriguez-Estevez, V.; Dominguez-Vidal, A.; Ayora-Canada, M.J.; Arce, L.; Valcarcel, M., Ion mobility spectrometry of volatile compounds from Iberian pig fat for fast feeding regime authentication, *Talanta* 2008, 76, 591–596.

61. Kolakowski, B.M.; D'Agostino, P.A.; Chenier, C.; Mester, Z., Analysis of chemical warfare agents in food products by atmospheric pressure ionization-high field asymmetric waveform ion mobility spectrometry—mass spectrometry, *Anal. Chem.* 2007, 79, 8257–8265.

62. Jafari, M.T.; Khayamian, T.; Shaer, V.; Zarei, N., Determination of veterinary drug residues in chicken meat using corona discharge ion mobility spectrometry, *Anal. Chim. Acta* 2007, 581, 147–183.

63. Bota, G.M.; Harrington, P.B., Direct detection of trimethylamine in meat food products using ion mobility spectrometry, *Talanta* 2006, 68, 629–635.

64. Purves, R.W.; Guevremont, R., Electrospray-ionization high-field asymmetric waveform ion mobility spectrometry–mass spectrometry, *Anal. Chem.* 1999, 71, 2346–2357.

65. Purves, R.W.; Barnett, D.A.; Ells, B.; Guevremont, R., Gas-phase conformers of the [M+2H]$^{2+}$ ion of bradykinin investigated by combining high-field asymmetric wave-form ion mobility spectrometry, hydrogen/deuterium exchange, and energy-loss measurements, *Rapid Commun. Mass Spectrom.* 2001, 18, 1453–1456.

66. Gabryelski, W.; Froese, K.L., Rapid and sensitive differentiation of anomers, linkage, and position isomers of disaccharides using high-field asymmetric waveform ion mobility spectrometry FAIMS, *J. Am. Soc. Mass Spectrom.* 2003, 14, 265–277.

67. Valentine, S.J.; Kulchania, M.; Barnes, C.A.S.; Clemmer, D.E., Multidimensional separations of complex peptide mixtures—a combined high-performance liquid chromatography/ion mobility/time-of-flight mass-spectrometry approach, *Int. J. Mass Spectrom.* 2001, 212, 97–109.

68. Hoaglund, C.S.; Valentine, S.J.; Clemmer, D.E., An ion-trap interface for ESI-ion mobility experiments, *Anal. Chem.* 1997, 69, 4186–4161.

69. Lee, D.S.; Wu, C.; Hill, H.H., Detection of carbohydrates by electrospray ionization ion mobility spectrometry following microbore high-performance liquid-chromatography, *J. Chromatogr. A* 1998, 822, 1–9.

70. Hoaglund, C.S.; Valentine, S.J.; Sporleder, C.R.; Reilly, J.P.; Clemmer, D.E., 3-Dimensional ion mobility TOFMS analysis of electro sprayed biomolecules, *Anal. Chem.* 1998, 70, 2236–2242.

71. Koomen, J.M.; Ruotolo, B.T.; Gillig, K.J.; McLean, J.A.; Russell, D.H.; Kang, M.J.; Dunbar, K. R.; Fuhrer, K.; Gonin, M.; Schultz, J.A., Oligonucleotide analysis with MALDI-ion-mobility-TOFMS, *Anal. Bioanal. Chem.* 2002, 373, 612–617.

72. Srebalus, C.A.; Clemmer, D.E., Assessment of purity and screening of peptide libraries by nested ion mobility TOFMS—identification of RNase S-protein binders, *Anal. Chem.* 2001, 73, 424–433.

73. Wyttenbach, T.; Kemper, P.R.; Bowers, M.T., Design of a new electrospray ion mobility mass spectrometer, *Int. J. Mass Spectrom.* 2001, 212, 13–23.

74. Counterman, A.E.; Clemmer, D.E., *Cis-trans* signatures of proline-containing tryptic peptides in the gas-phase, *Anal. Chem.* 2002, 74, 1946–1951.

75. Srebalus, C.A.; Li, J.W.; Marshall, W.S.; Clemmer, D.E., Gas-phase separations of electro sprayed peptide libraries, *Anal. Chem.* 1999, 71, 3918–3927.

76. Hudgins, R.R., Conformations of GlynH+ and ALAnH+ peptides in the gas phase, *Biophys. J.* 1999, 76, 1891–1897.

77. Beegle, L.W.; Kanik, I.; Matz, L.; Hill, H.H., Effects of drift-gas polarizability on glycine peptides in ion mobility spectrometry, *Int. J. Mass Spectrom.* 2002, 216, 257–268.

78. Wu, C.; Siems, W.F.; Klasmeier, J.; Hill, H.H., Separation of isomeric peptides using electrospray ionization/high-resolution ion mobility spectrometry, *Anal. Chem.* 2000, 72, 391–395.

79. Wu, C.; Klasmeier, J.; Hill, H.H., Atmospheric-pressure ion mobility spectrometry of protonated and sodiated peptides, *Rapid Commun. Mass Spectrom.* 1999, 13, 1138–1142.

80. Wyttenbach, T.; Vonhelden, G.; Bowers, M.T., Gas-phase conformation of biological molecules—bradykinin, *J. Am. Chem. Soc.* 1996, 118, 8355–8364.

81. Hudgins, R.R.; Woenckhaus, J.; Jarrold, M.F., High-resolution ion mobility measurements for gas-phase proteins—correlation between solution-phase and gas phase conformations, *Int. J. Mass Spectrom.* 1997, 165, 497–507.

82. Li, J.W.; Taraszka, J.A.; Counterman, A.E.; Clemmer, D.E., Influence of solvent composition and capillary temperature on the conformations of electro sprayed ions—unfolding of compact ubiquitin conformers from pseudo native and denatured solutions, *Int. J. Mass Spectrom.* 1999, 187, 37–47.

83. Wyttenbach, T.; Batka, J.J.; Gidden, J.; Bowers, M.T., Host/guest conformation of biological systems: valinomycin/alkali ions, *Int. J. Mass Spectrom.* 1999, 193, 143–182.

84. Cox, K.A.; Julian, R.K.; Cooks, R.G.; Kaiser, R.E., Conformer selection of protein ions by ion mobility in a triple quadrupole mass spectrometer, *J. Am. Soc. Mass Spectrom.* 1994, 5, 127–136.

85. Uetrecht, C.; Rose, R.J.; van Duijn, E.; Lorenzen, K.; Heck, A.J.R., Ion mobility mass spectrometry of proteins and protein assemblies, *Chem. Soc. Rev.* 2010, 39, 1633–1655.

86. Kaddis, C.S.; Loo, J.A., Native protein MS and ion mobility: large flying proteins with ESI, *Anal. Chem.* 2007, 79, 1778–1784.

87. Karpas, Z.; Berant, Z.; Stimac, R.M., An IMS/MS study of the site of protonation in anilines, *Struct. Chem.* 1990, 1, 201–204.

88. Karpas, Z., Evidence for proton-induced cyclization in α,w diamines from mobility measurements, *Int. J. Mass Spectrom. Ion Proc.* 1989, 93, 237–242.

89. Scarff, C.A.; Thalassinos, K.; Hilton, G.R.; Scrivens, J.H., Travelling wave ion mobility mass spectrometry studies of protein structure: biological significance and comparison with X-ray crystallography and nuclear magnetic resonance spectroscopy measurements, *Rapid Commun. Mass Spectrom.* 2008, 22, 3297–3304.

90. Jurneczko, E.; Barran, P.E., How useful is ion mobility mass spectrometry for structural biology? The relationship between protein crystal structures and their collision cross sections in gas phase, *Analyst* 2011, 136, 20–28.

91. Knapman, T.W.; Berryman, J.T.; Campuzano, I.; Harris, S.A.; Ashcroft, A.E., Considerations in experimental and theoretical collision cross-section measurements of small molecules using travelling wave ion mobility spectrometry-mass spectrometry, *Int. J. Mass Spectrom.* 2010, 298, 17–23.

92. Bush, M.F.; Hall, Z.; Giles, K.; Hoyes, J.; Robinson, C.V.; Ruotolo, B.T., Collision cross sections of proteins and their complexes: A calibration framework and database for gas-phase structural biology, *Anal. Chem.* 2010, 82, 9557–9565.

93. Hopper, J.T.S.; Oldham, N.J., Collision induced unfolding of protein ions in the gas phase studied by ion mobility-mass spectrometry: the effect of ligand binding on conformational stability, *J. Am. Soc. Mass Spectrom.* 2009, 20, 1851–1858.

94. Leary, J.A.; Schenauer, M.R.; Stefanescu, R.; Andaya, A.; Ruotolo, B.T.; Robinson, C.V.; Thalassinos, K.; Scrivens, J.H.; Sokabe, M.; Hershey, J.W.B., Methodology for measuring conformation of solvent-disrupted protein subunits using T-WAVE ion mobility MS: an investigation into eukaryotic initiation factors, *J. Am. Soc. Mass Spectrom.* 2009, 20, 1699–1706.

95. Pessoa, G.S.; Pilau, E.J.; Gozzo, F.C.; Arruda, M.A.Z., Ion mobility mass spectrometry: an elegant alternative focusing on speciation studies, *J. Anal. At. Spectrom.* 2011, 26, 201–206.

96. Fenn, L. S.; Kliman, M.; Mahsut, A.; Zhao, S.R.; McLean, J.A., Characterizing ion mobility-mass spectrometry conformation space for the analysis of complex biological samples, *Anal. Bioanal. Chem.* 2009, 394, 235–244.

97. Slaton, J.G.; Sawyer, H.A.; Russell, D.H., Low-pressure ion mobility-time-of-flight mass spectrometry for methalated peptide ion structural characterization: tyrosine-containing tripeptides and homologous septapeptides, *Int. J. Ion Mobil. Spectrom.* 2005, 8, 13–18.

98. Zhou, P.; Tian, F.; Li, Z., Quantitative structure-property relationship studies for collision cross sections of 579 singly protonated peptides based on a novel descriptor as molecular graph fingerprint (MoGF), *Anal. Chim. Acta* 2007, 597, 214–222.

99. Ruotolo, B.T.; McLean, J.A.; Gillig, K.J.; Russell, D.H., The influence and utility of varying field strength for the separation of tryptic peptides by ion mobility-mass spectrometry, *J. Am. Soc. Mass Spectrom.* 2005, 16, 158–165.

100. Stauber, J.; MacAleese, L.; Franck, J.; Claude, E.; Snel, M.; Kaletas, B.K.; Wiel, I.; Wisztorski, M.; Fournier, I.; Heeren, R.M.A., On-tissue protein identification and imaging by MALDI-ion mobility mass spectrometry, *J. Am. Soc. Mass Spectrom.* 2010, 21, 338–347.

101. Liu, X.; Valentine, S.J.; Plasencia, M.D.; Trimpin, S.; Naylor, S.; Clemmer, D.E., Mapping the human plasma proteome by SCX-LC-IMS-MS, *J. Am. Soc. Mass Spectrom.* 2007, 18, 1249–1264.

102. Shvartsburg, A.A.; Bryskiewicz, T.; Purves, R.; Tang, K.; Guevremont, R.; Smith, R.D., New directions for FAIMS and IMS opened by dipole alignment of macro ions in strong electric fields, *Int. J. Ion Mobil. Spectrom.* 2006, 9, 6–7.

103. Li, J.; Purves, R.W.; Richards, J.C., Coupling capillary electrophoresis and high-field asymmetric waveform ion mobility spectrometry mass spectrometry for the analysis of complex lipopolysaccharides, *Anal. Chem.* 2004, 76, 4676–4683.

104. Snyder, A.P.; Shoff, D.B.; Eiceman, G.A.; Blyth, D.A.; Parsons, J.A., Detection of bacteria by ion mobility spectrometry, *Anal. Chem.* 1991, 63, 526–529.

105. Snyder, A.P.; Blyth, D.A.; Parsons, J.A., Ion mobility spectrometry as an immunoassay detection technique, *J. Microbiol. Methods* 1996, 27, 81–88.
106. Snyder, A.P.; Harden, C.S.; Brittain, A.H.; Kim, M.G.; Arnold, N.S.; Meuzelaar, H.L.C., Portable hand-held gas chromatography/ion mobility spectrometry device, *Anal. Chem.* 1993, 65, 299–306.
107. Tripathi, A.; Maswadeh, W.M.; Snyder, A.P., Optimization of quartz tube pyrolysis atmospheric-pressure ionization mass spectrometry for the generation of bacterial biomarkers, *Rapid Commun. Mass Spectrom.* 2001, 18, 1672–1680.
108. Snyder, A.P.; Maswadeh, W.M.; Parsons, J.A.; Tripathi, A.; Meuzelaar, H.L.C., Dworzanski; J.P.; Kim, M.G., Field detection of bacillus spore aerosols with stand-alone pyrolysis-gas chromatography-ion mobility spectrometry, *Field Anal. Chem. Technol.* 1999, 3, 318–326.
109. Snyder, A.P.; Tripathi, A.; Maswadeh, W.M.; Ho, J.; Spence, M., Field detection and identification of a bioaerosol suite by pyrolysis-gas chromatography-ion mobility spectrometry, *Field Anal. Chem. Technol.* 2001, 5, 190–204.
110. Snyder, A.P.; Maswadeh, W. M.; Wick, C.H.; Dworzanski, J.P.; Tripathi, A., Comparison of the kinetics of thermal decomposition of biological substances between thermogravimetry and a fielded pyrolysis bioaerosol detector, *Thermochim. Acta* 2005, 437, 87–99.
111. Snyder, A.P.; Dworzanski, J.P.; Tripathi, A.; Maswadeh, W.M.; Wick, C.H., Correlation of mass spectrometry identified bacterial biomarkers from a fielded pyrolysis-gas chromatography-ion mobility spectrometry biodetector with the microbiological gram stain classification scheme, *Anal. Chem.* 2004, 76, 6492–6499.
112. Snyder, A.P.; Maswadeh, W.M.; Tripathi, A.; Dworzanski, J.P., Detection of gram-negative *Erwinia-Herbicola* outdoor aerosols with pyrolysis-gas chromatography-ion mobility spectrometry, *Field Anal. Chem. Technol.* 2000, 4, 111–126.
113. Snyder, A.P.; Maswadeh, W.M.; Tripathi, A.; Eversole, J.; Ho, J.; Spence, M., Orthogonal analysis of mass- and spectral-based technologies for the field detection of bioaerosols, *Anal. Chim. Acta* 2004, 513, 365–377.
114. Schmidt, H.; Tadjimukhamedov, F.; Mohrentz, I.V.; Smith, G.B.; Eiceman, G.A., Microfabricated differential mobility spectrometry with pryrolysis gas chromatography for chemical characterization of bacteria, *Anal. Chem.* 2004, 76, 5208–5217.
115. Cheung, W.; Yu Xu, Y; Thomas C.L.P.; Goodacre, R., Discrimination of bacteria using pyrolysis-gas chromatography-differential mobility spectrometry (Py-GC-DMS) and chemometrics, *Analyst* 2009, 134, 557–563
116. Vinopal, R.T.; Jadamec, J.R.; Defur, P.; Demars, A.L.; Jakubielski, S.; Green, C.; Anderson, C.P.; Dugas, J.E.; Debono, R.F., Fingerprinting bacterial strains using ion mobility spectrometry, *Anal. Chim. Acta* 2002, 457, 83–95.
117. Smith, G.B.; Eiceman, G.A.; Walsh, M.K.; Critz, S.A.; Andazola, E.; Ortega, E.; Cadena, F., Detection of *Salmonella typhimurium* by hand-held ion mobility spectrometer: a quantitative assessment of response characteristics, *Field Anal. Chem. Technol.* 1997, 4, 213–226.
118. Ochoa, M.L.; Harrington, P.B., Chemometric studies for the characterization and differentiation of microorganisms using in situ derivatization and thermal desorption ion mobility spectrometry, *Anal. Chem.* 2005, 77, 854–863.
119. Dwivedi, P.; Puzon, G.; Tam, M.; Langlais, D.; Jackson, S.; Kaplan, K.; Siems, W.F.J.; Schultz, A.J.; Xun, L.; Woods, A.; Hill, H.H., Metabolic profiling of *Escherichia coli* by ion mobility-mass spectrometry with MALDI ion source, *J. Mass. Spectrom.* 2010, 45, 1383–1393.
120. Thomas, J.J.; Bothner, B.; Traina, J.; Benner, W.H.; Siuzdak, G., Electrospray ion mobility spectrometry of intact viruses, *Spectroscopy* 2004, 18, 31–36.

121. Krebs, M.D.; Zapata, A.M.; Nazarov, E.G.; Miller, R.A.; Costa, I.S.; Sonenshein, A.L.; Davis, C.E., Detection of biological and chemical agents using differential mobility spectrometry (DMS) technology, *IEEE Sens. J.* 2005, 5, 696–703.

122. Clark, J.M.; Daum, K.A.; Kalivas, J.H., Demonstrated potential of ion mobility spectrometry for detection of adulterated perfumes and plant speciation, *Anal. Lett.* 2003, 36, 218–244.

123. Vautz, W.; Sielemann, S.; Baumbach, J.I., Qualitative detection of odours using ion mobility spectrometry, 12th International Conference on Ion Mobility Spectrometry, July 27–31, 2003, Umeå, Sweden.

124. Statheropoulos, M.; Agapiou, A.; Georgiadou, A. Analysis of expired air of fasting male monks at Mount Athos, *J. Chromatogr. B: Analytical Technologies in the Biomedical and Life Sciences* 2006, 832, 274–279.

125. Rudnicka, J.; Mochalski, P.; Agapiou, A.; Statheropoulos, M.; Amann, A.; Buszewski, B., Application of ion mobility spectrometry for the detection of human urine, *Anal. Bioanal. Chem.* 2010, 398, 2031–2038.

126. Vautz, W.; Nolte, J.; Bufe, A.; Baumbach, J.I.; Peters, M., Analyses of mouse breath with ion mobility spectrometry: a feasibility study, *J. Appl. Physiol.* 2010, 108, 697–704.

127. Guamán, A.V.; Carreras, A.; Calvo, D.; Agudo, I.; Navajas, D.; Pardo, A.; Marco, S.; Farré, R., Rapid detection of sepsis in rats through volatile organic compounds in breath, *J. Chromatogr. B. Anal. Technol. Biomed. Life Sci.* 2012, 881–882, 76–82.

128. Marcus, S.; Menda, A.; Shore, L.; Cohen, G.; Atweh, E.; Friedman, N.; Karpas, Z., A novel method for the diagnosis of bacterial contamination in the anterior vagina of sows based on measurement of biogenic amines by ion mobility spectrometry: a field trial, *Theriogenology* April 26, 2012 [Epub ahead of print].

129. Hashemian, Z.; Mardihallaj, A.; Khayamian, T., Analysis of biogenic amines using corona discharge ion mobility spectrometry, *Talanta* 2010, 81, 1081–1087.

130. Karpas, Z.; Litvin, O.; Cohen, G.; Mishin, J.; Atweh, E.; Burlakov, A., The reduced mobility of the biogenic amines: trimethylamine, putrescine, cadaverine, spermidine and spermine, *Int. J. Ion Mobil. Spectrom.* 2011, 14, 3–6.

131. Mäkinena, M.; Sillanpää, M.; Viitanene, A.K.; Knapd, A.; Mäkeläc, J.M.; Puton, J., The effect of humidity on sensitivity of amine detection in ion mobility spectrometry, *Talanta* 2011, 84, 116–121.

132. Steiner, W.E.; Clowers, B.H.; English, W. A.; Hill, H.H., Atmospheric pressure matrix-assisted laser desorption/ionization with analysis by ion mobility time-of-flight mass spectrometry, *Rapid Commun. Mass Spectrom.* 2004, 18, 882–888.

133. Bocos-Bintintan, V.; Moll, V.H.; Flanagan, R.J.; Thomas, C.L.P., Rapid determination of alcohols in human saliva by gas chromatography differential mobility spectrometry following selective membrane extraction, *Int. J. Ion Mobil. Spectrom.* 2010, 13, 55–63.

134. Dwivedi, P.; Hill, H.H., A rapid analytical method for hair analysis using ambient pressure ion mobility mass spectrometry with electrospray ionization (ESI-IMMS), *Int. J. Ion Mobil. Spectrom.* 2008, 11, 61–69.

19 Current Assessments and Future Developments in Ion Mobility Spectrometry

19.1 STATE OF THE SCIENCE AND TECHNOLOGY OF IMS

19.1.1 Ion Mobility Spectrometry Comes of Age

Ion mobility spectrometry (IMS) and related mobility methods are now recognized as beneficial and accepted within several communities, including analytical sciences, military establishments, security organizations, biomolecular researchers, clinicians, and pharmaceutical researchers. The developments in IMS today, including commercial, research, and other applications, are truly international, and this high level of activity is evident in the number of publications, which has risen sharply in the past 5 years and now surpasses 200 publications annually.

Unlike previous editions of this book, the vitality of this method is evident in an energetic community of instrument manufacturers, well-established and small startups, government employees (researchers, users, managers), academic research teams, and industries seeking solutions that IMS can provide. The atmosphere of innovations can be seen also in annual conferences, with the International Society for Ion Mobility Spectrometry (ISIMS) and the American Society of Mass Spectrometry, and finally in their respective publications, including the recent arrangement between ISIMS and Springer-Verlag. Advances in ion mobility are treated now in six books: *Plasma Chromatography*, the first and second editions of *Ion Mobility Spectrometry*, this third edition of the same title, *Field Asymmetric Ion Mobility Spectrometry*, and *Ion Mobility Mass Spectrometry*.[1-5] Four of these were published in the last decade.

During this past decade, knowledge of IMS emerged from the spheres of military and security use, for which efforts had been under way for 30 years, into a national and international awareness that IMS analyzers are the core technology to counter increased civilian vulnerability to unconventional warfare through their use in military preparedness, aviation security, and explosive detection. This, when supplemented with successful commercialization of ion mobility–mass spectrometry (IM–MS) and development of pharmaceutical and clinical applications have altered the presence of IMS in the scientific enterprise.

19.1.2 Assessment of Technology

The status of IMS is unusual since the technology has been available for several decades and is mature, yet innovations continue in all areas of engineering improvements, understandings of principles, and methodologies. This is unlike other analytical methods that may be considered mature methods, such as gas chromatography (GC). The multiple modes of operation of IMS are a major strength of versatility, enabling a broad range of applications. Developments in ion sources operating at ambient pressure have changed the value of IMS by changing the usefulness of the method. The fact that IMS devices generally operate at atmospheric pressure has encouraged the development of sample introduction techniques that can be readily interfaced with an IMS. Thus, gas, vapor, aerosol, liquid, and solid samples from a variety of matrices can be analyzed directly by IMS. Whenever a gas phase ion of reasonable stability can be made from a sample, a mobility measurement may be obtained. Amid all the possible variations of sizes, shapes, methods, and ion sources, there may be a sense of complexity; however, mobility measurements remain as practically uncomplicated, and the simplicity, size, and power demands ensure that IMS will have a large presence in field techniques far into the future, whereas other methods such as MS contend with requirements of vacuums.

The classic linear drift tube mobility spectrometers now are complemented by the orthogonal techniques of differential mobility, the growing promise for aspiration IMS, and the astonishing success of traveling-wave-based IMS. These have added layers of richness and possibilities in measurements and instrumentation. Interestingly, only differential mobility spectrometry (DMS) has emerged in combination with GC and IMS, once thought to be an inexpensive alternative to mass spectrometers (MSs) as advanced GC detectors had not emerged with either gas or liquid chromatography apart from IM–MS combinations.

While small drift tubes existed in 2004, as described in the last edition of this book, miniaturization and fabrication to a standard found in consumer electronics is occurring. Analyzers that are pocket size, unobtrusive, and lightweight are exemplified in the lightweight chemical detector (LCD) from Smiths Detection. Similarly, bench-scale explosive detectors are now available as handheld instruments. The field asymmetric IMS (FAIMS) from Owlstone NanoTech may have reached the ultimate dimensions, and yet questions still arise on the ultimate dimensions of a mobility drift tube. More than a decade ago, Bob Bradshaw of Bulstrode Technology Limited and formerly of Graseby Dynamics Limited spoke of practical limits on dimensions governed by the interface of technology, observing that fingers and eyesight are size-limiting factors in miniaturization of mobility instruments, not electronics or engineering. Can we expect further miniaturization and integration of mobility analyzers into Internet or other mass communication methods? Could mobility spectrometers enter the consumer world in home air quality monitoring?

19.1.3 A Missing Application: Environmental Monitoring and Testing

One of the large shibboleths in expanded use of mobility methods was the ever-present radioactive ion source, and this pattern of design and use of IMS is changed. The appearance of viable alternatives to radioactive sources, such as miniature pulsed corona discharges, photodischarge, electrospray ionization, and even paperspray portend uses of IMS previously hindered by the cost and logistics of instruments equipped with radioactive foils. Although the form, fit, and analytical properties of IMS should have made this method a preeminent tool in field analytical chemistry, a journal dedicated to this ceased publication in the past decade, and there is little visibility of IMS in the environmental community and no known ongoing applications. Certainly there is no description of IMS in testing methods certified by the Environmental Protection Agency (EPA). After a period of curiosity within the agency, largely led by a team in Nevada, little more has arisen, although Chapter 18 suggested this should be otherwise. Would a change in agency philosophy to performance-based technology rather than regulatory environments driven by legal descriptions of protocols open expanded environmental applications?

19.1.4 ASSESSMENT OF UNDERSTANDINGS IN THE PRINCIPLES OF ION MOBILITY MEASUREMENTS

At the release of the first edition of this book in 1994, there was no comprehensive scientific model for ion chemistry or ion behavior in drift tubes. These effects of kinetics and ion behavior on the appearance of mobility spectra are now well described and accepted. Still, new challenges are now apparent in refined understandings for the meaning of K_o (which reflects the averaged size of the collision cross section) and detail of ion structure or shape. This is not fully developed and is dependent on computational modeling. Spectral libraries are now available, although perhaps not well publicized or obtainable in searchable and user-friendly software packages.

Finally, the combination of mobility and MS is seen as a highly visible trend in this edition and was only barely discernible in the prior edition. This is surely one of the transformative influences today in IMS, making the combination of the two methods for ion measurement very much more than the sum of parts. We believe that advances in mass spectrometry will occur largely outside the vacuum chamber and with ion preparation and handling. Mobility spectrometers are ideally suited as an interface between ambient pressure ion sources, which can sometimes produce a complex mixture of ions from a sample, and the MS.

19.2 NEXT GENERATION IN ION MOBILITY METHODS

19.2.1 SOME FUNDAMENTAL CONSTRAINTS FOR IMS AT AMBIENT PRESSURE

The advantages of measuring ion mobility values at ambient pressure are manifest, and some of the challenges should be amplified here. Making ions at ambient pressure is associated with high collision frequencies, charge exchange based on well-known principles of reaction chemistry, and high sensitivity. In some instances, this is a strong advantage of IMS when the analyte of interest has the most favorable properties of ionization. In this instance, such as the historic uses of IMS in detecting chemical warfare agents and drugs, concerns over matrix interferences are substantially marginalized. Otherwise, analytical response with substances with less-favorable ionization properties is problematic at best and perhaps flawed in the absence of some prefractionation before sample enters the analyzer. Complication resides here, yet opportunities exist in practice of measurements and in fundamental engineering design of ion sources and sample introduction systems.

Quantitative response in IMS is today acceptable in applications where IMS has been successful yet unacceptable compared to other detector technologies, such as flame ionization detectors or MSs. This is limited by kinetics of ion formation at the low end of response and by ion source saturation at the top end of the response curve and by matrix effects. Other technologies, such as electron capture detectors and the ion trap MS, shared a similar history and were engineered free of the limitations. There is no such advancement under way today in IMS.

The Faraday plate detectors, the mainstay of detector technology in IMS from its inception as a modern analytical method, are seen as robust and effective for field instruments and relatively small ions. Nonetheless, the poor gain and susceptibility to microphonic noise can also be seen as disadvantages. Probably, these worries and wishes are small compared to the fundamental barrier to improved resolving power, which is established with the ion shutter. The Bradbury–Neilson (BN) or Tyndall–Powell (TP) shutters have and will certainly be the method of choice into the foreseeable future for IMS analyzers. The constraint is ambient pressure based. The limitation induced with field-mobility-based injections are large, yet no improved solution has been demonstrated.

19.2.2 SOME FUNDAMENTAL CONSTRAINTS FOR IMS AT 1 TO 10 TORR

A distinction of the past decade has been the emergence of analytical mobility with drift tubes operating in subambient pressures, usually in the range of 1 to 10 torr. These methods also face technical or operational challenges, including the scattering of ions and subsequent loss of ion signal if pressures exceed several torr. Similarly, these methods are dependently solely on the mobility of ions through nonclustering gases. Thus, separations are wholly physical and lack some of the nuance of the benefits from modifying drift gases with reagents or dopants.

19.2.3 CLARIFICATIONS IN EXPERIMENTAL PRACTICES IN IMS

The mobility of ions in gases is sensitive to moisture, gas temperature, analyte concentration, and residence time of ions (controlled by electric fields, drift tube length, and pressure). These are pivotal in the initial step of sample ionization and in the behavior of ions derived from a sample inside the drift tube. Although an understanding and acceptance of the importance of these parameters exists today, there is no recognized agreement on standard conditions for these parameters or any conformity on recording or reporting values. This casual approach to experimental controls observed over the history of modern analytical IMS may be associated with the precision of measurements; that is, reduced mobility values reported as three significant figures are amazingly insensitive to a large range for common parameters, and instruments continue to function reliably. Any aspirations to a new standard in measurements or instruments, where K_o values are reported to four or five significant figures, will be hindered by such diversity of experimental parameters. One intermediate solution is the calibration of instruments to chemical standards, and activity in this has already begun.

Chemical standards are used to test response in military field units, and the National Institute for Standards and Technology (NIST) has recently become active in establishing procedures to calibrate explosive detectors. Research teams have proposed substances that could provide a reference to calculate K_o values. Will these or any others be incorporated into IMS instruments or methods? Will the role of ion residence time in drift tubes or analyzers be widely recognized as a factor in the appearance of mobility spectra? How will we know that we have the same ion species? For the next-generation instruments as suggested, will the identity of the ion species be controlled and known sufficiently so mobility may be considered an intrinsic parameter of an gas ion? Until then, the method is disproportionately limited by control of experimental variables.

19.2.3 OTHER THOUGHTS

At one level, the analytical value of ion mobility depends only on a result that discloses details of the composition of a sample useful to a user. Mobility measurements today are being used to link ion structure and mobility of biomolecules through a blend of computational modeling and biological activity of the original substances. Quite apart from questions concerning the relationships between a molecule in solution and a gas phase ion derived from this molecule, the very first step of linking drift time and ion collision cross section to some structure must be advanced in concept, practice, and examples.

The comparison of mobility results between laboratories or between instruments is today not controlled, and this has animated the development of chemical standards as described. Establishing international protocols and then testing these protocols for qualitative and quantitative suitability is a distant thought in this generation of users or developers of ion mobility methods.

Finally, the long-standing limitation of ion mobility instruments as point sensors with all the limitations may be solvable with multiple instruments either distributed through some area or region or on airborne platforms in motion. This has been demonstrated with the unmanned airborne vehicle (UAV) work of C. S. Harden and colleagues and could develop with reduced size and cost of ion mobility analyzers and UAVs.

19.3 DIRECTIONS FOR IMS

19.3.1 EXPLOSIVE DETECTION AND APPLICATIONS WITH MILITARY AND LAW ENFORCEMENT

In the next decade, we anticipate that the capabilities of mobility spectrometers will be extended for detecting explosives, narcotics, and chemical weapons. This may occur through improvements in drift tube technology as in microfabricated drift tubes. Alternatively, we expect that advances in inlet and ion source will bring transformations to this application of IMS. In addition, the large number of analyzers already deployed in the field will continue in service, and yet new challenges with explosives will be introduced by terrorist organizations. This has been evident in recent and continued threats of new materials (at present, improvised explosives based on peroxides, liquids, and binary mixtures) and new methods for penetrating the defense lines (after the "shoe bomber" and the "exploding underwear," there have been innocuous liquids), and in the future, use of other types of devices can be anticipated. The explosive detectors, particularly the IMS-based instruments, will thus require continuous upgrades to meet these new challenges.

The promise of small MSs has been a consideration in the race for in-field analyzers of explosives and chemical warfare agents during the past two decades, and small analyzers have been developed and demonstrated. The same constraints of field use, such as low maintenance (ideally, low need for attention in any form from users) and low size and power apply equally to all methods of measurements. As MSs have been reduced in size, there has been no breakthrough yet seen on vacuum requirements and technology; simultaneously, mobility spectrometers have been reduced further in size and made, with the LCD, pocket size. After the attack on the World Trade Center in New York in September 11, 2001, vulnerabilities throughout infrastructures were assessed and efforts made to provide chemical detectors for toxic industrial chemicals (TICs), which could be poisons of opportunity for terrorists. Testing and development of IMS configurations suitable for detection of TICs have occurred, and a next generation of rugged hardware for these substances may emerge, although this is not guaranteed.

19.3.2 Ion Mobility with Mass Spectrometry

Mass spectrometry today stands as one of the most sophisticated and refined analytical instruments in history, and a position could be taken that there is little more inside the vacuum chamber to improve. The interface to ion sources operated at ambient pressure, however, is positioned for opportunity to innovate, create, and generally advance much of the practice and value of MS. We believe this will occur with ion mobility methods placed between these ion sources and the vacuum flange of the mass spectrometer, that is, mobility instruments operated at ambient pressure for IM–MS methods. While most IMS instruments will not be interfaced to MS, most MS instruments will be interfaced to IMS.

19.3.3 Tandem Ion Mobility Instruments at Ambient Pressure

An ion mobility instrument is comparatively small and simple enough that two or more instruments could be joined without difficulty or extra cost. Much as in two-dimensional GC, or GCxGC, IMSs could be combined without some of the constraints experienced in GCxGC. This is motivated by the low resolving power of aIMS, DMS, and IMS instruments, and IMSxIMS has already started with exploratory studies and demonstration. Indeed, one configuration of tandem ion mobility analyzers, a DMS-IMS$_2$ instrument, was part of a DHS (Department of Homeland Security) program of technology development.

19.3.4 Emerging Applications

Should our expectation that a large fraction of MSs will be fitted with mobility inlets, the world of applications is nearly as large as the current uses of the MS. While this expectation is not ensured, ion mobility could become a routine topic in future measurement sciences. This alone would transform practically all aspects of this subject of ion characterizations in electric fields by mobility in gases. Should this not occur, or even if IM–MS does develop, we expect that ion mobility methods will be expanded and adopted into other uses as stand-alone instruments of relatively low cost.

Metabolomics in diagnostics as pioneered by a few teams today is foremost among the exciting possibilities for IMS to enter medical or clinical service. This is already in practice in lung clinics in Germany and could become mainstream once accepted by clinicians and when technology is matured for this specific application. Diagnosis and treatment of lung disease is only one clinical use, and other applications are in development. These all feature the same attraction: high-speed diagnosis on-site so medication can be given at point of care. This results in prompt and proper treatment without repeat visits, which is attractive in face of trends in expenses of modern medicine. Some of these practices could jump the fence into the agricultural industry for health care of animals.

19.4 FINAL THOUGHTS

We have no way to anticipate unexpected events such as the attack on the World Trade Center and the impact this event had on surveillance technology and the visibility of IMS. This occurred because IMS provided a unique solution (or at least a more practical solution than the alternatives) and happened to be with the right degree of maturity to meet the technical requirements. An environmental accident that could lead to a human catastrophe, such as the release of methyl-isocyanate in Bophal, India, in 1984, may create a demand that can be met by IMS for on-site, real-time monitoring of a toxic compound. The existence of a large variety of IMS-based devices with the capability to handle many different types of samples and to provide results within seconds, with a high sensitivity and specificity in complex matrices for a suite of compounds, can be viewed as "a solution looking for (more) problems." When such a need will arise, it is hoped the IMS community of researchers and manufacturers will be in a position to provide a practical, competitive solution.

A comparison of this monograph to prior editions demonstrates that the world of IMS is now serving or in service to a larger community than previously understood and even anticipated. This is in large measure because of the combined effect of investigators and users at the extreme ends of measurement regimes. We anticipate that ion mobility will be understood and accepted as a general principle of measurement science rather than an assortment of technologies. More than 30 years ago, F. W. Karasek urged "there is much left to be discovered and developed in plasma chromatography." His words are as true today as then, and the horizon is bright with potential for discovery and enrichment to human endeavors as we enter into the fourth decade of modern analytical IMS.

REFERENCES

1. Carr, T.W., *Plasma Chromatography*, Plenum Press, New York, 1984.
2. Eiceman, G.A.; Karpas, Z., *Ion Mobility Spectrometry*, CRC Press, Boca Raton, FL, 1994.
3. Eiceman, G.A.; Karpas, Z., *Ion Mobility Spectrometry*, 2nd edition, CRC Press, Boca Raton, FL, 2005.
4. Shvartsburg, A.A., *Differential Ion Mobility Spectrometry*, CRC Press, Boca Raton, FL, 2008.
5. Trimpin, S.; Wilkins, C.L., *Ion Mobility Spectrometry-Mass Spectrometry*, Taylor & Francis, Boca Raton, FL, 2010.

Index

Page references in **bold** refer to tables.

Printed and bound by CPI Group (UK) Ltd, Croydon, CR0 4YY

21/10/2024

01777103-0008